Instrumental Methods for Analysis and Characterizations of Phytopharmaceuticals

This book focuses on phytopharmaceuticals and herbal drugs, effectively bridging traditional knowledge with modern sciences. It comprehensively discusses both conventional and novel drug delivery systems and addresses analytical and quality control challenges to ensure that herbal formulations comply with scientific and regulatory standards. The book elucidates the fundamentals of herbal drug delivery, rational strategies for dosage form selection, and key issues related to analysis and standardization. The chapters explore the instrumental methodologies for qualitative and quantitative analysis (HPLC, HPTLC, LC-MS), structural elucidation (NMR, FTIR, LC-MS), crystallinity evaluation (XRD, SAED, DVS, SEM), pharmaceutical property assessment (such as rheometer, pH, DLS), thermal behavior studies (DSC, TGA), estimation of volatile components and residual solvents (GC), morphological characterization (TEM, AFM), and elemental analysis (XPS, ICPMS). By integrating theoretical concepts with practical methodologies, this book serves as a valuable reference for students, researchers, and professionals of ethnopharmacology and pharmaceutical sciences.

Instrumental Methods for Analysis and Characterizations of Phytopharmaceuticals

Alakh Niranjan Sahu, Debadatta Mohapatra,
Soki Daeme Malang, and Pooja Kathait

CRC Press
Taylor & Francis Group
Boca Raton London New York

CRC Press is an imprint of the
Taylor & Francis Group, an **informa** business

[Designed cover image: The Cover Image has been created by the Authors]

First edition published 2026
by CRC Press
2385 NW Executive Center Drive, Suite 320, Boca Raton FL 33431

and by CRC Press
4 Park Square, Milton Park, Abingdon, Oxon, OX14 4RN

CRC Press is an imprint of Taylor & Francis Group, LLC

ISBN: 978-1-032-89020-3 (hbk)
ISBN: 978-1-032-89022-7 (pbk)
ISBN: 978-1-003-54082-3 (ebk)

DOI: 10.1201/9781003540823

Typeset in Times
by Apex CoVantage, LLC

Contents

About the Authors

Alakh Niranjan Sahu (B. Pharm., M. Pharm., Ph.D.) is a distinguished and acclaimed academician and researcher. He has 22 years of experience in teaching, research, administration, and outreach. He currently serves as a Professor in the Department of Pharmaceutical Engineering & Technology, Indian Institute of Technology (Banaras Hindu University), Varanasi, India. Dr. Sahu leads an independent research group at the Phytomedicine Research Laboratory with research interests in herbal drug formulation and standardization, phytopharmaceuticals, nutraceuticals, herbal cosmeceuticals, preparation of carbon dots for healthcare and environmental applications, classical pharmacognosy, and quality control of indigenous medicinal herbs. He is an active member of various professional bodies and a reviewer of various peer-reviewed international journals. His work in pharmaceutical engineering has been influential, impacting both the academic community and industry practices.

Debadatta Mohapatra (B. Pharm., M. Pharm., Ph.D.) is a postdoctoral researcher at the Leslie Dan Faculty of Pharmacy, University of Toronto, Canada. He earned his Ph.D. in Pharmaceutical Engineering & Technology from the Indian Institute of Technology (Banaras Hindu University), Varanasi, India. His research centers on quality-by-design (QbD)-based formulation development and novel drug delivery systems, such as solid dispersions, transferosomes, nanoemulsions, carbon dots, hydrogels, and phytoformulations. He has expertise in analytical and bioanalytical method development, validation, herbal drug standardization, pharmacokinetics, cancer therapeutics, and *in vitro* and *in vivo* evaluations. Dr. Mohapatra has extensively published in high-impact journals, has presented his work at international conferences, has led multidisciplinary research teams, and actively collaborates with academia, industry, and research institutions to drive advancements in pharmaceutical sciences and innovative therapeutic strategies.

Pooja Kathait (B. Pharm., M. Pharm.) is a Ph.D. research scholar at the Department of Pharmaceutical Engineering & Technology, Indian Institute of Technology (Banaras Hindu University), Varanasi, India. Her research covers phytopharmaceutical development, including extraction, marker-based standardization of medicinal plant compounds, quality control, and pharmacological evaluation. She is actively involved in assessing the *in vitro* and *in vivo* biological activities of plant-derived substances for potential therapeutic applications.

Soki Daeme Malang (B. Pharm., M. Pharm.) is a Ph.D. research scholar at the Department of Pharmaceutical Engineering & Technology, Indian Institute of Technology (Banaras Hindu University), Varanasi, India. Her research focuses on the development, optimization, and characterization of herbal formulations, alongside investigating their pharmacological properties. Her work involves designing novel herbal formulations, analyzing their chemical composition and evaluating their therapeutic potential through preclinical and clinical studies, with the goal of validating and enhancing the use of traditional herbal medicine.

Preface

Phytopharmaceuticals play an important role in global healthcare systems. Recent scientific advances in natural products have reshaped our overall understanding of herbal medicines. The growing recognition in their therapeutic potential has driven extensive research and development in recent times. However, the inherent complexity of plant-based extracts, formulations, polyherbal formulations, and multicomponent mixtures, coupled with the need for consistent quality, safety, and efficacy, poses significant challenges for their broader application. Addressing these challenges necessitates the use of precise, reliable, and sophisticated analytical and characterization tools.

This book aims to provide a comprehensive account of the advanced instrumental techniques used in the study of plant-derived medicines. It bridges traditional knowledge with modern scientific approaches, demonstrating how instrumental analysis highlights standardization, authentication, characterizations, and regulatory compliance in phytopharmaceutical research and industry. The content introduces fundamental principles, instrumentation, challenges, and applications of major instrumental methods, such as chromatography, spectroscopy, microscopy, and hyphenated techniques for the analysis of herbal drugs and formulations.

This book is intended for students, researchers, and professionals in pharmaceutical sciences, natural products chemistry, nutraceuticals, cosmeceuticals after pharmacognosy, and allied disciplines, as well as scholars engaged in herbal drugs and R&D, who may benefit from it. It aims to serve as both a theoretical framework and a practical tool for laboratory applications. We hope that the knowledge compiled herein will not only enrich readers' academic and professional pursuits but also contribute to the advancement of evidence-based natural medicine.

While every effort has been made to enhance the quality and accuracy of this work, some shortcomings may remain. We sincerely welcome constructive feedback and suggestions from researchers and experts to further enhance and refine future editions.

Acknowledgments

The authors express their sincere gratitude to the researchers whose dedicated work and valuable contributions have been instrumental in identifying the challenges addressed in this book, providing a strong foundation that allowed us to structure its content. The authors are also thankful to the researchers whose work provided valuable insights and inspiration for the ideas presented in this book.

We are deeply grateful to the Indian Institute of Technology (Banaras Hindu University, Varanasi, India) for providing the academic environment and infrastructure that greatly facilitated the advancement of this work. We also extend our heartfelt thanks to the staff of the Department of Pharmaceutical Engineering & Technology for their encouragement, guidance, and support throughout the development of this book.

We are particularly thankful to both the present and the former student members of the Phytomedicine Research Laboratory (PRL) for their unwavering support, insightful discussions, and collaborative spirit, which have been instrumental in shaping ideas and synthesizing the contents of this book. Their guidance and mentorship fostered a culture of innovation and rigorous scholarly review, providing the collaborative platform where many of the book's concepts took shape. The dedication of the PRL members and their commitment to advancing knowledge in phytomedicine have been a continual source of inspiration throughout the preparation of this work.

We are thankful to the Ministry of Education, Government of India, for providing support that contributed to this book. Their assistance has indirectly facilitated the pursuit of high-quality research and provided the resources necessary for the realization of this publication.

We would like to thank all colleagues, collaborators, and students who engaged in insightful discussions, offered constructive feedback, and shared their expertise, thereby enriching this work. We hope that this book serves as a valuable resource for the scientific community and encourages further research.

We are sincerely thankful to the publisher, the editor, and their dedicated staff members for their constant cooperation, guidance, and professional support in bringing out this book. Their timely assistance, valuable suggestions, and attention to detail have played a vital role in shaping the final outcome of this book.

Last but not least, we are grateful to all those guides of the Righteous Path because of whom the sources of actuality are alive even today. Gratitude is the fundamental value for progression toward awakening. Gratitude itself is the foundation and conserving value for culture and civilization. One who is not grateful cannot produce evidence of bearing culture and civilization. One who does not bear culture and civilization cannot follow norms and orderliness. Culture, Civilization, Norms, and Orderliness are mutually complementary. Without these, it is not possible to establish Undivided Society and Social Harmony. Therefore, without gratitude, honor is not; without honor, simplicity is not; without simplicity, coexistence is not; without coexistence, continuity of gratitude is not. The human being who lives with gratitude: only his conduct is educative and inspiring for future generations. This is achievable only in the purview of Humanness. Gratitude is for all those who have been helpful, in whichever way, toward Consciousness Development – Value Education.

1 Understanding Phytopharmaceuticals, Herbal Drugs, and Conventional and Novel Drug Delivery Systems

ABBREVIATIONS

Abbreviation	Description
AFLP	Amplified Fragment Length Polymorphism
AYUSH	Ayurveda, Yoga & Naturopathy, Unani, Siddha, Sowa Rigpa, and Homoeopathy
CE	Capillary Electrophoresis
D&C	Drug and Cosmetics
DSHEA	Dietary Supplement Health and Education Act
EMEA	European Medicines Agency
GACP	Good Agricultural and Collection Practices
GC	Gas Chromatography
GMP	Good Manufacturing Practice
HMPs	Herbal Medication Products
HPLC	High-Performance Liquid Chromatography
HPTLC	High-Performance Thin Layer Chromatography
IR	Infrared Spectroscopy
ISSR	Inter-Simple Sequence Repeats
LC	Liquid-Crystal Systems
MAO	Monoamino oxidase
matK	Maturase K
MS	Mass Spectrometry
PSLCs	Precursors for Liquid Crystals
RAPD	Random Amplified Polymorphic DNA
rbcL	Ribulose-1,5-biphosphate carboxylase
RCTs	Randomized Clinical Trials
SARS-CoV-2	Severe Acute Respiratory Syndrome
SLNs	Solid Lipid Nanoparticles
SSR	Simple Sequence Repeats
TLC	Thin-Layer Chromatography
UNICEF	United Nations Children's Fund
UV/Vis	Ultraviolet/visible spectroscopy
WHO	World Health Organization

DOI: 10.1201/9781003540823-1

1.1 INTRODUCTION

The traditional practices of utilizing plants for nutrition, therapy, and diagnostics have been followed by humans for centuries. Herbal medicines or herbal medicinal products (HMPs) are treatments derived from plants, including their parts or extracts, used to prevent, treat, and mitigate different diseases. They are complex mixtures of active compounds and other natural compounds that work together to provide their effects [1]. Many modern drugs, such as artemisinin and paclitaxel, are derived from plant sources, demonstrating the importance of plants in drug development. To ensure efficacy and safety, instrumental methods like chromatography and mass spectrometry are used to analyze and standardize these plant-based products. This ensures consistent quality in phytopharmaceuticals, making them reliable for clinical use. Additionally, both conventional and novel drug delivery systems (NDDS's) are critical in optimizing the therapeutic effects of phytopharmaceuticals. Conventional forms such as tablets and capsules remain common, while novel systems like nanoparticles, liposomes, and phytosomes improve bioavailability, stability, and targeted delivery of herbal compounds, enhancing their effectiveness. Understanding phytopharmaceuticals and their integration into conventional and NDDS's is essential for advancing modern healthcare and maximizing the benefits of natural medicines.

1.2 DEFINITIONS

1.2.1 Herbal Drugs

Herbal drug or phytomedicine or botanicals are those which are obtained from roots, stems, leaves, bark, fruit, seeds, or flowers of various medicinal plants and have medicinal properties to treat an illness, or to achieve good health. Phytochemically, herbal drugs are multicomponent mixtures. They are single herb preparations or polyherbal formulations.

It includes use of whole or part of medicinal plant, their secretions, excretions, exudations and whole extracts, purified extracts, enriched fractions, formulations, and so forth. As per World Health Organization (WHO), herbal drugs include herbs, herbal materials, herbal preparations, and finished herbal products that contain phytochemicals as active ingredients, parts of plants, or other plant materials, or combinations.

Herbs: Crude plant material such as leaves, flowers, fruit, seed, stems, wood, bark, roots, rhizomes or other plant parts, which may be entire, fragmented, or powdered.

Herbal materials: In addition to herbs, fresh juices, gums, fixed oils, essential oils, resins, and dry powders of herbs. In some countries, these materials may be processed by various local procedures, such as steaming, roasting, or stir-baking with honey, alcoholic beverages, or other materials.

Herbal preparations: The basis for finished herbal products and may include comminuted or powdered herbal materials, or extracts, tinctures, and fatty oils of herbal materials. They are produced by extraction, fractionation, purification, concentration, or other physical or biological processes. They also include preparations made by steeping or heating herbal materials in alcoholic beverages and/or honey, or in other materials.

Finished herbal products: Herbal preparations made from one or more herbs. If more than one herb is used, the term "mixture herbal product" can also be used. Finished herbal products and mixture herbal products may contain excipients in addition to the active ingredients. However, finished products or mixture products to which chemically defined active substances have been added, including synthetic compounds and/or isolated constituents from herbal materials, are not considered to be herbal.

Herbal drugs can be broadly categorized into the following four main groups:

- **Indigenous herbal medicines (HMs):** These are well established within local communities and known for their composition, treatment efficacy, and dosage due to their long history of use.
- **HMs in systems:** This category includes traditional practices like Ayurveda, Unani, and Siddha, which have been utilized for extensive periods. Efficacy assessments for local use are often not required for these medicines.
- **Modified HMs:** These represent adaptations of indigenous or system-based herbal medicines, altering their form, dosage, mode of administration, ingredients, preparation methods, and medical indications. They must comply with national regulatory safety and efficacy standards, with preclinical and clinical data requirements varying based on the modifications made.
- **Imported HMs:** This includes both raw materials and finished products, which must be registered in their countries of origin. Safety and efficacy data need to be submitted to the importing country's national authority, and adherence to Good Manufacturing Practices (GMPs) is critical for these products.

1.2.2 Herbal Medication Products

Herbal medication products (HMPs), also known as herbal remedies or botanical medicines, are medicinal products derived from plants or plant extracts derived from different parts of the plant, including leaves, roots, stems, flowers, seeds, and fruits [2]. They can be made from combinations of various plants (formulas) or single plant species (monographs). They have been used for centuries in various traditional healing systems around the world and continue to be popular due to their perceived natural origins and potential health benefits. The preparation of the HMPs can be done by taking a part of the plant or the whole herb made into infusions, decoctions, capsules, tinctures, ointments, and creams for traditional use purposes [3].

1.2.3 Traditional Medicines

Traditional medicine includes the collective knowledge, skills, and practices developed over generations within various cultures. Rooted in indigenous theories, beliefs, and experiences – whether scientifically explicable or not – it plays a significant role in maintaining health, preventing diseases, and diagnosing, treating, or improving physical and mental conditions. This system reflects the cultural heritage and holistic approaches of communities worldwide, addressing health through natural remedies, spiritual practices, and traditional therapies [4].

1.2.4 Ayurveda, Siddha, Unani Drugs, or ASU Drugs

It includes all medicines, for internal or external use, aimed at diagnosing, treating, preventing, or mitigating diseases in humans or animals, and are prepared strictly following the formulations outlined in the authoritative texts of these traditional systems, as listed in the First Schedule [4].

1.2.5 Phytopharmaceutical Drugs

As defined by the Drug and Cosmetic (D&C) Act of 1940, they are purified and standardized extracts from medicinal plants containing at least four identified bioactive or phytochemical compounds (qualitatively and quantitatively assessed). They are intended for internal or external use in humans or animals for diagnosing, treating, preventing, or mitigating diseases, but they exclude parenteral administration [4].

1.2.6 Chemical Markers

Marker substances are chemically defined constituents of herbal drugs, herbal drug preparations, and HMPs [4]. Chemical markers generally refer to phytoconstituents, including primary and secondary metabolites and other macromolecules. The chemical markers are used for (i) authentication of genuine species & identification of adulterants, (ii) differentiation of herbal medicines with multiple sources, (iii) consideration of the best harvesting time, (iv) identification of best collection sites, (v) selection of herbal part, (vi) identification and quantitative evaluation of proprietary products, (vii) stability testing and shelf life determination of proprietary products, (viii) assessment of optimum processing methods, quality control of herbal drugs, (ix) diagnosis of herbal intoxication, and (x) lead compounds development for drug discovery.

Chemical markers are further classified into active markers and analytical markers [5].

- Active markers or active principles: These are the phytoconstituents or groups of constituents responsible for therapeutic property.
- Analytical markers: These are the constituents or groups of constituents that serve solely for analytical quality control purposes.

1.3 THERAPEUTICALLY SIGNIFICANT BIOACTIVE COMPOUNDS IN HERBAL DRUGS

Plants with high medicinal value are rich in different bioactive compounds, including classes of carbohydrates, flavonoids, steroids, tannins, alkaloids, and terpenoids, which are very much necessary in promoting human health. These bioactive compounds are synthesized through both primary and secondary metabolic pathways in plants. The secondary metabolites, in particular, are essential for maintaining health and managing diseases. A thorough understanding of the chemical profiles, pharmacological effects, safety, dosing, and possible interactions of these bioactive compounds is required for the development of safe and effective therapeutic interventions. The following points highlight the significance of secondary metabolites.

- **Flavonoids**: A major class of phenolic secondary metabolites widely found in fruits, vegetables, mushrooms, etc. They are generally low in toxicity and classified into flavanones, flavan-3-ols, anthocyanins, flavones, isoflavones, and flavonols. Flavonoids, along with phenolic acids, play a key role in managing chronic and degenerative diseases, aid in wound healing, protect against inflammatory skin conditions, and act as prebiotics by regulating intestinal microflora, although only a small fraction of dietary polyphenols is absorbed in the small intestine [6, 7].
- **Terpenoids**: Naturally derived from mevalonic acid and composed of isoprene units, terpenoids are classified into monoterpenes, diterpenes, triterpenes, sesquiterpenes, pentaterpenes, and polyterpenes. They are widely distributed in nature, commonly present in medicinal plants as volatile oils. Terpenoids play vital roles in plant growth and development, while therapeutically they downregulate cytokine expression, reduce tissue damage, and exhibit antioxidant activity, contributing to their effectiveness against infections [8].
- **Quinones**: Secondary metabolites are classified into benzoquinones, naphthoquinones, anthraquinones, and phenanthrenequinones. They are known to inhibit inflammatory mediators, suppress cytokine expression, and reduce oxidative stress. Due to these

properties, quinones demonstrate therapeutic potential in controlling inflammation and related disorders, highlighting their importance as bioactive compounds.

- **Alkaloids**: Plant-derived nitrogen-containing compounds formed from amino acids, typically with a heterocyclic ring structure. While toxic at high doses, they possess significant therapeutic activity at low doses. Morphine, the first alkaloid isolated from the poppy plant, is a classic example, widely used for analgesia and sleep disorders. Alkaloids also provide plants with defense against herbivores and pathogens. Their diverse biological actions make them valuable in pharmaceuticals, particularly as central nervous system stimulants, analgesics, and anesthetics [7, 9].
- **Saponins and glycosides**: Saponins are glycosides represented by their soap-forming ability and belong to the steroidal or triterpenoid class. Saponins are broadly classified into triterpenoid saponin, steroid saponin, and alkaloid saponin [10]. Hydrolysis of glycoside molecule produces two portions: aglycone (genin) and a glycone (sugar moiety). According to the types of glycosidic linkages, they are classified into O-glycosides, C-glycosides, S-glycosides, and N-glycosides [11]. They possess anti-inflammatory, anticancer, antimicrobial, immune-modulatory, and cholesterol-lowering activity [12].

1.4 MAJOR CHALLENGES WITH HERBAL DRUGS

Herbal drugs are often faced with multiple challenges that can affects their efficacy and potency, which are often caused by a lack of a fixed standard for preparation, standardization, and quality control. Sometimes, safety becomes a concern due to contamination or mixing with other substances. Since there aren't enough scientific studies, and rules differ across regions and countries, it's hard to fully believe and use them in modern healthcare. The following points outline the key challenges associated with herbal drugs, their preparation, and the final products.

- **Quality control challenges**: Variability in raw materials due to genetics, ecology, and cultivation practices complicates quality assurance. Issues include adulteration (intentional or unintentional), contamination with heavy metals or synthetic drugs, and substitution with inferior materials, all of which compromise safety and trust [13].
- **Stability challenges**: Herbal drugs are prone to physical, environmental, and chemical instability. Factors like microbial growth, poor storage, temperature, humidity, and oxidative or enzymatic degradation reduce efficacy and shelf life, while complex multi-component formulations complicate stability testing [14].
- **Monitoring challenges**: Safety assessment is difficult due to misidentification of plant species, contamination, overdose, and interactions with prescription drugs. Multiple common names and improper taxonomy increase confusion; for instance, *Artemisia absinthium* has several common names that differ widely from its scientific name. Effective monitoring requires collaboration among experts and standardized procedures [15].
- **Herb–drug interactions**: Herbal medicines may interact with allopathic drugs, especially those with narrow therapeutic windows. For example, garlic and ginger can enhance bleeding risk in patients taking warfarin [12].
- **Clinical research challenges**: Conducting large-scale trials faces various challenges, including ethical, economic, and methodological. Blinding is difficult due to herbal characteristics (e.g., odor, holistic treatments), while appropriate controls, sample size calculation, and therapist consistency remain key obstacles [12].
- **Regulatory and safety challenges**: Regulations differ widely across countries, with many herbs marketed as supplements requiring minimal safety evidence. Frameworks like the US DSHEA (1994) and India's AYUSH system highlight variability. Stronger monitoring, quality assurance, and harmonized global standards are needed to ensure safety and prevent adulteration [12].

1.5 QUALITY CONTROL OF HERBAL DRUGS

With environmental and agricultural practices significantly influencing the quality of herbal materials, quality control of herbal medicines is crucial for ensuring their safety and efficacy [16].

1.5.1 Identification

Proper identification and authentication of botanical raw materials are fundamental to ensuring product quality. This involves botanical examination, morphological and microscopic analysis, and chemical fingerprinting to confirm the identity of plant species and detect any adulteration [3].

1.5.2 Physicochemical Parameters

This test involves evaluating the product's physical and chemical properties. These include assessing water content to prevent spoilage, determining inorganic impurities through ash values, and checking for foreign matter. For finished herbal preparations, additional tests like disintegration, dissolution, hardness, and dosage uniformity are performed to guarantee consistent performance. Safety is further verified by testing for microbial limits and mycotoxins.

1.5.3 Chromatographic and Spectroscopic Analysis

Advanced chromatographic and spectroscopic techniques are crucial for establishing a detailed chemical profile of the product. Methods like thin layer chromatography (TLC), high-performance liquid chromatography (HPLC), and gas chromatography (GC) are used to separate, identify, and quantify active ingredients or chemical markers. This multi-technique approach, known as "phytoequivalence," helps match the product's profile to a clinically proven reference, ensuring consistency despite the natural variability of herbal sources. These modern analytical methods, often combined with spectroscopic tools like ultraviolet-visible spectroscopy (UV/VIS), infrared spectroscopy (IR), and mass spectroscopy (MS) are essential for thorough phytochemical analysis and for quantifying specific components, such as essential oils via steam distillation, to guarantee both efficacy and safety.

1.5.4 Safety Assessment of Herbal Drugs

Plants contain various chemical compounds, some of which may be toxic. Therefore, phytopharmaceuticals derived from plants are not essentially safe. For instance, compounds such as digitalis, anti-cancer agents, and pyrrolizidine alkaloids can exhibit high toxicity. Therefore, it is important to conduct thorough safety assessments, including toxicity testing, to identify potential harmful effects, drug interactions, and side effects. These adverse effects can range from mild to severe, necessitating stringent safety evaluations by regulatory authorities. Toxicity studies, clinical trials with placebos, and post-marketing surveillance play a significant role in determining the safety of these compounds. Safety considerations include factors such as dosage, exposure duration, molecular reactivity, and patient-specific factors like coexisting medical conditions [17, 18].

1.5.5 Efficacy Assessment of Herbal Drugs

The efficacy of phytopharmaceuticals is determined by their ability to produce therapeutic effects, influenced by the concentration of bioactive compounds and their activity. Rigorous studies, including preclinical and clinical trials are essential to establish optimal dosages and administration routes. In vitro studies assess biological effects under controlled conditions, such as curcumin's antioxidant and anti-inflammatory properties. In vivo studies, using animal models and human trials, examine therapeutic efficacy, pharmacokinetics, and safety [17, 18].

1.6 STANDARDIZATION OF HERBAL DRUGS

Herbal drug evaluation is crucial for confirming the identity, quality, and purity of herbal medicines, and addressing issues like adulteration, deterioration, and variability. It involves several methods, which ensure the standardization, quality, efficacy, and safety, relying on two primary aspects: purity and content/assay of herbal drugs [18]. Standardization is needed to (i) provide guarantees on safety, purity, potency and efficacy, (ii) maintain sufficient quality profile, (iii) maintain batch to batch consistency and dosing uniformity, (iv) avoid the use of adulterated or substituted drugs for therapeutic use [19].

Standardization of herbal medicines is a tool in the quality control process, which involves prescribing a set of standards or inherent characteristics, constant parameters, and definitive qualitative and quantitative values that carry an assurance of safety, efficacy, quality, and reproducibility. It is the process of developing and agreeing upon technical standards.

According to WHO, standardization and quality control of herbals is a process involved in the physicochemical evaluation of crude drug covering aspects, such as selection and handling of crude material, safety, efficacy and stability assessment of finished product, documentation of safety and risk based on experience, provision of product information to consumer, and product promotion.

As per the American Herbal Product Association, "standardization" represents a body of information and control necessary to product material of reasonable consistency. This achieved through minimizing the inherent variation of natural product composition through quality assurance practices applied to agricultural and manufacturing processes.

- **Organoleptic evaluation:** Sensory characteristics like color, taste, and texture; microscopic evaluation analyses cellular structures are evaluated to identify plant materials [20].
- **Quantitative analysis:** It involves the quantification of specific active compounds or marker compounds in herbal extracts using analytical techniques such as HPLC, GC, or MS, hyphenated techniques like liquid chromatography-mass spectrometry (LC-MS), gas chromatography-mass spectrometry (GC-MS) [3].
- **Biological assays**: Assessing the biological activity or pharmacological effects of herbal medicines through *in vitro* or *in vivo* assays to determine their potency and efficacy [3]. It also involves microbiological (total viable count, pathogen determination, aflatoxins, etc.) and toxicological evaluations (heavy metal determination, estimation of pesticides, radioactive contaminants, etc.) [19].
- **Chemical fingerprints:** Creating characteristic chromatographic fingerprints of herbal extracts using techniques such as high-performance thin-layer chromatography (HPTLC) or HPLC with diode array detection (HPLC-DAD). These fingerprints serve as reference profiles for batch-to-batch consistency and authentication [3, 21].
- **Physical evaluation:** It involves the estimation of moisture content, ash values (total ash value, acid-insoluble ash, and water-soluble ash), extractive values, and foreign matter (harmful and poisonous foreign matter, insects or molds, stones, sand, microbial contaminants and chemical residues). The moisture content of the herbal crude drug should be minimized in order to prevent degradation of drug either due to chemical modification or microbial contamination. Ash value is used to examine the quality and purity of the herbal drug and to establish its identity. Determining the total extractable content of herbal materials to ensure that the extraction process efficiently extracts the desired constituents from the plant material [3]. Foreign matters are evaluated by microscopic and macroscopic techniques [19].

1.7 STANDARDIZATION FOR POLYHERBAL FORMULATIONS

Polyherbal formulation standardization is crucial to ensure the quality, safety, and consistency of formulation, as these are combinations of more than one herb to achieve the desired therapeutic effect, through scientific assessment of organoleptic, physical, and physicochemical properties. Techniques like TLC and HPTLC are used to create fingerprint profiles and validate formulations such as Artrex® (arthritis treatment), Madhumehari Churna, which contains eight herbal drugs (antidiabetics), and Dashamularishta (postpartum care). Minimizing batch-to-batch variation and providing robust identity, purity, and strength verification, standardization helps transform traditional herbal medicines from ethnobotanical practices to scientifically validated therapeutic interventions, addressing the current challenge that many polyherbal formulations lack, that is, comprehensive standardization [18, 22].

1.8 QUALITY ASSESSMENT OF HERBAL DRUGS

1.8.1 Identification and Authentication of Herbal Drugs

Authenticating medicinal plants can be challenging, especially if their purpose is therapeutic. In the drug discovery process, the authentication of plant material is a critical stage since the plant's identity is essential for determining patient safety, as incorrect identification of the plant might even result in patient mortality. Additionally, the widespread usage of medicinal plants has increased demand for them, ultimately leading to the adulteration of plant material to satisfy the high demand. Previously, authentication of the plant was carried out based on the morphological attributes, that is, confirming the identity of the plant through the color, shape, appearance, aroma, or taste. Then, authentication advanced to analyzing the form and contents of plant cells under a microscope. Although these organoleptic and anatomical methods provided useful insights, they often lacked precision. Therefore, herb authentication further relied heavily on analytical methods like liquid chromatography, HPLC, and TLC. But the overall identity of medicinal plants cannot be confirmed by taxonomical authentication and chromatographic methods alone; that's why molecular methods are required to verify the plant's authenticity [23, 24]. Molecular markers are biological elements, such as primary and secondary plant metabolites, proteins, DNA, and other macromolecules. The limitations of conventional authentication techniques are addressed by these markers using sophisticated techniques such as DNA barcoding, which uses gene regions like matK and rbcL, Amplified fragment length polymorphism (AFLP) for high sensitivity, random amplified polymorphic DNA (RAPD) for genetic variation detection, simple sequence repeats (SSR) for polymorphism analysis, and inter-simple sequence repeats (ISSR) for targeting regions between microsatellites [24]. These techniques guarantee the quality and authenticity of herbal drugs and their products [25, 26].

1.8.2 Phytochemical Composition/Marker-Based Analysis of Herbal Drugs

HMPs contain a mixture of constituents or compounds, which causes a great challenge in marker-based standardization of plants [27]. Various markers, including taxonomic, chemical, genomic, and proteomic, aid in identifying herbal drug components. Common identification techniques include macroscopic, microscopic, and chemical analyses like TLC, HPLC, LC/MS, and protein analysis [24, 28]. HMP quality control is frequently achieved through the use of chemical marker-based techniques [27]. Chemical markers are specific constituents or groups in herbal medicinal products essential for quality control, regardless of their therapeutic properties. They are categorized as analytical markers, which serve analytical purposes, and active markers, which contribute to therapeutic effects. In cases where no specific active constituent can be identified, the percentage of extractable matter with a solvent may be used. The European Medicines Agency (EMEA) emphasizes the importance of chemical markers in maintaining the quality of herbal medicines through methods like chemical fingerprinting, which creates distinct patterns of various markers. Although markers are associated with therapeutic effects, the absence of pharmacological data for certain compounds frequently necessitates the use of alternative markers [29]. The quality of herbal products can be influenced by factors such as seasonal variations, harvesting time, cultivation sites, and processing methods [30]. Marker-based standardization using advanced chromatographic techniques, such as HPTLC and HPLC, has gained attraction. Secondary metabolites are widely used for standardization, and DNA-based markers help identify variations within and between species [24, 29].

1.8.3 Chromatographic Fingerprinting Techniques

Chromatographic techniques, particularly hyphenated with spectroscopic methods like HPLC-DAD, GC-MS, and LC-MS, have become essential in the quality control and research of herbal products. These methods provide precise separation, detection, and identification of both known and unknown compounds in complex herbal formulations. TLC is widely used for initial screening due to its simplicity and cost effectiveness, while HPTLC improves this with automation and selective detection. GC, ideal for volatile compounds like essential oils, offers characteristic "fingerprints" for plant identification, and when coupled with mass spectrometry, GC-MS, provides a powerful tool for detailed qualitative and quantitative analysis [16, 24]. HPLC is versatile for separating polar and non-polar compounds, enhancing sensitivity when paired with UV, MS, or ELSD detectors. Capillary electrophoresis (CE) is effective in separating charged molecules for chemical fingerprinting of herbal extracts, though it faces reproducibility challenges [31].

Advanced hyphenated techniques such as liquid chromatography-infrared spectroscopy (LC-IR), LC-MS, and liquid chromatography-nuclear magnetic resonance (LC-NMR) combine liquid chromatography with infrared, mass, or nuclear magnetic resonance spectroscopy for enhanced structural identification and trace-level analysis of complex samples. LC-MS, using electrospray ionization (ESI) and atmospheric pressure chemical ionization (APCI), is highly sensitive, making it essential for bioactive compound detection. GC-MS remains one of the most sensitive methods for analyzing volatile compounds, providing detailed chemical fingerprints. HPLC-DAD allows for qualitative assessment of complex herbal samples, while HPLC-MS facilitates rapid phytochemical identification. CE, particularly when hyphenated with MS or NMR, offers high-sensitivity and fully automated methods, improving the efficiency and accuracy of herbal medicine analysis [21].

1.9 PHYTOPHARMACEUTICALS

Phyto-pharmaceuticals, derived from the Greek words "**phytón**" (plant) and "**pharmakon**" (medicine), are plant-based remedies made from dried parts like leaves, flowers, bark, or roots. Modern phytopharmaceuticals are optimized extracts with controlled drug-extract ratios and precise extraction methods, resulting in pharmacologically active substances [17]. Supported by clinical trials and scientific evidence, they are increasingly integrated into medical guidelines due to their favorable safety profiles and efficacy. Compared to synthetic drugs, they offer better metabolic compatibility, fewer side effects, and more efficient absorption. However, interactions with other medications are still possible, though their standardized forms enhance consistency and ease of use in modern drug delivery systems [7, 32].

The Ministry of Health and Family Welfare (Department of Health and Family Welfare), Government of India, New Delhi, vide the Gazette Notification G.S.R. 918(E) on 30 November 2015, and GSR 227(E) (2019), under the CDSCO, has introduced phytopharmaceutical drugs in rule 2 of the Drugs and Cosmetics Rules, 1945 clause (eb). "'Phytopharmaceutical drug' includes purified and standardised fraction with defined minimum four bio-active or phyto-chemical compounds (qualitatively and quantitatively assessed) of an extract of a medicinal plant or its part, for internal or external use of human beings or animals for diagnosis, treatment, mitigation or prevention of any disease or disorder but does not include administration by parenteral route" [4].

It is guided by separate regulations by Central Drugs Standard Control Organisation (CDSCO). This gazette notification defines regulatory provisions for phytopharmaceuticals and regulatory submission requirements for scientific data on quality, safety, and efficacy to evaluate and permit marketing for an herbal drug on similar lines to synthetic, chemical moieties.

1.9.1 Obstacles to Phytopharmaceutical Delivery

The effective delivery of phytopharmaceuticals requires addressing challenges related to consistent dosing, patient compatibility, stability, and maximizing therapeutic outcomes. This involves incorporating herbal drugs into well-designed formulations using both traditional and innovative methods.

- **Pharmaceutical challenges**: While developing phytopharmaceuticals, issues like poor solubility in both water and lipids and reduced bioavailability arise. Additional challenges include formulation complexity, pre-formulation issues, shelf life determination, instability, and pharmacokinetic variability. Effective strategies include solubility enhancement, stability testing, precise formulation, and pharmacokinetic optimization [17].
- **Pre-formulation challenges**: Herbal drugs face poor stability when exposed to light, heat, and moisture, and their low solubility often reduces bioavailability. Their complex composition with multiple active constituents makes individual analysis, standardization, and reproducibility challenging. Furthermore, characterizing parameters such as dissociation constant (pKa), partition coefficient, and dissolution is difficult, frequently resulting in batch-to-batch inconsistencies [17].
- **Challenges with formulation development**: Herbal dosage forms include liquids, solids, and semi-solids, but extracts pose problems due to hygroscopicity, viscosity, and poor flowability. They often absorb moisture, leading to stickiness and poor physico-mechanical properties. Selecting suitable solvents and excipients adds further difficulty. Regulatory and scientific hurdles, such as standardization, toxicity data, and pharmacovigilance, complicate development [17].
- **Challenges with bioavailability**: Poor solubility in water or lipids limits absorption and therapeutic effects. Larger molecular size and complex compounds face difficulty in crossing cell membranes. Techniques like micronization, nanosizing, complexation, and lipid-based systems improve solubility. Prodrugs and novel delivery approaches also enhance bioavailability [17].
- **Challenges with pharmacokinetics**: Phytochemicals show complex absorption, distribution, metabolism, and excretion (ADME) profiles. Limited data on pharmacokinetics, pharmacodynamics, and metabolomics reduces regulatory acceptance. The WHO highlights insufficient information on Indian medicinal plants, hindering global recognition and safe clinical use [33].
- **Stability challenges**: Herbal drugs are sensitive to temperature, humidity, and light, leading to chemical degradation and potency loss. Physical changes like stickiness or discoloration, and microbial contamination affect quality and safety. Stability testing with analytical tools (HPLC, IR) ensures efficacy. Variations in raw material quality and batch consistency remain major hurdles [14, 34].

1.9.2 Classification and Types of Herbal Formulations

Different types of plant-based medical systems use different forms of herbal preparations. The herbal formulations may be broadly classified as follows:

- **Single-herb formulations:** These consist of extracts or preparations from a single medicinal plant, offering targeted therapeutic benefits.

- **Polyherbal formulations:** These combine multiple herbs to create a synergistic effect, enhancing efficacy and minimizing side effects.

Based on various systems of medicine, the herbal formulations are classified into traditional herbal formulations, conventional herbal formulations, and modern/novel herbal formulations.

1.9.3 Traditional Herbal Preparations

These include age-old remedies like decoctions, infusions, powders, and pastes made using traditional methods/authentic literature. These formulations, rooted in authentic literature, are classified primarily on the basis of their physical state, though the specific types differ across systems [19]. In Ayurveda, formulations are categorized into liquids (*Swarasa, Kwaatha, Sneha, Asava, Arishta*), semisolids (*Kalka, Rasakriya, Lepa, Lehya, Avaleha, Prasha*), and solids (*Churna, Vati, Gutika, Modaka, Bhasma, Anjana*). The Siddha system similarly recognizes liquids (*Surasam, Kudineer, Ennai*), semisolids (*Karkam, Nei, Rasayanam*), and solids (*Chooranam, Mathirai, Chenduram*). The Unani system adopts a broader scheme, classifying remedies into gaseous forms (*Bukhoor, Inkabab, LakhLakh, Shamoom*), liquids (*Ma-ul-Jubn, Sharbat, Joshanda, Araq, Roghan*), semisolids (*Majoon, Khamira, Itrifal, Marham, Laooq*), and solids (*Sufoof, Sunoon, Kohal, Habb, Qurs, Kushta, Hamool*). In Traditional Chinese medicine (TCM), preparations are grouped as liquids (*decoctions, liniments, medicated wines*), semisolids (*pastes, ointments, coating agents, jellies*), and solids (*powders, moxa, pastilles, suppositories, medicated rolls*).

1.9.4 Conventional Herbal Formulations

Conventional herbal formulations are similar to allopathic formulations; however, the active pharmaceutical ingredient (API) is replaced by a multicomponent plant extract or enriched fraction. With the passage of time, people have understood and explored more about the newer excipients and novel techniques for large-scale production. The use of newer excipients and techniques has developed a conventional formulation which further improves the stability, palatability, and performance of the finished product. Herbal medicines are prepared in different types of formulations, categorized into solid, liquid, and semi-solid forms. Some of the solid formulations commonly prepared include herbal powders (Triphala or Ashwagandha powder), tablets and capsules (Brahmi tablets or Turmeric capsules), and traditional pills (Chandraprabha vati). Herbal teas (Tulsi tea) are also widely consumed for their therapeutic and wellness benefits. Also, liquid herbal formulations include syrups (*Tulsi cough syrup)*, decoctions (*Guduchi kashayam)*, juices (*Amla juice)*, herbal oils (*Neem oil)*, and concentrated forms (*Echinacea tincture or Garlic oxymel)*. Semi-solid preparations are mainly for external use, including creams (*Aloe vera cream)*, ointments and balms (*Neem ointment or herbal pain relief balm)*, and liniments (*Mahanarayan oil)* for muscle pain. These varied formulations make herbal remedies versatile for both traditional and modern therapeutic needs. These include tablets, capsules, syrups, or topical formulations, and so forth [35, 36].

1.9.5 Modern/Novel Herbal Formulations

Novel herbal formulations are reported to possess advantages such as improvement of solubility, absorption, bioavailability, protection from gastric degradation, improvement of physical and chemical stability, enhanced tissue distribution, reduction of the required dose, targeted delivery, reduction of side effects, control over the release of incorporated medicament, and enhancement of therapeutic activity over conventional herbal preparations. A wide variety of novel formulations, such as liposomes, solid dispersion, transferosomes, phytosomes, ethosomes, microspheres, polymeric nanoparticles, nanoemulsions, solid lipid nanoparticles (SLNs), nanocapsules, and so forth, have been formulated utilizing plant extracts and bioactive components [19].

1.9.6 Characterization and Evaluation

Herbal drugs and isolated phytoconstituents are characterized for their chemical integrity by NMR, FTIR, single crystal XRD, LC-MS, and GC-MS. Phytopharmaceuticals are characterized by general appearance, size and shape, organoleptic properties (color, odor, taste), yield, and so forth. Physicochemical properties are evaluated by analyzing saturation solubility, melt rheology, surface area, surface free energy, micromeritics properties (flow property, porosity, density, etc.), and moisture content. Thermal properties are evaluated by differential scanning calorimetry (DSC), thermogravimetric analysis (TGA), thermomechanical analysis (TMA), evolved gas analysis (EGA), and hot stage microscopy. Functional group and drug-excipient compatibility are studied by Fourier transform infrared spectroscopy (FT-IR), DSC, and so forth. Crystallinity analysis is carried out by powder X-ray diffraction (PXRD), polarized light microscopy (PLM), dynamic vapor sorption (DVS), DSC, microcalorimetry, selected area electron diffraction (SAED), scanning electron microscopy (SEM), Raman spectroscopy, solid-state nuclear magnetic resonance (ssNMR), and hot stage microscopy. Morphology is evaluated by optical microscopy, scanning electron microscopy (SEM), atomic force microscopy (AFM), and transmission electron microscopy (TEM). Elemental analysis is carried out by inductively coupled plasma-mass spectrometry (ICP-MS), X-ray photoelectron spectroscopy (XPS), and atomic absorption spectroscopy (AAS). Pharmaceutical characterizations, such as phytochemical content, content uniformity, dissolution profile, disintegration time, permeability, oral bioavailability, and stability studies, are carried out to check the quality and performance of the formulation. In addition to these, microbial contamination, pesticide residue, heavy metal estimation, preservative challenge tests, toxicity/safety, and therapeutic efficacy are studied for phytopharmaceuticals.

1.10 REGULATION OF HERBAL DRUGS

The WHO has conducted global surveys on national policies concerning traditional medicine and herbal regulation, revealing that as of 2003, only 37% of member states had established laws governing herbal medicines. Many countries sell these products as over-the-counter (OTC) drugs, with a smaller percentage requiring prescriptions. Key parameters for regulatory development include drug registration systems, pharmacopoeia development, and inclusion in essential medicine lists [29].

In various regions, including Southeast Asia and the Western Pacific, herbal drug regulations are becoming more standardized. For instance, the EMEA has a simplified registration procedure for traditional herbal medicines, while the US allows botanical products to be marketed under specific regulations. In India, the AYUSH department oversees herbal medicine regulation under the Drugs and Cosmetics Act, with many products not requiring clinical trials for safety and efficacy. Latin American countries like Brazil, Argentina, and Mexico have established regulations for herbal products, with requirements for efficacy, safety, and quality documentation. In the Eastern Mediterranean region, where herbal medicines are widely used, regulatory frameworks are still developing, and there is a pressing need for improved quality control and safety monitoring systems. Despite efforts to create global standards for herbal medicines, inconsistencies among different national pharmacopoeias create challenges for manufacturers, as monographs for the same plant can vary significantly [29].

The WHO has established guidelines for the safety and quality assurance of medicinal plants, including "Quality Control Methods for Medicinal Plant Materials" and "Guidelines on Good Agricultural and Collection Practices (GACP)." Regional guidelines from the European Union, China, and Japan also ensure that cultivation practices are free from harmful substances like heavy metals and pesticides [29].

Regulatory frameworks, such as India's Drugs and Cosmetics Act of 1940, govern the production and sale of herbal products, with amendments ensuring the safety and standardization of Ayurveda, Siddha, and Unani medicines. Institutions like the Pharmacopoeial Laboratory of Indian Medicine and various research councils are responsible for setting standards and ensuring the quality of traditional medicines [29].

1.11 REGULATIONS ASSOCIATED WITH PHYTOPHARMACEUTICALS

Phytopharmaceuticals are subject to varying legal classifications across different jurisdictions, making regulation crucial for ensuring their safety and efficacy. In some countries, these products are recognized and regulated, while in others, they are categorized as food, limiting therapeutic claims and potentially putting users at risk due to inadequate oversight. This lack of regulation can lead to harmful consumption patterns, raising public health concerns such as the risk of drug resistance.

Various organizations, including the Codex Alimentarius Commission, WHO, and UNICEF, work diligently to establish processing standards and guidelines for the safe use of phytopharmaceuticals. Additionally, organizations like Médecins Sans Frontières (MSF) contribute significantly to promoting responsible use, particularly in therapeutic food contexts. Governments are increasingly integrating phytopharmaceuticals into their regulatory frameworks, with entities like the US FDA expanding their oversight to include dietary supplements. Similar regulations are also present at regional levels, such as within the EU, SADC, and EAC, ensuring a comprehensive approach to monitoring these products. Regulating phytopharmaceuticals in India is crucial for ensuring their safety, efficacy, and quality. The CDSCO oversees these processes, ensuring that plant-based medicines undergo rigorous safety and efficacy evaluations. This includes toxicity studies, clinical trials, and post-marketing surveillance to monitor adverse effects and interactions. Phytopharmaceutical Reference Standards help ensure the quality and potency of plant compounds, while GMPs enforce strict controls over production and quality to safeguard consumer safety.

1.12 CONCLUSION

The introduction of phytopharmaceuticals and herbal medicines into contemporary therapeutics is a promising horizon in medicine. Traditional herbal medicines have been in use for millennia, but their science-based validation and standardization using phytopharmaceutical technologies are a novel step toward evidence-based medicine. Traditional drug delivery systems, although good, are limited by poor bioavailability, stability, and site-specific delivery. NDDS's like nanoparticles, liposomes, polymeric nanoparticles, and so forth, provide new solutions to these problems by improving the efficacy and safety profiles of both synthetic drugs and plant drugs.

REFERENCES

1. Kuruvilla, A., Herbal formulations as pharmacotherapeutic agents. *Indian Journal of Experimental Biology*, 2002. 40.
2. Kamboj, V.P., Herbal medicine. *Current Science*, 2000. 78(1): p. 35–39.
3. Jain, S.D., et al., WHO guidelines for quality control of herbal medicines: From cultivation to consumption. *Journal of Pharmaceufical and Chemical Analysis*, 2024. 11.
4. Nooreen, Z., V.K. Rai, and N.P. Yadav, Phytopharmaceuticals: A new class of drug in India. *Annals of Phytomedicine*, 2018. 7(1): p. 27–37.
5. Kushwaha, S.K., et al., Role of markers in the standardization of herbal drugs: A review. *Archives of Applied Science Research*, 2010. 2(1): p. 225–229.
6. Papoutsis, K., et al., Fruit, vegetables, and mushrooms for the preparation of extracts with α-amylase and α-glucosidase inhibition properties: A review. *Food Chemistry*, 2021. 338: p. 128119.
7. Afzal, I., U. Habiba, and H. Yasmeen, Review on therapeutic potential of phytochemicals from medicinal plants. *Journal of Bioresource Management*, 2023. 10(4): p. 7.
8. Yang, W., et al., Advances in pharmacological activities of terpenoids. *Natural Product Communications*, 2020. 15(3): p. 1934578X20903555.
9. Gutiérrez-Grijalva, E.P., et al., Plant alkaloids: Structures and bioactive properties. *Plant-Derived Bioactives: Chemistry and Mode of Action*, 2020: p. 85–117.

10. El Aziz, M.M.A., A.S. Ashour, and A.S.G. Melad, A review on saponins from medicinal plants: Chemistry, isolation, and determination. *Nanomedicine Research*, 2019. 8(1): p. 282–288.
11. Bartnik, M. and P. Facey, Glycosides, in *Pharmacognosy*. 2024, Elsevier. p. 103–165.
12. Desai, S.D., D.G. Desai, and H. Kaur, Saponins and their biological activities. *Pharma Times*, 2009. 41(3): p. 13–16.
13. Chandel, H.S., A. Pathak, and M. Tailang, Standardization of some herbal antidiabetic drugs in polyherbal formulation. *Pharmacognosy Research*, 2011. 3(1): p. 49.
14. Thakur, L., et al., Novel approaches for stability improvement in natural medicines. *Pharmacognosy Reviews*, 2011. 5(9): p. 48.
15. Saggar, S., et al., Traditional and herbal medicines: Opportunities and challenges. *Pharmacognosy Research*, 2022. 14(2).
16. Shulammithi, R., et al., Standardization and quality evaluation of herbal drugs. IOSR *Journal of Pharmacy and Biological Sciences*, 2016. 11(5–1): p. 89–100.
17. Pavithra, K. and V. Manimaran, A review of safety, quality, regulation, and delivery approaches for phytopharmaceuticals. *Jordan Journal of Pharmaceutical Sciences*, 2024. 17(2): p. 316–332.
18. Kumari, R. and M. Kotecha, A review on the standardization of herbal medicines. *International Journal of Pharma Sciences and Research*, 2016. 7(2): p. 97–106.
19. Sahu, A.N. and D. Mohapatra, *Herbal Drug Formulation and Standardization*. 2021, New Delhi: Ane Books Pvt. Ltd. ISBN-10: 9390658365, ISBN-13: 978-93-90658-36-7.
20. Kunle, O.F., H.O. Egharevba, and P.O. Ahmadu, Standardization of herbal medicines-A review. *International Journal of Biodiversity and Conservation*, 2012. 4(3): p. 101–112.
21. Olufunke, M.D., Developments in phytochemistry. *Drug Discovery Research in Pharmacognosy*, 2012: p. 12–23.
22. Choudhary, N. and B.S. Sekhon, An overview of advances in the standardization of herbal drugs. *Journal of Pharmaceutical Education and Research*, 2011. 2(2): p. 55.
23. Kiran, U., et al., SCAR markers: A potential tool for authentication of herbal drugs. *Fitoterapia*, 2010. 81(8): p. 969–976.
24. Noviana, E., G. Indrayanto, and A. Rohman, Advances in fingerprint analysis for standardization and quality control of herbal medicines. *Frontiers in Pharmacology*, 2022. 13: p. 853023.
25. Ganie, S.H., et al., Authentication of medicinal plants by DNA markers. *Plant Gene*, 2015. 4: p. 83–99.
26. Tharachand, C., C.I. Selvaraj, and M. Mythili, Molecular markers in characterization of medicinal plants: An overview. *Research in Plant Biology*, 2012. 2(2): p. 01–12.
27. Shi, Z.-Q., et al., Identification of effective combinatorial markers for quality standardization of herbal medicines. *Journal of Chromatography A*, 2014. 1345: p. 78–85.
28. Zhang, T., et al., The method of quality marker research and quality evaluation of traditional Chinese medicine based on drug properties and effect characteristics. *Phytomedicine*, 2018. 44: p. 204–211.
29. Sahoo, N., P. Manchikanti, and S. Dey, Herbal drugs: Standards and regulation. *Fitoterapia*, 2010. 81(6): p. 462–471.
30. Li, S., et al., Chemical markers for the quality control of herbal medicines: An overview. *Chinese Medicine*, 2008. 3: p. 1–16.
31. Yadav, N. and V. Dixit, Recent approaches in herbal drug standardization. *International Journal of Integrative Biology*, 2008. 2(3): p. 195–203.
32. Bhusnure, O.G., et al., Phytopharmaceuticals: An emerging platform for innovation and development of new drugs from botanicals. *Journal of Drug Delivery and Therapeutics*, 2019. 9(3-s): p. 1046–1057.
33. Mehta, P., et al., Pharmacokinetic profile of phytoconstituent (s) isolated from medicinal plants – a comprehensive review. *Journal of Traditional and Complementary Medicine*, 2015. 5(4): p. 207–227.
34. Ali, H., et al., Novel Drug delivery approaches in formulation development; stability considerations and quality features of herbal products. *RADS Journal of Pharmacy and Pharmaceutical Sciences*, 2019. 7(3): p. 155–164.
35. Elkordy, A.A., et al., An overview on natural product drug formulations from conventional medicines to nanomedicines: Past, present and future. *Journal of Drug Delivery Science and Technology*, 2021. 63: p. 102459.
36. Kushare, V.N., et al., Review of herbal formulations. *World Journal of Pharmaceutical Research*, 2023. 12(15): p. 436–446

2 Choosing a Rational Dosage Form for Herbal Drugs

ABBREVIATIONS

Abbreviation	Description
HPLC	High-Performance Liquid Chromatography
GC	Gas Chromatography
MS	Mass Spectrometry
GC-MS	Gas Chromatography-Mass Spectrometry
GACP	Good Agricultural and Collection Practices
GMP	Good Manufacturing Practice
IPR	Intellectual Property Rights
TLC	Thin Layer Chromatography
ISM	Indian System of Medicine
TCM	Traditional Chinese Medicine
AYUSH	Ayurveda, Yoga & Naturopathy, Unani, Siddha, and Homoeopathy
WHO	World Health Organization
FDA	Food and Drug Administration

2.1 UNDERSTANDING HERBAL DRUGS, THEIR ACTIVE CONSTITUENTS, AND CHALLENGES FOR FORMULATION DEVELOPMENT

Herbal drugs are derived from plants and contain a variety of phytochemical substances that can exert therapeutic effects. These include alkaloids, glycosides, terpenoids, flavonoids, and other secondary metabolites. Unlike conventional pharmaceuticals, herbal medicines often consist of a mixture of bioactive compounds, which can vary in concentration based on factors such as plant species, growing conditions, harvest time, and extraction methods.

The challenges of herbal drugs are broadly classified into the following [1]:

- Plant-related challenges
- Pharmaceutical challenges
- Regulatory challenges

Plant-related challenges include the following:

- Phytochemical diversity
- Standardization constraints
- Diversity in the nature and distribution of chemical constituents
- Incorrect authentication
- Quality deterioration due to adulteration or substitution
- Insufficient scientific validation

DOI: 10.1201/9781003540823-2

- Poor quality clinical trials with inadequate reporting
- Non-compliance with good agricultural and collection practices (GACP) and good manufacturing practices (GMPs)
- Presence of extraneous material
- Biopiracy concerns and Intellectual Property Rights (IPR) protection

Pharmaceutical challenges include the following:

- Challenges in pre-formulation study
- Challenges in formulation and development
- Poor aqueous and lipid solubility leads to low bioavailability
- Stability issues
- Pharmacokinetics alterations
- Complexities in characterizations of formulations
- Challenges in assessing shelf life

Regulatory challenges include the following:

- Incorrect classification of products
- Inconsistencies in regulatory requirements

2.1.1 Phytochemical Diversity

Herbal products contain a mixture of multiple phytoconstituents, which makes it challenging to characterize each phytoconstituent and to establish an appropriate standardization method [1].

2.1.2 Standardization Constraints

One of the main difficulties in the formulation of herbal drugs is the standardization of various bioactive phytoconstituents. The standardized extracts are high-quality preparations that contain a consistent amount of various specified phytoconstituents. This can be achieved by implementing appropriate quality control and standardization protocols. However, it is challenging to adhere to all standardization parameters, including marker-based standardization, macroscopy, microscopy, and physicochemical evaluations. Due to the inherent complexity of crude herbal drugs, sophisticated analytical techniques, such as high-performance liquid chromatography (HPLC), gas chromatography (GC), mass spectrometry (MS), or hyphenated techniques such as LC-MS, GC-MS, are required for phytochemical profiling and also for extract standardization before formulation and product development [1].

2.1.3 Diversity in the Nature and Distribution of Chemical Constituents

In contrast to synthetic drugs, which contain well-defined and consistent active ingredients, herbal drugs often exhibit variability in their phytochemical composition. This variability can affect the bioavailability and therapeutic activity of the medicinal herb, therefore highlighting the need to choose a rational dosage form that enhances its therapeutic potential while minimizing the variations in therapeutic outcomes [1].

2.1.4 Incorrect Authentication

Incorrect authentication of unprocessed herbal materials can lead to serious public health issues. The complexity of plant nomenclature often leads to misidentification. A single plant can have diverse names. Similarly, the same name can be referred to different plant species, as well as closely related or entirely unrelated plant species. Moreover, alongside taxonomical authentication, DNA-based molecular authentication can be utilized for the authentication of herbal drugs [1].

2.1.5 Quality Deterioration Due to Adulteration or Substitution

Herbal raw materials are frequently found to be adulterated with parts of other plant materials or different parts of the same plant. In several cases, the finished herbal products may also be adulterated deliberately with synthetic chemical drugs. Such adulteration of the crude raw material may lead to deteriorated finished herbal products posing serious public health issues [1].

2.1.6 Insufficient Scientific Validation

Herbal products are generally assumed to be safe as they are obtained from natural sources. However, intake of herbal products may cause several toxic and adverse effects. The traditional use is considered an alternative or proxy form of evidence for the safety and efficacy of herbal drugs. Hence, in the present era, preclinical and clinical scientific evidence, experiments, and reports are required to guarantee the safety of herbal products [1].

2.1.7 Poor Quality Clinical Trials with Inadequate Reporting

The available data from medicinal plant-based medications are mostly obtained from *in vitro* experiments or animal models, while only a limited portion of the thousands of available plants utilized globally have been verified cautiously through controlled randomized trials. Despite having a historical background of traditional use passed down over generations often cited as supporting validation for their safety and efficacy, extensive evidence-based studies remain limited. Evidence-based studies on safety and efficacy remain inadequate. Therefore, well-planned scientific research is required to generate further evidence on the safe and effective use of the herbal drugs [1].

2.1.8 Non-Compliance with Good Agricultural and Collection Practices and Good Manufacturing Practices

The quality of a herbal product is generally determined by certain aspects such as the quality of the crude raw materials, GACP, and GMPs. Several other factors, including the state of plant material (fresh or dried), plant part collected, plant age, season, duration and method of harvesting, processing temperature, exposure to sunlight, drying condition, packing, transportation and storage of raw material, can largely influence the quality, consequently the therapeutic effectiveness of herbal products. Moreover, during the manufacturing of herbal formulation, various processes, such as drying, milling, and mixing, may produce heat, which can degrade the active phytoconstituents, while also posing a chance of contamination by external chemicals or airborne impurities [1].

2.1.9 Presence of Contaminants

The presence of unwanted materials, such as foreign organic matter, fumigants, pesticides, microorganisms, radioactive substances, and heavy metals, can change the quality of herbal drugs, thus compromising the safety and efficacy of herbal products [1].

2.1.10 Biopiracy and Intellectual Property Rights

Biopiracy is a major issue in the advancement of herbal medicines. Therefore, systematic documentation of traditional knowledge is essential for future research on herbal drugs and herbal products [1].

2.1.11 Challenges in Preformulation Study

The preformulation study involves solubility study (e.g., pH solubility profile, pKa determination, partition coefficient, dissolution, etc.), bulk characterization (e.g., crystallinity and polymorphism, micromeritics properties, hygroscopicity, etc.), and stability analysis (solution stability, solid state stability and drug excipient compatibility). In herbal formulations, API use is often a multicomponent plant extract or an enriched fraction. Performing pre-formulation studies on each single constituent is challenging; therefore, the entire plant extract is usually considered as the API, and as many pre-formulation parameters as possible are evaluated [1].

2.1.12 Challenges in Formulation and Development

The primary challenges of plant extracts are hygroscopic, viscous, and amorphous in nature and possess poor flow ability and physico-mechanical properties. The viscous and hygroscopic nature of herbal extracts makes them difficult to handle and to measure accurately. In addition, during formulation development, identifying a suitable common solvent for dissolving both the extract and excipients can be challenging. Also, certain phytoconstituents are thermolabile and may undergo degradation when exposed to high temperatures. Several phytoconstituents also show poor aqueous solubility and low bioavailability, which limits therapeutic effectiveness and complicates dosage form design. Due to the complex nature of the plant extracts, it is difficult to analyze the excipient compatibility [1].

2.1.13 Poor Aqueous and Lipid Solubility Leading to Low Bioavailability

Poor aqueous solubility limits dissolution and absorption, resulting in poor bioavailability. Likewise, poor lipid solubility reduces membrane permeability, leading to poor absorption. Furthermore, the bulky molecular size and structural complexity of many phytoconstituents hinder passive diffusion across biological membranes, thereby lowering systemic absorption and overall bioavailability [1].

2.1.14 Stability Issues

Some specific phytoconstituents tend to become unstable either during storage or within the gastric environment when administered orally. Herbal products are particularly prone to denaturation at the time of storage due to processes such as hydrolysis, oxidation, crystallization, enzymatic degradation, emulsion breakdown, and interactions with excipients. Temperature and moisture are among the key factors that affect their stability and quality. Therefore, special consideration must be given to the proper storage of raw materials and careful handling during processing. Moreover, incorporation of an appropriate formulation strategy can help protect these constituents from gastric degradation and preserve their therapeutic activity [1]. Herbal product stability is also affected by external factors such as soil, climate, harvesting methods, drying, and extraction techniques, as well as genetic variability of plants. Since formulations contain multiple constituents with different shelf lives, predicting overall stability is difficult [2].

2.1.15 Pharmacokinetics Alterations

The bioavailability of plant extract constituents can be influenced by the presence of other phytoconstituents, often referred to as coeffectors. Although these coeffectors may not exhibit direct pharmacological activity, they can enhance the solubility, absorption rate, half-life, and stability of the primary active constituents, thereby improving overall bioavailability. Variations in the chemical composition and proportion of constituents within the extract may alter the pharmacokinetic profile of the main active components. Such changes can either improve therapeutic efficacy or reduce bioavailability, highlighting the need for systematic evaluation in the rational design of safe and effective herbal formulations [1]. In addition, herbal medicines may alter the pharmacokinetics of co-administered drugs by influencing their absorption, metabolism, bioavailability, and half-life. These interactions can lead to therapeutic failure or drug toxicity. Although in vitro studies are useful in identifying potential targets and mechanisms of interaction, differences in extract composition and study design often limit their predictive value for clinical outcomes [3].

2.1.16 Complexities in Characterizations of Formulations

In the case of herbal products, due to the presence of numerous phytoconstituents, it is very difficult to evaluate the finished herbal formulation with respect to each and every component. For this instance, the finished herbal product should be evaluated with respect to known marker compounds (active marker and analytical markers). For example, the drug content should be evaluated by estimating the number of active markers and analytical markers. Furthermore, since herbal products contain multiple constituents, the concentration of any single constituent in the final formulation is very low, and the corresponding plasma concentration is very low, which requires sophisticated analytical techniques during an *in vivo* bioavailability study [1].

2.1.17 Challenges in Assessing Shelf Life

Shelf life refers to the duration for which the active pharmaceutical ingredient (API) in a formulation is expected to remain within acceptable stability limits, provided it is stored under the recommended conditions specified on the product label. At present, stability studies of APIs in allopathic medicines are conducted in accordance with guidelines issued by the International Council for Harmonisation (ICH) and the World Health Organization (WHO). For herbal medicines, stability evaluation follows the AYUSH guidelines. Parameters commonly considered include assay values of chemical markers (both active and analytical), the appearance or disappearance of spots in thin-layer chromatography (TLC), physicochemical factors such as moisture content, ash value, and particle size, as well as physical and functional attributes like color, caking, hardness, and phase separation. While using marker compounds for stability and shelf life determination is an effective strategy, many traditionally used plants lack identified markers. In some cases, markers are reported but without sufficient evidence regarding their bioactivity. Under such circumstances, analytical markers can serve as indicators for predicting shelf life. Furthermore, in polyherbal formulations, each component may have a different shelf life, which complicates the estimation of the overall shelf life of the final product and the recommendation of appropriate storage conditions [1].

2.1.18 Improper Classification of Products

Clear differentiation between herbal food supplements and medicinal products is important to ensure appropriate regulatory control regarding their purity, safety, and efficacy. Proper classification helps determine the extent of regulatory requirements. Depending on national legislation and definitions, herbal products may be regulated either as food, dietary supplements, and nutraceuticals or as medicinal products. Generally, food regulations are less stringent compared to those governing medicines. Products that do not claim to prevent, manage, or treat diseases should fall under food regulations. For example, garlic is considered a food when marketed as a spice, but it is classified as a medicinal product when promoted for its ability to lower blood cholesterol [1]. Improper classification of products can lead to various problems like regulatory gaps, public health risks, compliance issues, and market confusion.

2.1.19 Inconsistencies in Regulatory Requirements

Regulations governing herbal medicines vary across countries, with differing levels of control. For example, in the United States, herbal products are typically classified as dietary supplements and are therefore subject to less rigorous regulation than conventional medicines. In contrast, China treats herbal products as medicinal products, applying the same stringent regulatory standards as those used for conventional therapeutics. To support the global trade and safe use of herbal products, harmonization of regulatory frameworks is essential [1]. Regulatory bodies operate a two-tier control system consisting of pre-market and post-market control. Pre-market control involves the evaluation of herbal products before they are allowed to enter the market, while post-market control ensures ongoing monitoring and assessment once the products are available to consumers. Typically, finished herbal formulations are subject to stricter regulations than raw herbs. In many regions, stricter requirements are applied to pre-market assessments, while post-market surveillance may be less rigorous. This imbalance may lead to shortcomings in ensuring product safety and quality, emphasizing the necessity for uniform regulatory standards across both stages. Hence, maintaining equal regulatory rigor in pre-market and post-market controls is essential [1].

2.2 REQUIREMENT OF HERBAL FORMULATION

Besides providing better patient compatibility, uniform dosing, and safe and convenient delivery, various formulation/dosage forms are needed for additional reasons:

- To protect the drug from the destructive influence of gastric acid after oral ingestion (e.g., enteric-coated tablets).
- To accurately administer multiple drugs at a time (multi-drug formulation).
- To protect the drug substance from the deteriorative effect of humidity or atmospheric oxygen (e.g., coated tablets, sealed ampoules).
- To mask the bitter, salty, or offensive taste or disagreeable odor of a drug substance (e.g., coated tablets, capsules, flavored syrups).
- To provide control over the release of drugs at the administration site (e.g., various controlled-release nanoformulations, capsules, tablets, and suspensions).
- To administer insoluble drugs by suspending them in suitable vehicles (e.g., suspensions).
- To improve the therapeutic efficacy of a poorly soluble or permeable drug by increasing its bioavailability (e.g., Novel drug delivery systems).
- To provide for the insertion of a drug substance into one of the body cavities for local effects (e.g., rectal or vaginal suppositories).
- To target a particular drug to a specific site/organ of the body. (e.g., targeted drug delivery).
- To get optimal therapeutic responses of drugs from topical administration sites by improving the transdermal permeability (e.g., transdermal patches, creams, ointments, ophthalmic, ear, and nasal formulations).
- To produce optimal drug action through inhalation therapy (e.g., aerosols, and inhalers).

2.3 COMMON DOSAGE FORMS FOR HERBAL DRUGS

There are various dosage forms used for herbal drug delivery, each with its unique advantages and limitations. Herbal formulations are dosage forms composed of one or more herbs, intended to treat, prevent, or alleviate various diseases effectively. Herbal materials should be incorporated into appropriate formulations for consistent dosing, better patient compatibility, stability, and maximum therapeutic effect. While herbal drug formulation has ancient roots, modern approaches now influence a deeper understanding of plant constituents. Over time, advancements in excipients and production techniques have enhanced the stability, palatability, and efficacy of herbal products. Modern formulations often incorporate multicomponent plant extracts or enriched fractions rather than isolated constituents to preserve synergy and reduce the complexity of isolation processes [4]. These extracts should be standardized, ideally using active or analytical markers, to ensure consistent therapeutic responses and dosage uniformity across batches. The herbal formulations can be broadly classified into traditional formulations, conventional formulations, and novel formulations.

2.3.1 Traditional Formulation

The existence of diverse formulations of plant material can be understood from various systems of medicine that have been practiced since ancient times all over the world. Out of various traditional systems of medicine, the Indian system of medicine (ISM) and traditional Chinese medicine (TCM) have gained tremendous popularity throughout the world. The classification of various traditional herbal formulations is shown in Figures 2.1–2.5. ISM is the culmination of Indian thought of medicine, which is available in Scriptures, such as Ayurveda, Siddha, Yoga, and Naturopathy, and in combination with other civilizations, such as Greece (Unani Medicine) or Germany (Homoeopathy). Combining all the systems of medicine, they are called Ayurveda,

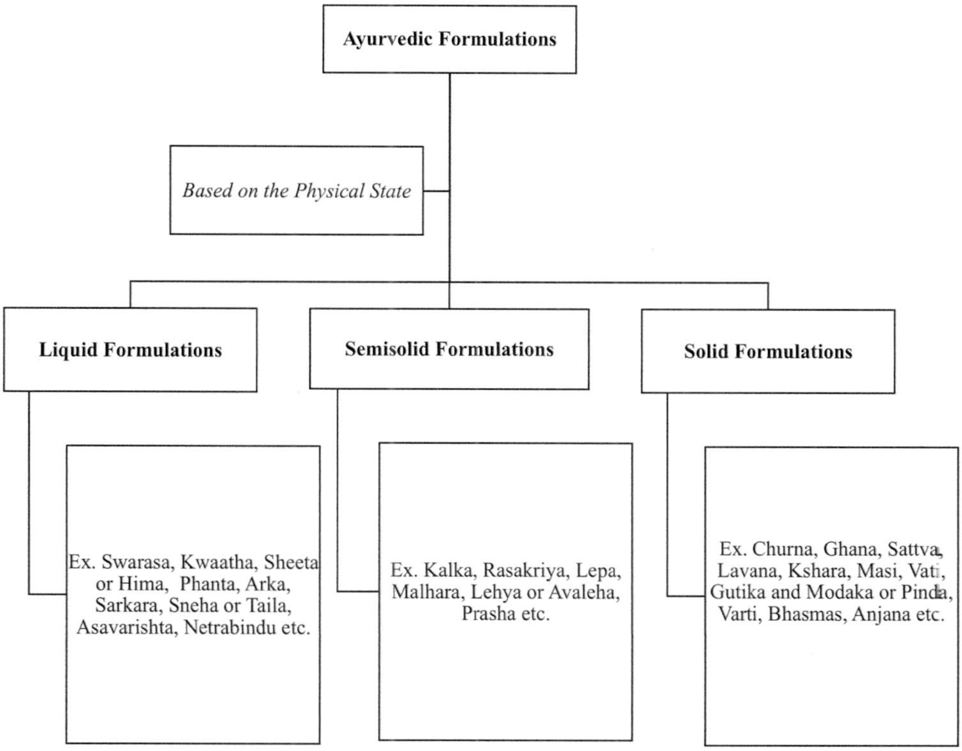

FIGURE 2.1 Major formulations in Ayurvedic system of medicine

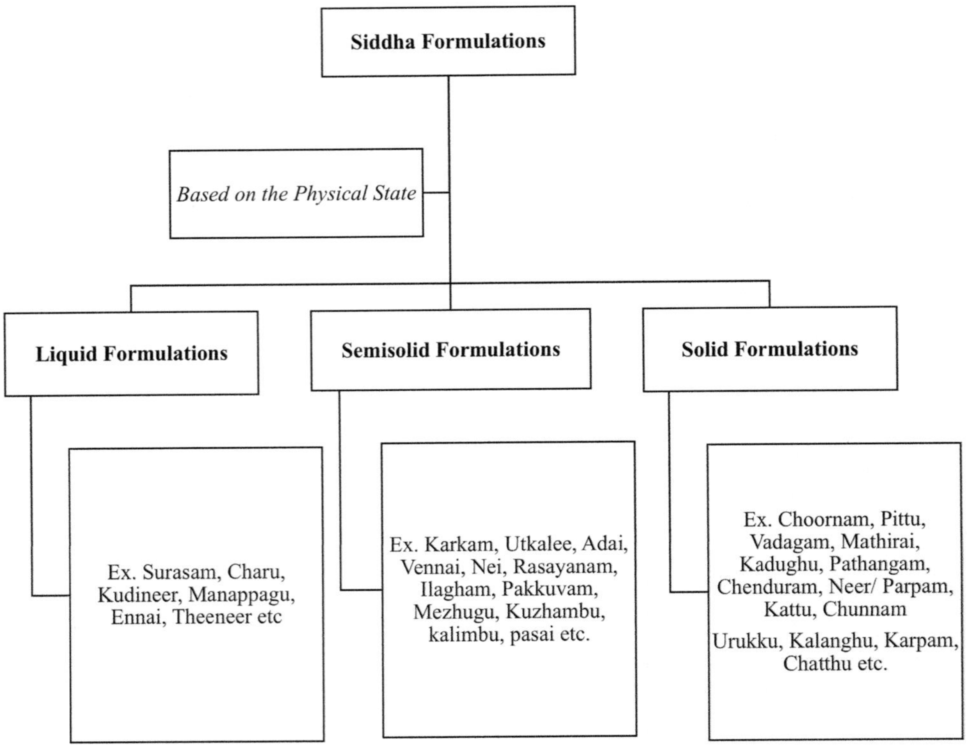

FIGURE 2.2 Major formulations in the Siddha system of medicine

Yoga & Naturopathy, Unani, Siddha, and Homoeopathy (AYUSH) [5–7]. Other Indian traditional medicines include Amchi or Sowa-Rigpa and Folk medicine [7]. As per Rule 3 a of the Drugs and Cosmetics Act, 1940 "Ayurvedic, Siddha or Unani Drug" includes all medicines intended for internal or external use in the diagnosis, treatment, mitigation, or prevention of disease or disorder in human beings or animals and manufactured exclusively in accordance with the formulae described in the authoritative books of Ayurveda, Siddha, Unani systems of medicine, specified in the First Schedule [8].

2.3.2 Conventional Herbal Formulation

Some of the conventional herbal formulations are shown in Figures 2.6 and 2.7 [9, 10]. Formulating plant extracts into conventional dosage forms has provided significant benefits in the fields of pharmaceutical and biomedical sciences. Conventional herbal formulations offer uniform dosing, improved palatability, stability, and improved therapeutic efficacy compared to traditional herbal formulations due to the use of various pharmaceutical excipients. However, such drug delivery often leads to systemic side effects, which can be attributed to inconsistent drug release and non-targeted biodistribution [11]. Further, such formulations can't address the low solubility, absorption, and bioavailability issues of poorly soluble/bioavailable phytocomolecules. Additionally, issues like instability in the body and lack of targeted delivery may lead to unwanted side effects [9]. Incorporating plant extracts/herbal drugs into novel dosage forms can offer improved bioavailability and therapeutic benefits with fewer side effects.

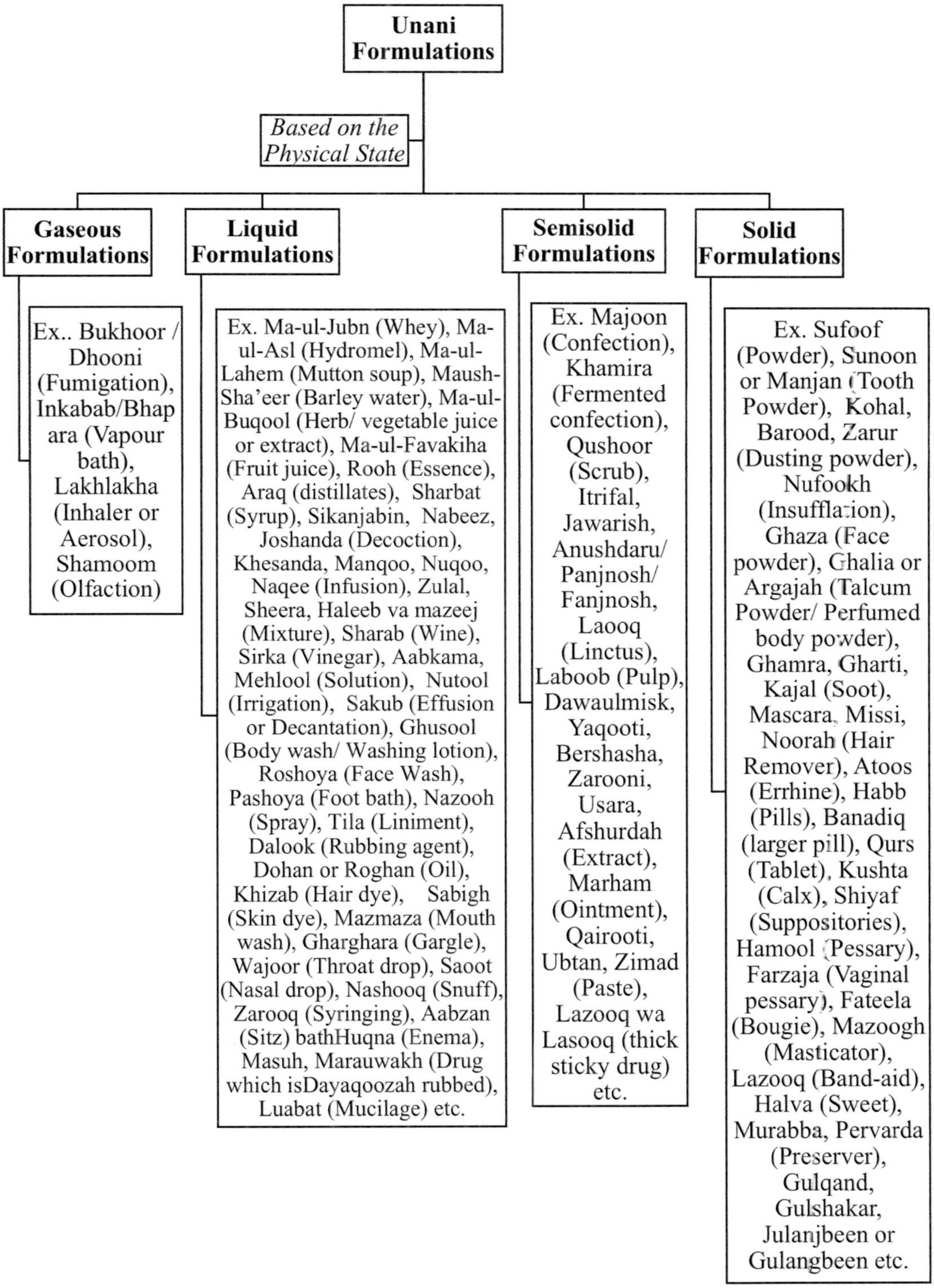

FIGURE 2.3 Major formulations in Unani system of medicine

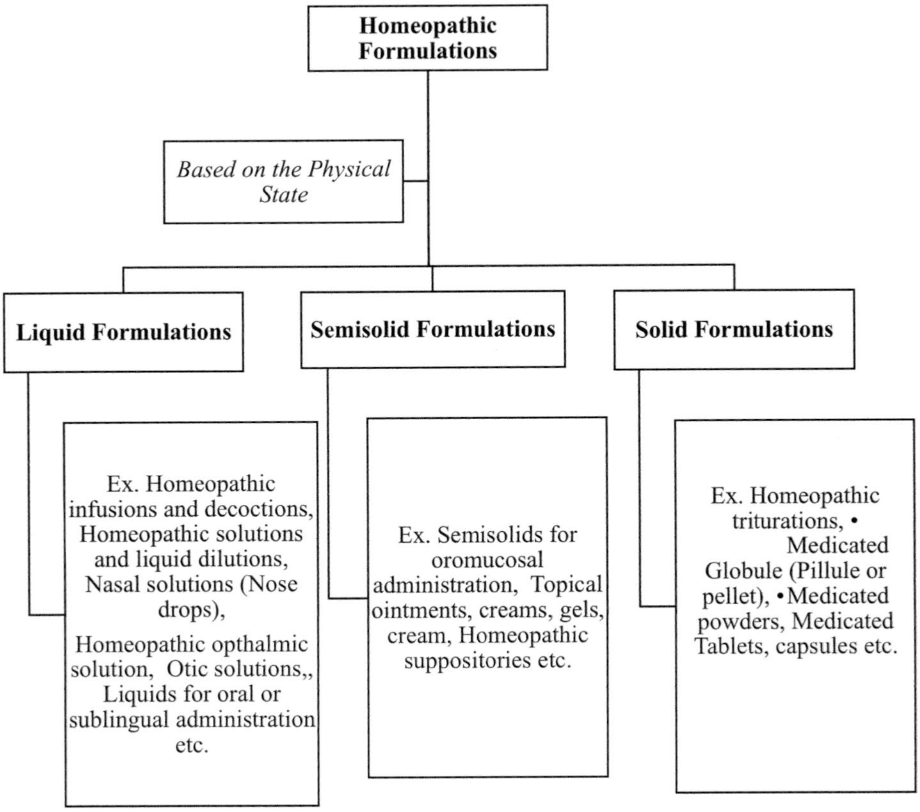

FIGURE 2.4 Major formulations in homeopathic system of medicine

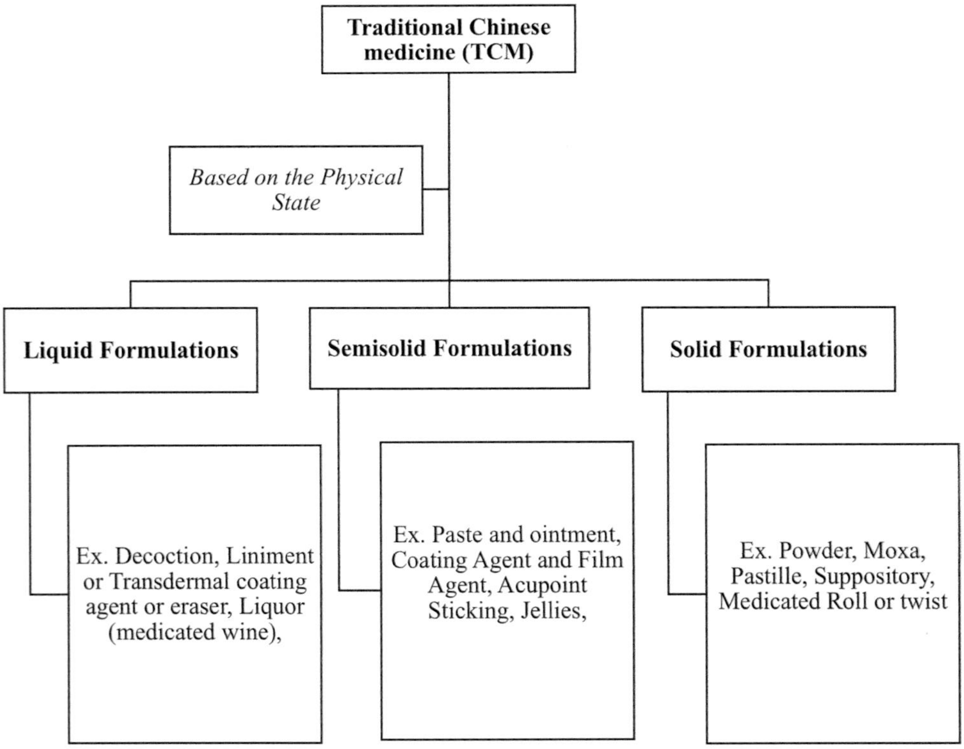

FIGURE 2.5 Major formulations in traditional Chinese medicine

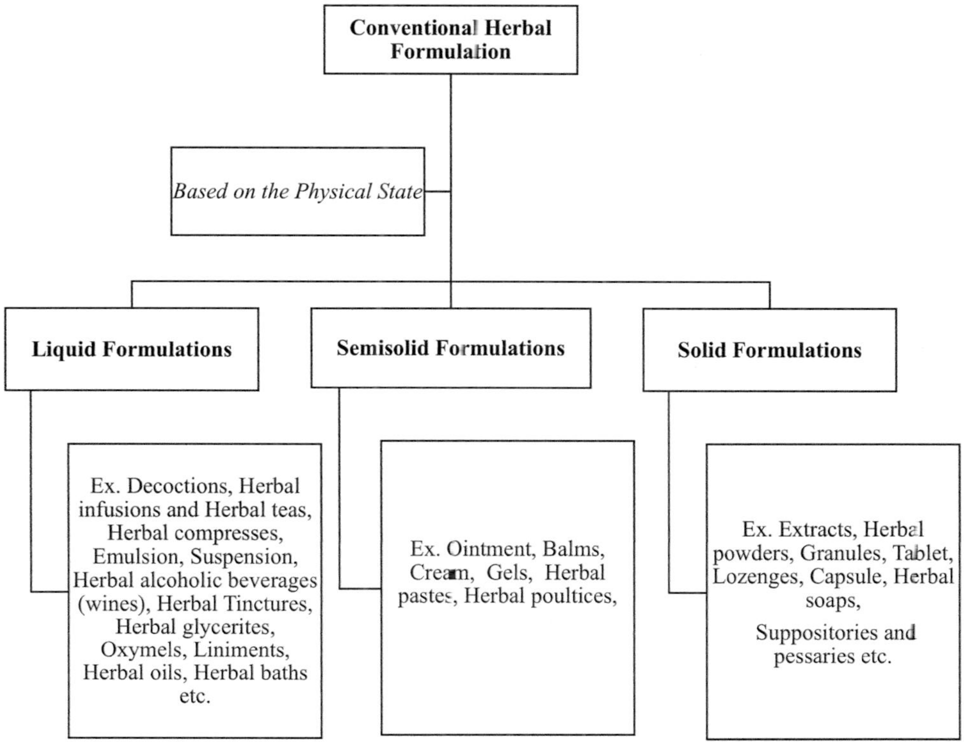

FIGURE 2.6 Various conventional herbal formulations

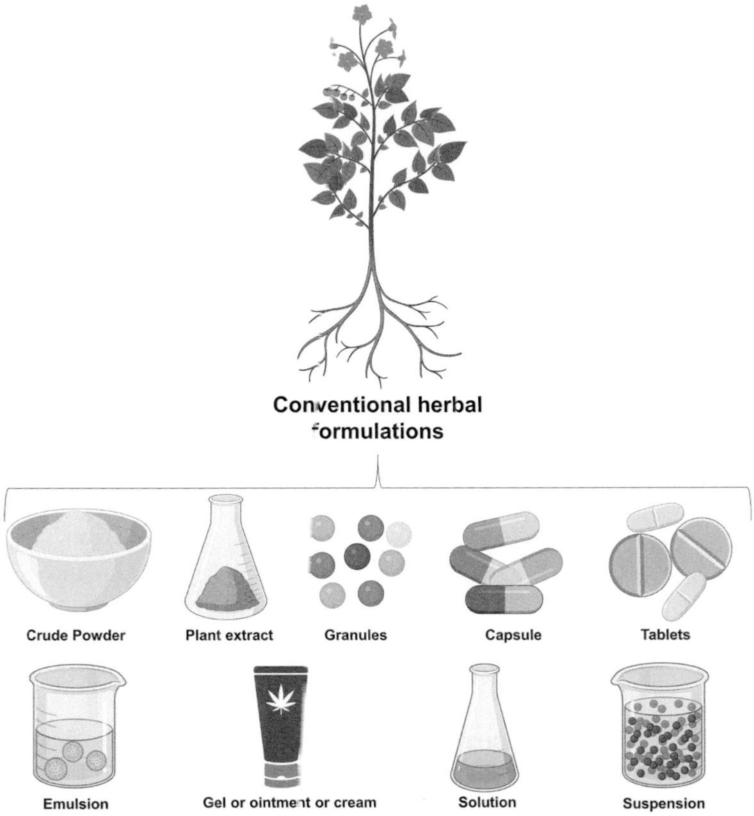

FIGURE 2.7 Conventional herbal formulations

2.3.2.1 Need for Conventional Herbal Formulations

- Herbal formulations allow accurate combinations of extracts to maximize the potency of the bioactive compounds.
- Most of the bioactive compounds have poor solubility, permeability, and absorption. So making formulations can enhance the solubility and facilitate the efficient delivery of these compounds.
- The majority of plant extracts' active ingredients are highly unstable. Degradation occurs when exposed to elements including heat, light, and oxygen. The bioactive is preserved and the shelf life is increased when extracts are formulated into stable dosage forms (such as tablets, capsules, or solutions).
- Plant extracts generally have very bitter and strong flavors with unpleasant odors, which lowers patient compliance. These challenges can be addressed by formulating these extracts using coating, sweetening agents, or encapsulation techniques.
- With the advances in targeted delivery techniques, incorporating the plant extracts in these modern herbal formulations enhances a controlled or sustained release of the active compounds, which leads to a steady therapeutic effect over time.
- Formulating the plant extracts in the form of tablets, capsules, syrups, or creams for topical use makes it easy for administration.
- While most of the plant extracts or products fail to meet the regulatory requirements, formulating them into conventional formulations provides stringent quality control measures, ensuring the safety, efficacy, and uniformity of the product [10].

2.3.3 Novel Drug Delivery Systems of Herbal Drugs

Recent advances in phytochemical and phytopharmacological research have highlighted the composition and biological activities of various medicinal plants. However, many active compounds in these plants, such as phenolics, flavonoids, tannins, and terpenoids, face challenges in drug delivery due to their poor lipid membrane permeability, large molecular sizes, and low absorption rates. As a result, their bioavailability and therapeutic effectiveness are limited.

To address these challenges, novel drug delivery systems (NDDSs) have been developed (Table 2.1). These include advanced formulations like solid lipid nanoparticles (SLNs), liquid-crystal systems (LC), nano-emulsions, and lipid-based carriers such as liposomes, niosomes, and transferosomes (Figures 2.8 and 2.9) [12]. These innovative delivery methods, when combined with herbal drugs and phytoconstituents, enhance drug delivery by increasing bioavailability and effectiveness.

For example, the lipid-based drug delivery system has shown promise for controlled and targeted delivery. Pharmacosomes enhance drug bioavailability through amphiphilic phospholipid complexes. Phytosomes, which consist of lipophilic plant-derived components like *Silybum marianum* and *Ginkgo biloba* bound to phospholipids, also improve bioavailability and therapeutic effects. They are created through a patented process that links herbal compounds to phosphatidylcholine. Phytosomes are beneficial for treating acute liver diseases and have applications in cosmetology, representing a bridge between conventional and NDDS's [13].

Nanotechnology plays a crucial role in this transformation, improving the solubility, stability, and duration of action of active compounds. It also facilitates the targeted delivery of substances to specific tissues, shields them from degradation, and reduces side effects. These advancements allow for more precise and efficient therapeutic applications, making these systems ideal for both market introduction and therapeutic use, although challenges remain in scaling production and ensuring stability.

Innovative delivery strategies address these issues by utilizing nanotechnology for targeted delivery with specific ligands, employing nano-sized carriers to protect phytochemicals and allow controlled release, and using transdermal methods such as patches and gels to bypass the

TABLE 2.1
Novel Drug Delivery Systems of Herbal Drugs

Category	Delivery System	Description	Advantages	Applications with Examples	Reference
Emulsifying Systems	**Microemulsions**	• Stable systems of water, oil, and surfactants with enhanced solubility and bioavailability.	• Increased solubility • Thermodynamic stability • Enhanced therapeutic effects	• Babchi oil microemulsions for anti-inflammatory effects	20, 21
	Nanoemulsions	• It is an emulsion system where two immiscible liquids, oil and water, are mixed and stabilized using surfactants, with droplet sizes ranging from 20 to 200 nanometers.	• Enhanced bioavailability • Small droplet size • Improved stability	• Curcumin nanoemulsions for anti-inflammatory activity	20
Lipid-Based Systems	**Liposomes**	• Liposomes are spherical, bilayered vesicular systems composed of biodegradable phospholipids, the primary structural units. Phospholipids are amphipathic molecules with both hydrophilic and hydrophobic regions. Upon hydration, they form a spherical structure with an aqueous core. • This dual nature allows liposomes to encapsulate both hydrophilic drugs in the aqueous core and lipophilic drugs within the lipid bilayer, making them versatile carriers for drug delivery.	• Improved bioavailability • Reduced toxicity, • Enhanced therapeutic activity • Localized delivery	• Deliver curcumin to macrophages for anti-inflammatory effects • Enhance tea catechins' antioxidant effects	20–22
	Niosomes	• Niosomes are multilamellar vesicles composed of cholesterol and non-ionic surfactants.	• Cost effective • Biologically stable • Variable purity		22, 23
	Proliposomes	• Dry particles form liposomes upon contact with water, overcoming stability issues.	Enhanced stability, ease of storage, and increased absorption	Deliver silymarin for oral bioavailability improvement.	21
	Phytosomes	• Phytosomes are bilayered systems of lipid-compatible complexes made from water-soluble plant components, mainly polyphenols, and phospholipids.	• Improved drug absorption, stability • Therapeutic efficacy, particularly for polar drugs like flavonoids	• Enhance flavonoid absorption and stability • Solubility enhancement of embelin • Prolonged action and enhanced bioavailability of naringenin	20, 23, 24

(Continued)

TABLE 2.1 (*Continued*)
Novel Drug Delivery Systems of Herbal Drugs

Category	Delivery System	Description	Advantages	Applications with Examples	Reference
	Solid Lipid Nanoparticles	• Submicron particles (50–1000 nm) made from solid lipids stabilized by surfactants.	Biodegradable, non-toxic, controlled release, suitable for large-scale production, and compatible with both hydrophilic and lipophilic drugs	Formulations include mouthwashes, gargles, and inhalations	20, 21
	Transferosomes	• Vesicles enhance skin hydration and penetration for transdermal drug delivery.	Improved skin penetration, targeted upper-layer drug delivery, and better patient compliance	• Deliver colchicine and curcumin for localized and systemic effects	20, 24
	Ethosomes	• Ethosomes, composed of water, phospholipids, and high ethanol content, are non-invasive systems that enhance lipid fluidity, enabling deep skin penetration and efficient transdermal or systemic drug delivery.	• Deeper skin penetration • Systemic absorption • Improved patient compliance	• Enhanced systemic absorption of drugs like curcumin • Enhances drug penetration for topical delivery of *Sophora alopencerides*	20, 24
Polymer-Based Systems	**Microspheres**	• Microspheres are spherical particles with an average size of 1 to 50 microns, with drugs dispersed in a polymer matrix. • They release drugs via diffusion, dissolution, or surface erosion, and follow first-order release kinetics.	• Sustained release, targeted delivery • Ease of administration and reduced side effects	• Zedoar oil, rutin, camptothecin, tetradrin, quercetin, and *Cynara scolymus* extract are used in microsphere formulations	20, 23, 24
	Solid Dispersions	• Drugs dispersed in a carrier matrix to enhance dissolution and bioavailability.	Faster drug release, improved absorption, and therapeutic efficacy for water-insoluble drugs	• Poorly soluble drugs in oral formulations	20, 21
	Nanoparticles	• Nano-sized polymer-based particles for improved delivery of poorly soluble drugs.	Enhanced stability, bioavailability, and targeted delivery; protection against degradation	• Silver nanoparticles of *Ocimum sanctum* for antibacterial activity	20, 22

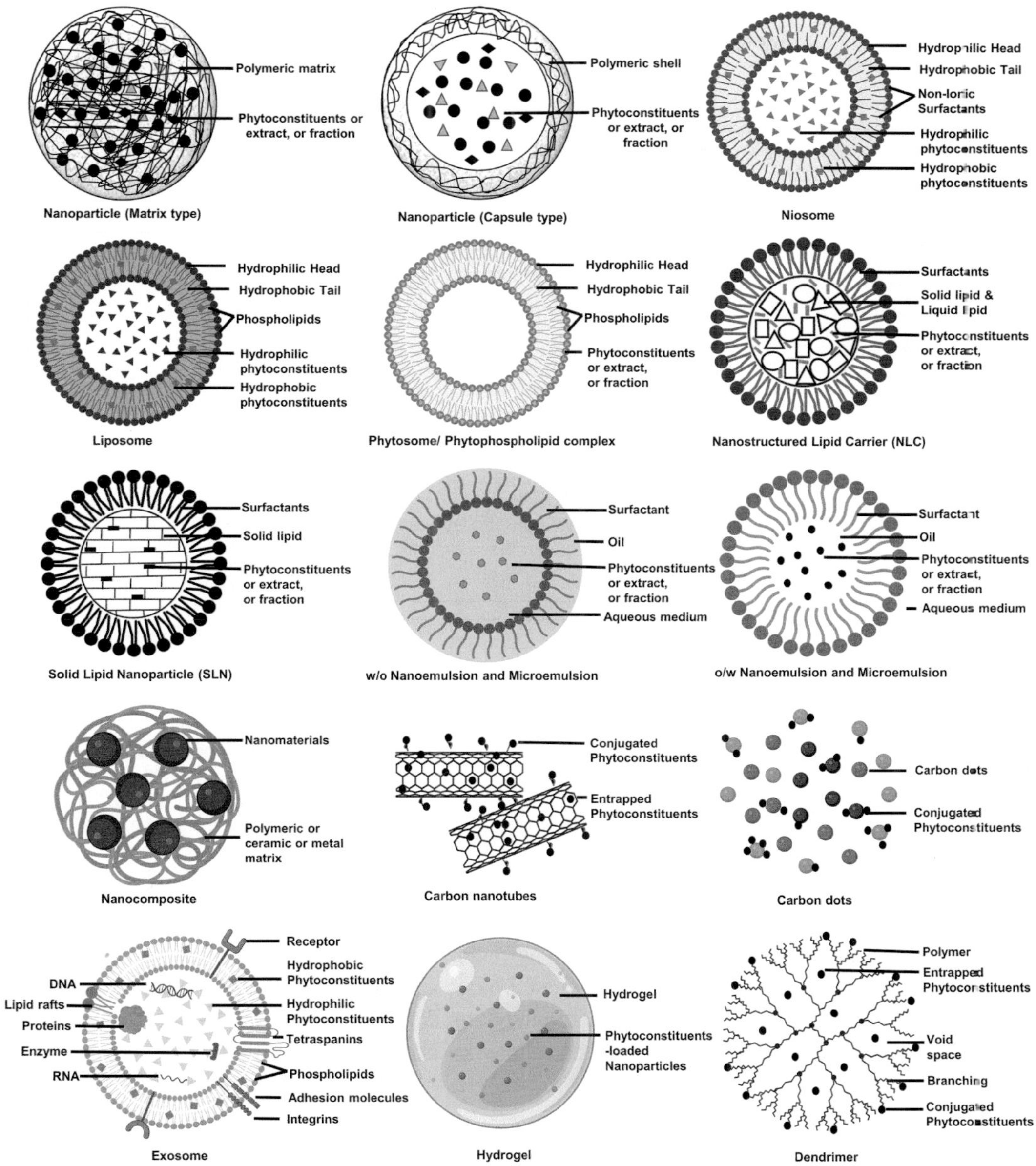

FIGURE 2.8 Schematic representation of novel drug delivery systems of herbal drugs

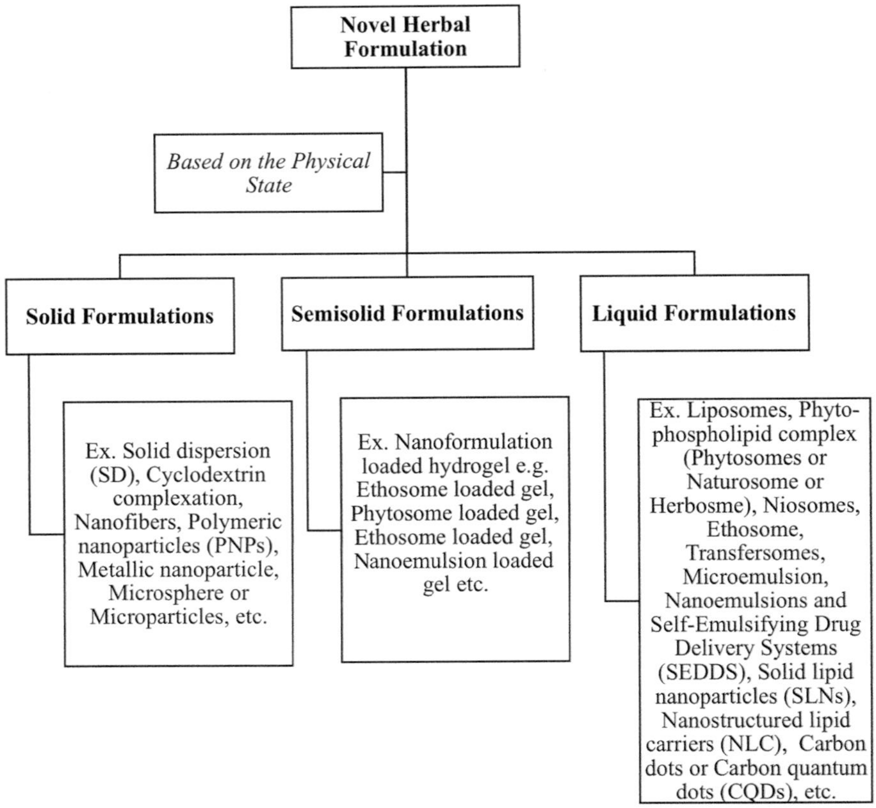

FIGURE 2.9 Classifications of novel herbal formulations

gastrointestinal tract. Additionally, advanced systems facilitate the intracellular delivery of bioactive compounds, while protective coatings and gastro-resistant formulations ensure stability in harsh environments. Finally, combining multiple phytopharmaceuticals or standard drugs into single formulations enhances therapeutic effects and simplifies dosing for patients.

2.3.3.1 Need for NDDS

- **To improve solubility, permeability, absorption and bioavailability**: Some of the phytoconstituents are poorly water soluble, hence possessing poor dissolution and bioavailability. Most of the bioactive components of plant extracts, such as flavonoids, tannins, and terpenoids, are highly water soluble and have low oral absorption due to their low tendency to cross biological lipid membranes. Most of the phytoconstituents have high molecular sizes and low lipophilicity; hence, their permeation across the bio-membrane is a rate-limiting step in systemic absorption. Due to these factors, plant constituents possess poor oral bioavailability. Due to these reasons, herbal medicaments have good *in vitro* activity, which is not reproducible in *in vivo* experiments due to their poor dissolution, absorption and bioavailability [14–16].
- **Targeted/site-specific drug delivery**: Novel formulation can be designed by attaching ligands that enable them to target the formulation to a specific site.
- **Reduction of therapeutic dose**: Improved bioavailability and site-specific drug delivery, allowing novel formulation to achieved the same therapeutic effect with a lesser dose. Hence, the therapeutic dose of novel formulations is generally low compared to crude extracts and traditional and conventional formulations.

- **To control the release of medicaments**: Novel formulations have the capacity to release the incorporated phytoconstituents in a controlled or sustained manner, thus prolonging the duration of action, reducing the frequency of administration, and improving patient compliance.
- **To mask the disagreeable taste and odor**: Some of the plant extracts are bitter and possess unpleasant odor. Novel formulations have the capacity to mask their taste and odor.
- **To develop a compatible formulation:** Some essential phytoconstituents are very rarely used during formulation, because they are incompatible with other excipients [15]. Hence, such constituents can be coated or encapsulated in novel formulations to make them compatible with other constituents or excipients.
- **To protect the phytoconstituents from degradation**: Some of the constituents of herbal extract get degraded when exposed to the gastric medium, subsequently losing their stability as well as therapeutic value [16, 17].
- **For increasing the intracellular uptake:** Most of the nanoparticles are passively taken up by cancer cells through the enhanced permeation and retention effect (EPR effect). Cancerous cells possess leaky vasculature and low lymphatic drainage. The leaky vasculature near the cancerous cell allows the permeation of nanoparticles from the blood. Again, the lower lymphatic drainage prevents the earlier clearance and better accumulation of the permeated nanoparticles at the tumor site. Hence, most of the intracellular uptake of nano formulation occurs through the EPR mechanism [18, 19].

2.4 FACTORS INFLUENCING THE CHOICE OF DOSAGE FORM

The selection of an appropriate herbal dosage form is a critical aspect of herbal medicine, ensuring both safety and therapeutic efficacy. A rational approach to selecting herbal dosage forms involves a careful evaluation of several key factors, including the properties of the herb, its intended therapeutic effect, the bioavailability of active compounds, safety, patient compliance, and the stability of the formulation. Several factors must be considered when selecting an appropriate dosage form for a herbal drug.

2.4.1 Nature of the Herbal Drug/Type of Active Constituents

The chemical composition and the bioactive compounds of the herb largely determine which dosage form is most appropriate. Some herbs contain volatile compounds (e.g., limonene, linalool, myrcene, pinene, terpinolene) [25], while others have compounds that are more stable in concentrated forms (e.g., alkaloids, glycosides) [26]. A rational dosage form must preserve the bioactive components of the herb and deliver them effectively to the body. Herbs rich in volatile oils are often formulated as essential oils, tinctures, or infusions. On the other hand, herbs with more stable constituents are commonly formulated into extracts, tablets or capsules for better stability and controlled dosage.

2.4.2 Bioavailability and Absorption

Bioavailability refers to the rate and extent to which a drug reaches to the systemic circulation. The bioavailability of herbal drugs is influenced by the following.

2.4.2.1 Molecular Size and Lipophilicity

Herbal drugs contain a diverse range of molecules, which may vary in terms of their solubility, molecular size, and lipophilicity. Larger molecules suffer from permeability challenges. Lipophilic compounds may be better absorbed when formulated into oil-based dosage forms, whereas hydrophilic compounds and aqueous extracts are mostly suited to aqueous formulations. Lipophilic phytomolecules suffer from solubility issues, whereas hydrophilic phytomolecules suffer from permeability issues. Hence, a certain degree of hydrophilicity and lipophilicity is required for better absorption and bioavailability of herbal drugs. Suitable novel formulations should be chosen to improve the bioavailability [27–29].

2.4.2.2 First-pass Metabolism/Presystemic Metabolism

First-pass metabolism is a phenomenon where a drug is largely metabolized in the gastrointestinal tract and liver before it reaches the systemic circulation [30]. Many herbal drugs undergo significant first-pass metabolism in the liver, which reduces their bioavailability. For example, herbal drugs taken orally may be subjected to enzymatic breakdown, reducing the amount of active drug that reaches the bloodstream [30]. After oral administration, some phenylpropanoids, like ferulic acid and chlorogenic acid, are metabolized by CYP3A and monoamine oxidase in the liver before the drugs reach systemic circulation. Some flavonoids, such as scutellarein, undergo first-pass metabolism by hepatic CYP450 enzymes [30]. This factor often dictates the choice of an alternative route of administration (sublingual and transdermal routes) via a suitable formulation, to bypass the first-pass effect [31].

2.4.2.3 Extraction Method and Dosage Form

The method of extraction can affect the bioavailability of herbal drugs. For example, some herbal compounds may be better absorbed in their alcohol-based tincture forms, while others may benefit from solid dosage forms like capsules, which can provide a controlled release of the active ingredients. Solid dosage forms, such as tablets and capsules, are considered one of the most reliable forms for ensuring uniform dosage. While liquid formulations like tinctures and syrups may offer rapid absorption, they are harder to standardize and may introduce variability in dosing.

2.4.3 Stability of Herbal Drugs

Herbal drugs often contain compounds that are sensitive to environmental factors like heat, light, moisture, and oxygen. The degradation of active ingredients over time can compromise the efficacy of the drug. They are often prone to degradation, especially during storage, leading to loss of active components, generation of inactive metabolites, and, in extreme cases, generation of toxic metabolites.

2.4.3.1 Preservation of Potency

The shelf life of the dosage form and its ability to maintain the efficacy of active ingredients are important considerations. Some herbal preparations, especially those in liquid form, may degrade over time or require specific storage conditions. Therefore, selecting a dosage form that enhances the stability of the active ingredients is paramount. Various conventional dosage forms, like capsules and coated tablets, and novel formulations, such as polymeric nanoparticles, solid lipid nanoparticles, liposomes, and nanoemulsions, are reported to improve the stability of herbal drugs [32, 33]. For example, tinctures typically have a longer shelf life compared to aqueous extracts due to the alcohol content, which acts as a preservative. Powders, when stored in airtight containers, also tend to maintain their potency for extended periods.

2.4.3.2 Storage Conditions

During the storage of herbal products, the chemical nature of phytoconstituents is altered by various environmental factors. The phytoconstituents may change into other unstable constituents. Sometimes additional compounds may be produced due to the incompatibility with formulation excipients or with the container or closure system. Phytoconstituents belonging to different chemical classes in a herbal formulation may undergo varied inter-molecular or intra-molecular reactions under the influence of temperature, humidity, and/or light experienced during its manufacture, transportation, and storage. Such interactions may produce products less or more active and/or toxic [34]. Degradation of phytoconstituents inside the herbal product leads to a quantitative reduction of active makers and finally the loss of therapeutic value. The various pathways for chemical degradation include the following: hydrolysis (ester, amide), dehydration, isomerization, racemization, decarboxylation, elimination, oxidation, photodegradation, drug/phytoconstituent-excipient interactions, and interaction among phytoconstituents in a polyherbal product, and so forth [35, 36]. The phytoconstituents are mostly secondary metabolites. The quantities of a particular plant are regulated by harmonized activities of various plant metabolizing enzymes (transferases, hydrolases, oxidoreductases, lyases, isomerases, and ligases). Before the collection of crude plant materials, the enzyme activities are controlled by a feedback mechanism. After harvesting, the enzyme activities inside the crude plant material depend on the processing and/or storage conditions. Various enzymes, such as catalase, peroxidase, chlorophyllase, lipoxygenase, esterase, ascorbate oxidase, galactolipase, pectin, polygalacturonase, and phospholipase, remain active during post-harvest time. They continuously metabolize the phytoconstituents, degrade them, and reduce their levels [34]. Some plant extracts and their finished herbal product (liquid/semisolid formulations) may require refrigeration or special packaging to maintain stability. On the other hand, solid dosage forms (tablets, capsules) are generally more stable and less susceptible to environmental degradation. So herbal formulations may be designed in such a way to achieve sufficient storage stability over the shelf life. Stability data is required to support the proposed shelf life for the finished herbal formulation [34].

2.4.3.3 Coating and Encapsulation

Encapsulating herbal extracts in soft or hard gel capsules can protect sensitive compounds from degradation. Additionally, coated tablets or capsules can shield the contents from moisture and light, enhancing shelf life [37]. Multiple reports demonstrated the extension of shelf life after encapsulation in polymeric nanoparticles [38, 39], liposomes [40], and microparticles [41].

2.4.4 Dosage Precision/Dosage Uniformity

Herbal drugs are often prepared as mixtures of various plant parts, leading to significant variability in the concentration of active ingredients. This variability can make precise dosing difficult. A rational dosage form should ensure uniformity and consistency in the delivery of active compounds. The extracts or fractions should be standardized before formulation and development to achieve dosage uniformity [42].

2.4.5 Patient Compliance and Convenience

Patient compliance plays a critical role in the therapeutic success of any medication, including herbal drugs. The choice of dosage form should take into consideration factors that may impact patient acceptance and adherence to the regimen.

2.4.5.1 Ease of Use

Tablets and capsules are often preferred by patients because they are easy to swallow, especially for those with chronic conditions who need to take medication regularly. Liquid dosage forms are more convenient for pediatric and geriatric patients.

2.4.5.2 Dosing Frequency

The herbal formulation should be designed in such a way as to reduce the dosing frequency and to achieve better patient compliance. Sustained-release and controlled-release herbal formulations should be used to reduce the dosing frequency.

2.4.5.3 Taste and Odor

Many herbal extracts have unpleasant tastes or odors that can discourage patient use. In such cases, taste masking agents or alternative dosage forms like soft gels or enteric-coated tablets may be used to improve palatability.

2.4.5.4 Age and Condition Considerations

The choice of dosage form may also depend on the patient's age or medical condition. Individuals with swallowing difficulties may benefit from liquid or dissolvable forms. For instance, children and elderly individuals may prefer chewable tablets or syrups and other liquid dosage forms, whereas tablets and capsules are convenient for the younger age group.

2.4.6 Regulatory and Quality Control Considerations

The rational selection of a dosage form must also comply with regulatory standards, which ensure that the herbal drug meets safety, efficacy, and quality benchmarks. Quality control measures such as testing for uniformity, stability, and bioavailability are essential to ensuring that the herbal product performs as intended.

Standardization of herbal products: Regulatory bodies, such as the World Health Organization (WHO) and the U.S. Food and Drug Administration (FDA), provide guidelines for the standardization and quality control of herbal medicines [43]. These include guidelines for herbal monographs, extraction methods, and acceptable limits for active ingredient concentrations.

Good manufacturing practices (GMPs): Ensuring that herbal drugs are produced according to GMPs is critical for ensuring product safety and efficacy. Adherence to GMP guidelines helps minimize contamination, improper handling, and variability in the final product. Poor agriculture and propagation methods, poor pre- and post-harvesting practices, and the adaptation of incorrect processing techniques lead to substandard herbal drugs. Good agricultural and collection practices (GACP) can guarantee the quality and stability of herbal formulations.

2.4.7 Safety Considerations

Safety is paramount when choosing a herbal dosage form. Some forms, like concentrated tinctures, can pose risks of overdose if not used appropriately. Liquid dosage forms might require accurate dosing, while tablets or capsules offer more controlled dosages. Additionally, certain herbs might interact with medications or cause allergic reactions, so it's important to ensure that the dosage form allows for ease of control and monitoring.

2.4.7.1 Toxicity Considerations

The safety of the herbal remedy is paramount. Some herbs may have toxic effects in large doses. They may produce acute and chronic toxicity [44, 45]. Their dosage form with a consistent dose must be carefully selected to control the amount of active ingredients consumed. Administering herbal formulations through the appropriate route, considering dosing uniformity, may avoid toxicity.

2.4.7.2 Standardization and Quality Control

Standardized extracts, where the concentration of active compounds is fixed, help minimize the risk of overdose and ensure consistent therapeutic effects [46]. The form in which the herb is consumed must be accurately dosed to avoid variations in potency and toxicity. Marker-based standardization with the use of suitable chromatographic techniques or spectroscopy methods ensures the content uniformity of the plant extract or fraction. This reduces the risk of unexpected side effects and toxicity.

2.4.8 Therapeutic Action

The desired therapeutic outcome should guide the type of dosage form chosen.

2.4.8.1 Immediate Versus Sustained Release

The therapeutic objectives dictate whether a fast-acting or slow-release dosage form is preferable. If rapid absorption is necessary (e.g., for acute conditions), liquid formulations (tincture, syrup, liquid extract) may be preferred due to their faster bioavailability. In contrast, for sustained action or long-term use (e.g., for chronic conditions), capsules, tablets, and novel sustained dosage forms (nanoparticles, phytosomes) might be more appropriate as they provide a slow, steady release [31, 47].

2.4.8.2 Localized Versus Systemic Effects

Some herbal formulations are intended to have localized effects, while others are meant for systemic treatment. This is crucial in determining the appropriate dosage form. Topical preparations like ointments, creams, films, fibers, nano-particles, micro-particles, and vesicular systems (phytosomes, transferosome, ethosome, niosomes) are used for local effects [48, 49], while oral preparations like capsules, tablets, novel dosage forms (solid dispersion, nano and microparticles, emulsions) are for systemic effects [31].

2.4.8.3 Targeted Versus Nontargeted Effects

Targeted delivery refers to the precise and controlled release of active herbal compounds at specific sites in the body where they can exert their therapeutic effects [50]. This approach aims to increase the efficacy of the herbal remedy while minimizing side effects. In contrast, nontargeted herbal formulations deliver the drug to the entire body. When selecting a suitable herbal dosage form for targeted delivery, several critical factors must be considered, including the physicochemical properties of the herbal active compounds, the desired site of action, the delivery system, and the pharmacokinetics of the herb. Novel herbal formulations can be designed with suitable ligands, antibody, peptides, and aptamers to achieve targeted drug delivery [50]. Targeted delivery often requires advanced delivery systems to control the release and location of the active herbal ingredients. Some of these systems include nanoparticles, liposomes, microparticles, and microspheres, micelles, dendrimers, and phytosomes, all of which can protect herbal compounds from degradation, enhance absorption, and direct them to specific sites in the body [51]. Understanding the pharmacokinetics (absorption, distribution, metabolism, and excretion) of the herbal compounds is essential when selecting a targeted delivery system. The goal is to enhance the compound's bioavailability and ensure its therapeutic concentration at the desired site of action.

2.4.9 Cost and Affordability

The cost of herbal preparations and their availability in specific dosage forms are important factors for both practitioners and patients. While concentrated extracts or specialized delivery systems may offer higher therapeutic efficacy, they might be more expensive than simpler forms like teas or bulk powders.

2.5 CONCLUSION

A rational approach to dosage form selection ensures that patients receive the optimal balance of effectiveness and safety, which is essential for achieving desired health outcomes. The rational selection of herbal dosage forms involves a holistic approach that considers the drug's bioavailability, stability, patient compliance, and manufacturing feasibility. By carefully considering these factors, herbal medicines can be effectively formulated into dosage forms that maximize their therapeutic potential while minimizing variability and ensuring consistency. By choosing the appropriate form, healthcare providers can maximize the therapeutic benefits of herbal remedies while minimizing risks and enhancing patient compliance. With ongoing research and advancements in formulation technology, the future of herbal drug dosage forms holds great promise in making these traditional remedies more accessible and reliable for modern medical use.

REFERENCES

1. Sahu, A.N. and D. Mohapatra, *Herbal Drug Formulation and Standardization*. 2021, New Delhi: Ane Books Pvt. Ltd. ISBN-10: 9390658365, ISBN-13: 978–9390658367.
2. Ansari, F.A., et al., Novel decontamination approaches for stability and shelf-life improvement of herbal drugs: A concise review. *The Microbe*, 2024. 3: p. 100070.
3. Rombolà, L., et al., Pharmacokinetic interactions between herbal medicines and drugs: Their mechanisms and clinical relevance. *Life*, 2020. 10(7): p. 106.
4. Noviana, E., G. Indrayanto, and A. Rohman, Advances in fingerprint analysis for standardization and quality control of herbal medicines. *Frontiers in Pharmacology*, 2022. 13: p. 853023.
5. Nooreen, Z., V.K. Rai, and N.P. Yadav, Phytopharmaceuticals: A new class of drug in India. *Annals of Phytomedicine*, 2018. 7(1): p. 27–37.
6. Sen, S., and R. Chakraborty, Toward the integration and advancement of herbal medicine: A focus on traditional Indian medicine. *Botanics: Targets and Therapy*, 2015. 5: p. 33–44.
7. Sen, S., and R. Chakraborty, Revival, modernization and integration of Indian traditional herbal medicine in clinical practice: Importance, challenges and future. *Journal of Traditional and Complementary Medicine*, 2017. 7(2): p. 234–244.
8. Raval, N., D. Khunt, and M. Misra, Microemulsion-based delivery of triamcinolone acetonide to posterior segment of eye using chitosan and butter oil as permeation enhancer: An in vitro and in vivo investigation. *Journal of Microencapsulation*, 2018. 35(1): p. 62–77.
9. Elkordy, A.A., et al., An overview on natural product drug formulations from conventional medicines to nanomedicines: Past, present and future. *Journal of Drug Delivery Science and Technology*, 2021. 63: p. 102459.
10. Kushare, V.N., et al., Review of herbal formulations. *World Journal of Pharmaceutical Research*, 2023. 12(15): p. 436–446.
11. Singh, H., Recent development of novel drug delivery of herbal drugs. *RPS Pharmacy and Pharmacology Reports*, 2023. 2(4): p. rqad028.
12. Pavithra, K. and V. Manimaran, A review of safety, quality, regulation, and delivery approaches for phytopharmaceuticals. *Jordan Journal of Pharmaceutical Sciences*, 2024. 17(2): p. 316–332.
13. Devi, V.K., N. Jain, and K.S. Valli, Importance of novel drug delivery systems in herbal medicines. *Pharmacognosy Reviews*, 2010. 4(7): p. 27.
14. Alexander, A., et al., Recent expansion of pharmaceutical nanotechnologies and targeting strategies in the field of phytopharmaceuticals for the delivery of herbal extracts and bioactives. *Journal of Controlled Release*, 2016. 241: p. 110–124.
15. Bonifacio, B.V., et al., Nanotechnology-based drug delivery systems and herbal medicines: A review. *International Journal of Nanomedicine*, 2014. 9: p. 1.
16. Saraf, S., Applications of novel drug delivery system for herbal formulations. *Fitoterapia*, 2010. 81(7): p. 680–689.
17. Bhattacharya, S., Phytosomes: The new technology for enhancement of bioavailability of botanicals and nutraceuticals. *International Journal of Health Research*, 2009. 2(3): p. 225–232.
18. Kratz, F., P. Senter, and H. Steinhagen, *Drug Delivery in Oncology: From Basic Research to Cancer Therapy*. 2013, John Wiley & Sons.
19. Li, J., et al., Recent advances in targeted nanoparticles drug delivery to melanoma. *Nanomedicine: Nanotechnology, Biology and Medicine*, 2015. 11(3): p. 769–794.
20. Verma, H., S.B. Prasad, and S.H. Yashwant, Herbal drug delivery system: A modern era prospective. *International Journal of Current Pharmaceutical Review and Research*, 2013. 4: p. 88–101.
21. Musthaba, S.M., et al., Status of novel drug delivery technology for phytotherapeutics. *Expert Opinion on Drug Delivery*, 2009. 6(6): p. 625–637.
22. Zahid, S.A., et al., A review on herbal drug excipient used in NDDS. *International Journal of Research in Pharmacy and Allied Science (IJRPAS)*, 2023. 2(4): p. 124–135.
23. Lavande, J., S. Bais, and G. Pawar, A review on advanced herbal drug technology. *International Journal*, 2024. 2(1): p. 1099–1117.
24. Ali, H., et al., Novel drug delivery approaches in formulation development; stability considerations and quality features of herbal products. *RADS Journal of Pharmacy and Pharmaceutical Sciences*, 2019. 7(3): p. 155–164.
25. Kim, M.-H., et al., Usage of natural volatile organic compounds as biological modulators of disease. *International Journal of Molecular Sciences*, 2021. 22(17): p. 9421.
26. Velu, G., V. Palanichamy, and A.P. Rajan, Phytochemical and pharmacological importance of plant secondary metabolites in modern medicine, in *Bioorganic Phase in Natural Food: An Overview*. 2018, Springer. p. 135–156.

27. Chavan, A.K. and A.U. Tatiya, Development of novel carrier system: A key approach to enhance bioavailability of herbal medicines. *Pharmacognosy Research*, 2025. 17(1).
28. Ahmed, H.M., S. Nabavi, and S. Behzad, Herbal drugs and natural products in the light of nanotechnology and nanomedicine for developing drug formulations. *Mini Reviews in Medicinal Chemistry*, 2021. 21(3): p. 302–313.
29. Saraf, S., Applications of novel drug delivery system for herbal formulations. *Fitoterapia*, 2010. 81(7): p. 680–689.
30. Zeng, M., et al., Metabolic pathways and pharmacokinetics of natural medicines with low permeability. *Drug Metabolism Reviews*, 2017. 49(4): p. 464–476.
31. Byeon, J.C., et al., Recent formulation approaches to oral delivery of herbal medicines. *Journal of Pharmaceutical Investigation*, 2019. 49: p. 17–26.
32. Thakur, L., et al., Novel approaches for stability improvement in natural medicines. *Pharmacognosy Reviews*, 2011. 5(9): p. 48.
33. Kumadoh, D. and K. Ofori-Kwakye, Dosage forms of herbal medicinal products and their stability considerations-an overview. *Journal of Critical Reviews*, 2017. 4(4): p. 1–8.
34. Bansal, G., et al., Stability testing of herbal drugs: Challenges, regulatory compliance and perspectives. *Phytotherapy Research*, 2016. 30(7): p. 1046–1058.
35. Yoshioka, S. and V.J. Stella, *Stability of Drugs and Dosage Forms*. 2000, Springer Science & Business Media.
36. Sachan, A.K. and A. Kumar, Stability testing of herbal products. *Journal of Chemical and Pharmaceutical Research*, 2015. 7(12): p. 511–514.
37. Thakur, L., et al., Novel approaches for stability improvement in natural medicines. *Pharmacognosy Reviews*, 2011. 5(9): p. 48.
38. Hassan, F., et al., Shelf-life extension of sweet basil leaves by edible coating with thyme volatile oil encapsulated chitosan nanoparticles. *International Journal of Biological Macromolecules*, 2021. 177: p. 517–525.
39. Das, S., et al., Chitosan based encapsulation of Valeriana officinalis essential oil as edible coating for inhibition of fungi and aflatoxin B1 contamination, nutritional quality improvement, and shelf life extension of Citrus sinensis fruits. *International Journal of Biological Macromolecules*, 2023. 233: p. 123565.
40. Al-Moghazy, M., et al., Edible packaging coating of encapsulated thyme essential oil in liposomal chitosan emulsions to improve the shelf life of Karish cheese. *Food Bioscience*, 2021. 43: p. 101230.
41. Wilar, G., et al., Microparticles of herbal extracts with antioxidant activity. *Pharmacognosy Journal*, 2021. 13(1).
42. Kunle, O.F., H.O. Egharevba, and P.O. Ahmadu, Standardization of herbal medicines-A review. *International Journal of Biodiversity and Conservation*, 2012. 4(3): p. 101–112.
43. Wah, C.L., S.C. Hock, and T.K. Yun, Current scientific status and regulatory control of traditional/herbal medicinal products: Globalization challenges. *Pharmaceutical Engineering*, 2012. 32.
44. Mensah, M.L., et al., Toxicity and safety implications of herbal medicines. *Herbal Medicines*, 2019: p. 63.
45. Knöss, W., Toxicity of herbal medicines: From past to present to future, in *Toxicology of Herbal Products*. 2017, Springer. p. 1–9.
46. Kumari, R. and M. Kotecha, A review on the standardization of herbal medicines. *International Journal of Pharma Sciences and Research*, 2016. 7(2): p. 97–106.
47. Chen, Z., et al., Sustained and controlled release of herbal medicines: The concept of synchronized release. *International Journal of Pharmaceutics*, 2019. 560: p. 116–125.
48. Rangrej, U., et al., Evidence based review on herbal local drug delivery. *IOSR Journal of Dental and Medical Sciences (IOSR-JDMS)*, 2017. 16(8): p. 77–85.
49. Khogta, S., et al., Herbal nano-formulations for topical delivery. *Journal of Herbal Medicine*, 2020. 20: p. 100300.
50. Khan, R.U., et al., Targeted delivery systems for herbal drugs, in *Herbal Pharmacopeia*. 2025, CRC Press. p. 345–366.
51. Kumari, S., et al., Bioactive loaded novel nano-formulations for targeted drug delivery and their therapeutic potential. *Pharmaceutics*, 2022. 14(5): p. 1091.

3 Analytical Challenges of Herbal Drugs/Dosage Forms

ABBREVIATIONS

Abbreviation	Description
AAS	Atomic Absorption Spectroscopy
AES	Atomic Emission Spectroscopy
API	Active Pharmaceutical Ingredients
ATR-FTIR	Attenuated Total Reflectance-Fourier Transform Infrared Spectroscopy
AYUSH	Ayurveda, Yoga & Naturopathy, Unani, Siddha and Homeopathy
BLAST	Basic Local Alignment Search Tool
FT-IR	Fourier Transform Infrared Spectroscopy
GACP	Good Agricultural and Collection Practices
GC	Gas Chromatography
GC-MS	Gas Chromatography-Mass Spectrometry
HPLC	High-Performance Liquid Chromatography
HPTLC	High-Performance Thin-Layer Chromatography
ICH	International Council for Harmonization
ICP-MS	Inductively Coupled Plasma Atomic Emission Spectroscopy
LC-MS	Liquid Chromatography-Mass Spectrometry
MALDI-TOF MS	Matrix-Assisted Laser Desorption Ionization Time-Of-Flight Mass Spectrometry
matK	Maturase K Gene
MS	Mass Spectrometry
NMR	Nuclear Magnetic Resonance
RbcL	Ribulose-Bisphosphate Carboxylase Gene
TLC	Thin-Layer Chromatography
trnH-psbA	Transfer RNA histidine-photosystem II protein D1 gene
UNODC	United Nations Office on Drugs and Crime
UPLC	Ultra-Performance Liquid Chromatography
WHO	World Health Organization

3.1 INTRODUCTION

Herbal drugs have long been a necessity in healthcare systems, valued for their therapeutic potential and cultural acceptance. The complexity of herbal drugs creates significant analytical challenges in ensuring their safety, consistency, and standardization, with important issues like authentication of raw materials, detection of adulterants, quantification of bioactive constituents, and identification of contaminants. The conversion of herbal extracts into dosage forms such as tablets, capsules, syrups, or advanced delivery systems further complicates the evaluation, as stability, bioavailability, and pharmacokinetics must also be addressed. While modern analytical tools – chromatography,

DOI: 10.1201/9781003540823-3

spectroscopy, DNA barcoding, and chemometrics – offer powerful solutions, variability within herbal matrices often limits their application.

This chapter explores the key analytical challenges associated with the evaluation of herbal drugs and their various dosage forms. It highlights the complexities researchers face in authentication, standardization, and quality assessment of plant-based formulations. Furthermore, it highlights the fundamental role of modern analytical science in integrating traditional herbal knowledge with contemporary pharmaceutical frameworks, with the overarching aim of ensuring that herbal medicines are safe, effective, and of consistent quality.

3.1.1 Difficulty in Following the Multistep Standardization Protocol

Plant extracts are typically prepared directly from crude plant raw materials such as leaves, roots, stems, and bark. These extracts exhibit significant variation in phytoconstituent composition, quality, and therapeutic effects. Standardized plant extracts, which maintain consistent levels of specified constituents, represent high-quality extracts. Achieving this consistency requires the adoption of appropriate standardization and quality control protocols. Standardization and quality control of plant extracts are typically performed through various methods, including macro and microscopic examination, estimation of foreign organic matter, determination of ash values, moisture content, extractive values, crude fiber, volatile oil content, bitterness value, hemolytic index, astringent property, swelling index, foaming index, as well as qualitative and quantitative chemical evaluation, chromatographic examination, microbial assays, and identification of pesticide residues and toxic elements. However, it is challenging to adhere to every standardization parameter. Some medicinal plants still lack identified marker components. Even when chemical marker(s) are known, there may be limited understanding of the present bioactive markers. Additionally, acquiring standard markers is not always feasible. Isolating and purifying these markers from plant extracts can be difficult due to their trace quantities [1].

3.1.2 Hurdles with Carrying Out Chemical Marker-Based Standardization

Chemical marker-based standardization is essential to achieve consistent levels of phytochemical composition in extracts. Chemical markers are primarily of two types: active markers or active principles, which have therapeutic activity, and analytical markers, which are used solely for quality control and standardization purposes [2]. For marker-based standardization, markers with known therapeutic activity (active principles) are generally preferred. However, when active markers are unknown, analytical markers should be selected for standardization. In some cases, both active and analytical markers are used. These markers, in purified form (i.e., as reference standards), should be available to compare their presence (qualitative estimation) and quantity (quantitative estimation) in the complex herbal extract or enriched fraction. The difficulty in obtaining reference standards can pose challenges in the standardization process [3]. Improper standardization due to a lack of chemical markers, technical challenges in marker production, and insufficient pharmacological data on chemical markers lead to substantial fluctuations in herbal product constituents, quality, and therapeutic effects [2]. In addition to that, since herbal raw materials are complex, sophisticated analytical instruments, such as high-performance liquid chromatography (HPLC), gas chromatography (GC), mass spectrometry (MS), liquid chromatography-mass spectrometry (LC-MS), and gas chromatography-mass spectrometry (GC-MS), are needed to analyze the constituents and standardize the plant extract before formulation and development [4].

3.1.3 Improper Authentication

Serious public health issues arise when herbal raw materials are not authenticated correctly [3]. Improper authentication of herbal ingredients can lead to the production of modified herbal products, and consuming the incorrect herbal product may pose serious health risks. The complexity of nomenclature leads to misidentification, as a single plant may have multiple names, and one name might be used for different plants, whether they are closely related or entirely unrelated species. Therefore, it is crucial to accurately authenticate plants through macroscopic and microscopic comparison with authentic material, and they should be referred to by their biological Latin names (italicized form of genus and species)/scientific/international nomenclature instead of synonyms. Establishing a correspondence between vernacular names with scientific nomenclature presents a challenge, as a single species can be identified by multiple vernacular names across various regions and languages. Conversely, a single vernacular name might be used to refer to multiple species, sometimes even belonging to different genera or families [5].

For many years, species and their synonyms were recorded in "The Plant List" at Kew Garden's website (www.theplantlist.org/). These records are now maintained and updated on the World Flora Online website (http://about.worldfloraonline.org/). Voucher specimens preserved in official herbaria serve as the authoritative references that connect scientific names to botanical species, and they should be well documented in all published research involving plant specimens [5].

Besides binomial nomenclature, herbal drugs are also identified by their pharmaceutical names as listed in official pharmacopoeias. These names are derived from a Latin term indicating the plant part (such as radix for root, semen for seed, caulis for stem, cortex for bark, fructus for fruit, flos for flower, folium for leaves, herba for aerial parts) combined with the Latin name of the plant, usually the genus. In some instances, the pharmaceutical name appears as the Latin name (in genitive form) followed by the plant part. These names are never italicized. For example, "*Rhizoma Alimatis*" or "*Alimatis rhizoma*" refers to the rhizome of Alisma, and "Radix Angelicae pubescentis" or "Angelicae pubescentis radix" refers to the root of *Angelica pubescens* [5].

Macroscopy is often the first step toward the authentication of herbal drugs and their organs, followed by organoleptic evaluations and microscopy. However, distinguishing between related plant species, varieties of the same species, or identifying adulterations is challenging [5]. Faced with the limitations of macroscopic analyses, the microscopy of herbal drugs or powder of a plant part presents a significant advantage. It enables detailed examination at the cellular level, which is more effective for identifying species and verifying adulteration [5]. In addition to that, many analytical methods, including hyphenated tools such as chromatography, spectrometry, and DNA barcoding techniques, have been employed to identify herbal products. DNA-barcoding-based plant authentication and adulterant identification is recognized as a gold standard. However, DNA-based genetic profiling has notable limitations and accessibility challenges, especially in basic laboratory environments [5]. In many instances, identifying key chemical markers (bioactive or analytical) is a primary approach. This can be achieved by comparing with reference standards or using hyphenated techniques, such as LC-MS and GC-MS, with their compound database. However, many phytochemicals are not commercially available or listed in databases. Furthermore, the phytochemistry of herbal drugs is highly complex and largely unexplored [5]. Analytical profiling, along with the development of spectroscopic and chromatographic fingerprints of a given plant or extract/fraction, can be used to authenticate a plant. This method requires substantial effort to develop a verified library of acceptable plant samples and their potential adulterants [5]. Modifications to existing analytical methods are continuously being developed and improved, resulting in a broader range of analytical tools. However, an appropriate analytical method or combination of multiple methods should be selected with careful consideration.

3.1.4 Chemical Complexity

Herbal products are typically complex mixtures of numerous constituents, making it challenging to fully characterize every component and implement rigorous standardization protocols [3]. The inherent chemical complexity of medicinal plants, combined with the limited availability of straightforward analytical tools capable of identifying and characterizing active constituents solely by chemical or biological means, underscores the requirement for a robust quality assurance system.

3.1.5 Challenges with a Sampling of Herbal Drugs

Sampling of herbal drugs is a problem during collection, as the content of phytochemicals depends on the sampling process. For instance, sampling of roots of various thicknesses may lead to variations in the bark-to-wood ratios. Consequently, depending on the sampled root's diameter, the plant bioactives' concentration could differ significantly [5]. Further, the content of phytoconstituents also depends on geographical location, harvesting time, complexity of constituents, chances of microbial contamination, and adulteration. These also pose challenges for the sampling of authentic herbal drugs. Currently, the detailed sampling strategy and processing have also been mentioned in various Pharmacopoeias. However, implementing certain strategies, such as Good Agricultural and Collection Practices (GACP), rigorous quality control parameters, and regulatory harmonization, can significantly improve the proper sampling of herbal drugs, ensuring safety and efficacy.

3.1.6 Handling Issues of Herbal Drugs for Analysis

From extraction to formulation development, herbal drugs face several challenges. During extraction, the solubilities of the present phytocomponents, chemical nature of target compounds, sample to solvent ratio, the temperature of extraction, duration of extraction, type of extraction technique (conventional: maceration, infusion, decoction, etc.; or advanced: microwave-assisted extraction, supercritical fluid based-extraction, etc.), and polarity and toxicity of extracting solvent should be considered. During the standardization of herbal drugs with respect to bioactive and analytical chemical markers, the initial step is the weighing of extract and quantification of the present markers using a range of analytical techniques, including HPLC and LC-MS. However, due to the sticky/semisolid nature of most of the plant extracts, it is very difficult to precisely weigh a definite quantity of plant extract. Also, due to the semisolid nature of most of the extracts, their content uniformity is not maintained, which leads to variation in dose. During formulation and development of standardized herbal extract, the initial step is the herbal drug characterization via physicochemical properties evaluation and pre-formulation studies. However, due to the sticky/oily/semisolid nature of most of the herbal extracts, it is difficult to perform pre-formulation studies (e.g., solubility, dissolution, flow property, porosity, density, stability, drug-excipient compatibility, etc.) and formulation. However, as per the phytopharmaceutical guidelines, any four chemical markers can be chosen, and the aforementioned studies can be performed with respect to the markers. The herbal extract should be dried properly to remove the residual aqueous or organic solvent. The herbal extract can be evaporated under reduced pressure, vacuum dried, freeze dried, or spray dried to remove the solvent and avoid stickiness and the toxic effect of residual organic solvents.

3.1.7 In-Tube Reactions-Based Analysis of Phytochemicals

In laboratories with limited access to advanced instrumental methods, the identification of phytochemicals, such as alkaloids, glycosides, flavonoids, terpenoids, and tannins, is often conducted through in-tube reactions. These tests, including the Vitali-Morin, Keller-Killiani, Shinoda's, Salkowski, and goldbeater's skin tests, depend on the interaction with phytocompounds to produce observable changes such as color, precipitate formation, or fluorescence. However, these methods only allow the qualitative analysis of these phytochemicals with limited sensitivity and selectivity [5]. Currently, these techniques have been replaced by profiling techniques for plant materials, while test kits from the United Nations Office on Drugs and Crime (UNODC) are recommended and extensively used at border controls for the identification of substances such as cannabinoids, cocaine, and opiates (www.unodc.org/unodc/en/scientists/druge-kits-new.html). Advanced spectroscopic and chromatographic techniques can be used to understand the chemical profile and their exact quantity in a multiconstituent herbal extract or herbal formulation.

3.1.8 Challenges with Chromatography-Based Phytochemical Analysis

Chromatography allows the separation of a mixture of phytocompounds/herbal extract/fractions into individual components with the use of a set of stationary and mobile phases to yield a characteristic profile. This allows both qualitative and quantitative estimation of phytoconstituents. The selection of a chromatographic method is determined by the specific characteristics of the extract or fraction being analyzed [5]. Thin-layer chromatography (TLC) allows the separation using a planar stationary phase, such as silica, polyamide, cellulose, or chemically modified TLC plates. Detection can be performed directly or after treatment with a specific reagent, under ultraviolet or visible light. In contrast, high-performance thin-layer chromatography (HPTLC) utilizes silica plates with a higher number of theoretical plates, resulting in significantly improved resolution and reproducibility of the profiles. The data obtained from HPTLC can be utilized for both qualitative and quantitative analyses. Currently, HPTLC is used as a standard technique in modern pharmacopoeias to authenticate herbal drugs [5]. Additionally, chromatographic techniques, including high-performance liquid chromatography (HPLC), ultra-performance liquid chromatography (UPLC), and gas chromatography (GC), are also extensively employed for both qualitative phytochemical profiling and quantitative analysis [5]. These advanced chromatographic techniques offer a high level of selectivity and sensitivity and can be integrated with spectroscopic techniques, such as mass spectroscopy (MS), allowing the development of hyphenated techniques (LC-MS or GC-MS) that allow qualitative and quantitative phytochemical analysis with a high degree of selectivity, sensitivity, and reproducibility [5]. However, due to the complexity and the variety of phytochemicals, it can be quite challenging to prepare the sample and to determine the appropriate composition for the mobile phase, the stationary phase, and the type of detector. Further, the analysis and integration of complex chromatographic data is very difficult. The complexity makes it difficult to separate and analyze individual components. Co-elution of compounds can occur, where multiple substances elute simultaneously, complicating the identification and quantification of individual analytes. The presence of other components in the herbal matrix can cause background noise and interfere with the separation and detection of target analytes. Some active compounds are present in very low concentrations, requiring highly sensitive detection methods to quantify them accurately. The presence of many compounds can lead to complex chromatograms with overlapping peaks, making it difficult to identify and quantify individual substances. Chromatographic techniques generate large and complex data sets, requiring sophisticated software and analysis expertise. There may be a lack of standardized reference materials for many herbal compounds, complicating method validation. Ensuring reproducibility across different laboratories and batches of herbal material is challenging due to the inherent variability in plant-derived products. However, standardizing extraction protocols and ensuring consistent sample preparation can minimize variability and matrix effects. Using hyphenated separation techniques such as LC-MS and GC-MS can improve separation and detection capabilities. Employing advanced data analysis tools and software for peak deconvolution, pattern recognition, and multivariate analysis, complex chromatographic data can be analyzed and explained. By creating extensive chromatographic libraries of known compounds, the identification of unknown phytochemicals is possible.

3.1.9 Challenges with Spectroscopy-Based Phytochemical Analysis

Spectroscopy is a powerful analytical technique that identifies and quantifies substances based on their interaction with electromagnetic radiation. It is essential in the analysis of herbal drugs to provide extensive data to ensure their quality, efficacy, and safety. Techniques such as UV-visible spectroscopy and infrared spectroscopy are widely used to identify specific compounds in herbal extracts by their characteristic absorption spectra, types of functional groups, quantification of phytochemicals, metabolite profiling, and structural elucidation. These are also used to estimate pesticide residues, heavy metals, and other contaminants in herbal products. Such techniques are mostly non-destructive techniques that offer rapid detection, high sensitivity, specificity, and reproducibility. Numerous uses of metabolic fingerprinting, based on Fourier transform infrared spectroscopy (FT-IR) and nuclear magnetic resonance (NMR) spectrometry in conjunction with multivariate analysis, have been developed to distinguish between closely related plant species and have applications in the quality control of herbal drugs. For instance, the authentication of Phyllanthus species was made possible by the combination of FT-IR, ^{1}H NMR, and chemometric analysis [6]. Using cluster analysis via attenuated total reflectance-Fourier transform infrared (ATR-FTIR) spectroscopy, it was possible to distinguish between genuine and fake tobacco with high accuracy and precision [7]. Atomic absorption spectroscopy (AAS), atomic emission spectroscopy (AES), and inductively coupled plasma atomic emission spectroscopy (ICP-MS) are employed for both qualitative and quantitative estimation of mineral elements and contaminants in herbal drugs [5]. However, due to the complexity of phytoconstituents, it is difficult to analyze and interpret spectroscopic data. Components within the plant matrix can cause background noise and matrix effects that interfere with the spectroscopic signals of target analytes. Overlapping signals in complex spectra can obscure the presence of minor constituents, making it difficult to identify and quantify all components. Due to the interference among phytocomponents, it is difficult to identify individual components using standard spectroscopic techniques. Some bioactive compounds are present in very low concentrations, requiring highly sensitive spectroscopic techniques to detect and quantify them accurately. Spectroscopic techniques, especially when coupled with multivariate analysis, generate large and complex data sets that require sophisticated statistical tools and expertise to analyze. Selecting appropriate multivariate analysis methods and interpreting the results correctly can be challenging, especially when dealing with subtle differences in complex spectra. By improving sample preparation, employing advanced analytical techniques, developing comprehensive spectral libraries of known herbal compounds, and utilizing robust data analysis methods, it is possible to enhance the accuracy, sensitivity, and reliability of spectroscopic analysis of herbal drugs. A comprehensive and non-biased high-throughput analysis of complex metabolite mixtures in herbal drugs can be achieved through metabolomics studies. To obtain a broad overview of metabolites, a fully integrated strategy is necessary, encompassing sample extraction, metabolite separation, identification, automated data acquisition, processing, analysis, and quantitative estimation. Due to the diverse nature of phytoconstituents, no single solvent can dissolve the entire range of compounds, thereby limiting the choice of extracting solvents. A single analytical method cannot provide a complete visualization of the metabolome; hence, multiple techniques are required [8].

3.1.10 Challenges with the Size Analysis of Herbal Formulations

Accurate size analysis of herbal formulations remains a significant challenge due to their complex compositions and limitations of current analytical techniques. Direct methods such as transmission electron microscopy (TEM) and scanning electron microscopy (SEM) provide nanometer-scale resolution but require complex and time-consuming sample preparation, including drying, coating, and filtering. These steps may introduce artifacts, cause structural damage, or alter the natural state of herbal particles. Similarly, atomic force microscopy (AFM) offers nanometer resolution and can operate in both air and liquid, but it demands particle immobilization, is time consuming, and unsuitable for large-scale or routine studies. Moreover, all these microscopy-based methods are costly, require highly skilled operators, and mainly provide morphological information rather than chemical insights, which often necessitates their combination with complementary analytical techniques [9].

Indirect light scattering methods also present challenges. Small-angle X-ray scattering (SAXS) provides structural information and broad material applicability, yet it struggles with polydisperse samples, particles larger than 200 nm, and typically requires expensive synchrotron radiation facilities. Laser diffraction (LD) can measure a wide particle size range but cannot detect particles smaller than ~150 nm, requires large sample volumes, and relies on complex algorithms. Dynamic light scattering (DLS) is rapid and convenient, but its accuracy is compromised in polydisperse herbal formulations because larger particles dominate scattering signals, masking smaller ones. Multi-angle light scattering (MALS), particularly when combined with field-flow fractionation (FFF) or size-exclusion chromatography (SEC), enhances accuracy and extends applicability to polydisperse systems; however, it requires complicated experimental conditions, sample dilution, and may suffer from particle–surface interactions during separation [9].

In addition to these method-specific limitations, broader challenges affect the size analysis of herbal formulations. Herbal systems are chemically diverse and highly heterogeneous, making uniform measurement and reproducibility difficult. Instruments such as Zeta sizers, TEM, and SEM are expensive, limited in accessibility, and unsuitable for large-scale or routine testing. Many techniques are also time intensive, low throughput, and sensitive to environmental conditions such as vacuum or electron beams, which can degrade samples. Furthermore, while they provide valuable physical and morphological data, they often fail to capture complete chemical profiles of phytoconstituents, necessitating the integration of size analysis with complementary chemical characterization methods [9].

3.1.11 Challenges with DNA Profiling/DNA-Based Molecular Authentication

In certain cases, it is necessary to perform DNA-based molecular authentication of closely related plant species due to their similar macro-and micromorphological features [5]. Given the absence of a universal DNA profiling method for plant authentication, the process can differ on a case-by-case basis. Consequently, various molecular techniques have been suggested to tackle this issue, utilizing variations in the sequences of different nuclear DNA and chloroplast regions unique to each species [5]. The DNA barcode identifies unknown plants using a short DNA sequence, which is far shorter than the complete genome, which can be acquired easily and affordably [5]. Using Basic Local Alignment Search Tool (BLAST) analysis, matK (MaturaseK gene, 99%) gene was the most successful plant species-level assigner, followed by trnH-psbA (95%) and rbcL (ribulose-bisphosphate carboxylase gene, 75%) gene [5]. In one of our recent works, the fruit of *Piper longum* has been authenticated by BLAST analysis using rbcL gene in addition to taxonomical authentication [10]. Current advancements in next-generation sequencing techniques and metabarcoding have further brought down sequencing costs and enhanced the process of authentication of complex herbal products with multiple ingredients. Genetic-method-based authentication is considered a highly effective tool for identifying plant species, even within complex mixtures. Due to the high sensitivity of DNA-based molecular techniques, they can detect trace amounts of DNA in both liquid and dry plant extracts. However, it is important to note that DNA methods alone are insufficient for the quality control and standardization of herbal drugs. Additionally, DNA-based molecular analysis of a plant does not correlate with the content of its bioactive phytochemicals [5]. These techniques also face several other challenges, such as microbial contamination of plant material/specimen during collection, processing, or testing; chances of DNA degradation during drying, grinding, and extraction; and following a multistep DNA extraction, purification, and amplification protocol. The high genetic variability leads to difficulties in defining a standard DNA barcode. The interspecies hybridization leads to the development of offspring with mixed genetic makeup, which leads to difficulty in identification. The limitations of the reference database lead to misidentification of species. The complexity of plant material leads to difficulty in deconvoluting the DNA profiles and hindrance of DNA extraction and PCR amplification, leading to complications in the molecular authentication. Such techniques are costly and require specialized equipment and expertise. However, using improved techniques of DNA extraction, stringent contamination control, development of comprehensive databases, use of advanced analytical techniques and bioinformatics tools, establishment of standardized protocols, and validation can avoid the aforementioned challenges.

3.1.12 Challenges with Protein Profiling

Although matrix-assisted laser desorption ionization time-of-flight mass spectrometry (MALDI-TOF MS) protein profiling is frequently used to identify microbes, this technique has rarely been applied in plants. Thus, appropriate techniques, m/z ranges, and markers must still be developed. By integrating two-dimensional electrophoresis with mass spectrometry (MS), proteomics has been applied to address the taxonomic and phylogenetic intricacies of certain plant taxa. Despite the promise these studies hold for proteomics, the extensive effort required renders it impractical for regular use, and advancements in MS are still necessary [5]. However, certain challenges, such as the complexity and diversity of plant proteomes, abundance of secondary metabolites, variation in protein expression across developmental stages and tissue specificity, and alteration of protein expression, pose serious challenges in responses to biotic and abiotic stresses. Further, there is difficulty in the extraction and purification of protein because of the presence of cell walls and secondary metabolites. Additionally, challenges in detecting low-abundance proteins due to the presence of highly abundant proteins, difficulty in achieving comprehensive proteome coverage due to the wide range of protein concentrations and the complexity of plant proteomes, and profiling complexity due to post-translational modifications (phosphorylation, glycosylation, and ubiquitination) require sophisticated bioinformatics tools for analysis and interpretation. There may be inconsistent results due to variations in sample preparation, analytical techniques, and data analysis methods; limitations of availability and quality of protein databases; and difficulty in achieving accurate quantification of proteins. Despite these challenges, advancements in technology and methodology are continuously improving plant protein profiling. Innovations such as improved mass spectrometry, advanced bioinformatics tools, and more robust protein extraction methods are helping to address these issues. Additionally, collaborative efforts to build comprehensive protein databases and standardized protocols are essential for advancing plant proteomics research.

3.1.13 Challenges in the Identification of Adulterants, Impurities, and Contaminants

The presence of adulterants, contaminants, or impurities in herbal drugs alters their therapeutic effectiveness or causes toxicity. Plants may contain traces of pesticides, other species of plant, microorganisms and their metabolic byproducts, radioactive elements, and toxic heavy metals as impurities or contaminants during the process of harvesting and cultivation. [5]. Therefore, it is essential to follow GACP along with contaminant and impurity control to ensure the safety and effectiveness of herbal products [5]. The adulterants, contaminants, and impurities are usually found in very low concentrations, so efficient analytical techniques, such as microscopy (i.e., for particles, molds, insect parts), spectroscopy (FT-IR, NMR, MS, ICP-MS), chromatography (HPLC, GC), DNA-based methods, and combination of these techniques are needed for their estimation [5]. Some of the key challenges associated with the identification of adulterants, impurities, and contaminants include the complexity of herbal drugs, the presence of low levels of contaminants, diverse contaminants, variability in raw materials, a lack of a universal method, the presence of natural analogues, and limited availability of reference standards. The presence of very low levels of contaminants requires highly sensitive analytical techniques to detect and quantify them. Variability in raw materials complicates the establishment of standard profiles for comparison. The lack of universally accepted standardized methods and natural analogues with similar chemical profiles makes them difficult to distinguish using conventional analytical methods. The presence of diverse contaminants and the limited availability of reference standards for the vast array of possible contaminants and adulterants also create difficulty in identifying adulterants, impurities, and contaminants.

3.1.14 Analytical Challenges Related to Pre-formulation Study

The pre-formulation study of herbal drugs evaluates the physico-chemical properties of extracts, fractions, or bioactive compounds that affect formulation stability and efficacy. It includes solubility (pKa, pH profile, partition coefficient, dissolution), bulk characterization (crystallinity, polymorphism, hygroscopicity, micromeritic properties), and stability (solution, solid-state, drug-excipient compatibility). Herbal formulations usually use multicomponent extracts or fractions, making such studies challenging due to their chemical complexity [11].

3.1.15 Analytical Challenges During Formulation Development and Characterization

Plant extracts are often hygroscopic, viscous, and poorly flowable, making handling, accurate measurement, and solvent selection difficult in formulation development. Their chemical complexity further complicates characterization. Since analyzing every phytoconstituent is impractical, formulations are usually evaluated using representative bioactive or marker compounds. For example, evaluating drug content, content uniformity, diffusion, dissolution, bioavailability, and stability, major bioactive or analytical components should be considered instead of numerous phytoconstituents. Moreover, the low concentration of individual phytoconstituents results in very low plasma levels, requiring advanced analytical techniques for bioavailability and pharmacokinetic studies [12].

3.1.16 Challenges with Stability Study and Shelf life Prediction

Shelf life is the period for which a drug/phytochemical remains stable under recommended storage conditions. In Ayurveda, this is termed "Saviryata Awadhi," traditionally assessed through sensory attributes like taste, smell, and appearance, though these are insufficient by modern quality standards. Today, API stability in allopathic formulations follows ICH and WHO guidelines, while herbal formulations are evaluated under AYUSH regulations and the Ayurvedic Pharmacopoeia of India. Stability parameters include chemical marker assays, TLC profiling, physicochemical properties, and physical attributes [13]. As per G.S.R. 918(E) (2015), phytopharmaceuticals must be standardized to at least four marker compounds, with stability data generated accordingly [14]. Challenges in herbal stability testing arise from complex, variable compositions, marker selection difficulties, and enzymatic influences [1]. The type and amount of phytoconstituents in herbal raw materials are shaped by environmental conditions (air, humidity, rainfall), agricultural practices (seed quality, fertilizers, growth conditions, pesticide use), contamination (microbial or heavy metal), and collection techniques (harvesting, drying, and post-harvest handling). Consequently, herbal drugs from different sources often show variations in chemical composition. This leads to differences in the phytochemical profile of products across manufacturers, with similar formulations containing varying levels of marker compounds, which may alter required dosages, therapeutic outcomes, and shelf life [1].

Herbal formulations consist of complex, heterogeneous mixtures of phytoconstituents that span a wide spectrum of polarity, solubility, acidity/alkalinity, and molecular weight. This chemical diversity makes it difficult to predict the stability of individual components within the final product. Moreover, the presence of multiple chemical groups complicates the accurate estimation of the shelf life of herbal formulations [1].

Microbial contamination in herbal products can degrade certain phytoconstituents through enzymatic action, posing health risks. Therefore, monitoring microbial load is a critical aspect of stability studies. In addition, the presence of heavy metals and pesticide residues represents another major concern, as these contaminants not only introduce toxicity but may also interact with phytoconstituents, altering the phytochemical profile during storage and complicating stability evaluation [1].

While stability can be assessed using chemical markers, some medicinal plants still lack identified marker components. Acquiring standard markers is not always feasible [1]. In synthetic formulations, APIs are well defined and serve as direct markers for stability, with quantitative changes reflecting therapeutic efficacy. In contrast, herbal products contain complex mixtures of constituents, often with unidentified active components that may act independently, synergistically, or antagonistically [15]. Their therapeutic effects usually arise from the combined action of diverse phytoconstituents, making changes in individual markers less predictive of efficacy [1]. Hence, selecting suitable markers for stability testing is both critical and challenging, particularly in polyherbal formulations where complexity is further amplified [15].

In herbal formulations with multiple constituents, each component has a distinct shelf life, making it challenging to predict the final product's overall shelf life or expiry date and recommend appropriate storage conditions [16, 17]. The shelf life of a particular herbal substance or product may vary depending on the chemical markers considered, complicating the prediction of shelf life for herbal products [1]. Further, certain hurdles arise during stability predictions, such as (a) challenges in analyzing and developing analytical methods (b) possibility for chemical reactions between phytoconstituents, and (c) enzymatic activities during shelf life.

Since herbal drugs are complex, developing analytical methods for determining every chemical marker is very tedious. In the case of synthetic drug-based formulation, mostly one or three drugs are present, and quantification at various time points of stability study is quite easy compared to the stability study of the herbal product, in which simultaneous analysis of several phytoconstituents is required [15]. In many cases, a complete herbal product is often regarded as the active ingredient, rather than its separate constituents [18]. Chemical markers of herbal drugs can be selected to assess the quality, stability, and shelf life of the final herbal formulation. The amount of a chemical marker serves as an indicator of the herbal product's quality [2]. An ideal protocol for stability studies of herbal substances or products should include the quantitative estimation of markers, whose levels can be linked to their required pharmacological activities. However, identifying the active marker of a herbal drug remains challenging. In such cases, analytical markers may be used to estimate the shelf life of herbal drugs. In multicomponent herbal formulations, each component has a different shelf life, making it difficult to predict the shelf life of the final product and recommend appropriate storage conditions. [19].

During stability studies, exposing the herbal formulation to various environmental conditions, such as light, temperature, humidity, or air, can lead to potential chemical reactions among its phytoconstituents. Such reactions lead to chemical degradation, alteration of the number of bioactive constituents, and, ultimately, the therapeutic activities. This may lead to a false estimation of shelf life [1, 15].

Phytoconstituent levels in plants are regulated by enzymes, which remain active even after harvest and during storage, altering the chemical composition of herbal products. Enzymes such as catalase, peroxidase, lipoxygenase, and others significantly influence product stability and shelf life. Therefore, identifying and monitoring post-harvest enzyme activity through specific assays is essential for assessing the shelf life of herbal formulations [1].

3.2 CONCLUSION

Herbal formulations present numerous challenges that start from authentication and standardization to analytical evaluation and size characterization. The inherent chemical complexity, variability in raw materials, and presence of multiple bioactive compounds make it difficult to establish uniform quality, safety, and efficacy standards. Additionally hurdles arise from methodological limitations, such as difficulties in sample preparation, size analysis, and handling of heterogeneous phytoconstituents. Analytical and regulatory inconsistencies across regions further complicate the process, hindering reproducibility, large-scale application, and international acceptance. Addressing these challenges requires a multidisciplinary approach that integrates advanced analytical tools, harmonized regulatory frameworks, and innovative formulation strategies to ensure reliable development and global standardization of herbal medicines.

REFERENCES

1. Bansal, G., et al., Stability testing of herbal drugs: Challenges, regulatory compliance and perspectives. *Phytotherapy Research*, 2016. 30(7): p. 1046–1058.
2. Kushwaha, S.K., et al., Role of markers in the standardization of herbal drugs: A review. *Archives of Applied Science Research*, 2010. 2(1): p. 225–229.
3. Kunle, O.F., H.O. Egharevba, and P.O. Ahmadu, Standardization of herbal medicines-A review. *International Journal of Biodiversity and Conservation*, 2012. 4(3): p. 101–112.
4. Sen, S. and R. Chakraborty, Revival, modernization and integration of Indian traditional herbal medicine in clinical practice: Importance, challenges and future. *Journal of Traditional and Complementary Medicine*, 2017. 7(2): p. 234–244.
5. Muyumba, N., et al., Quality control of herbal drugs and preparations: The methods of analysis, their relevance and applications. *Talanta Open*, 2021. 4: p. 100070.
6. Santos, M.S., et al., Authenticity study of Phyllanthus species by NMR and FT-IR Techniques coupled with chemometric methods. *Química Nova*, 2012. 35: p. 2210–2217.
7. Assi, S., et al. Identification of counterfeit tobacco using Fourier transform infrared spectroscopy, in *Meeting of the London Toxicology Group*. 2014, London Toxicology Group London.
8. Booker, A., D. Johnston, and M. Heinrich, Value chains of herbal medicines – ethnopharmacological and analytical challenges in a globalizing world, in *Evidence-Based Validation of Herbal Medicine*. 2015, Elsevier. p. 29–44.
9. Takechi-Haraya, Y., et al., Current status and challenges of analytical methods for evaluation of size and surface modification of nanoparticle-based drug formulations. *AAPS PharmSciTech*, 2022. 23(5): p. 150.
10. Mohapatra, D., et al., Quality by design – based development and optimization of fourth-generation ternary solid dispersion of standardized Piper longum extract for melanoma therapy. *Drug Delivery and Translational Research*, 2023. 13(12): p. 3094–3131.
11. Lachman, L., H.A. Lieberman, and J.L. Kanig, *The Theory and Practice of Industrial Pharmacy*. 1976, Lea & Febiger Philadelphia.
12. Bhattaram, V.A., et al., Pharmacokinetics and bioavailability of herbal medicinal products. *Phytomedicine*, 2002. 9: p. 1–33.
13. *General Guidelines for Drug Development of Ayurvedic Formulations, g.s.-i., Central Council for Research in Ayurvedic Sciences Ministry of AYUSH*. Government of India, New Delhi. 2018.
14. Government of India. *Drugs and Cosmetics (Eighth Amendment) Rules, 2015, G.S.R. 918(E), The Gazette of India, Extraordinary, Part II, Section 3, Sub-section (i)*. Ministry of Health and Family Welfare. 30 Nov 2015.
15. Sachan, A.K. and A. Kumar, Stability testing of herbal products. *Journal of Chemical and Pharmaceutical Research,* 2015. 7(12): p. 511–514.
16. Gupta, A., M. Jaiswal, and K. Pradeep Prajapati, Shelf life of Ayurvedic dosage forms-Traditional view, current status and prospective need. *Indian Journal of Traditional Knowledge*, 2011. 10.
17. Thakur, L., et al., Novel approaches for stability improvement in natural medicines. *Pharmacognosy Reviews*, 2011. 5(9): p. 48.
18. Kumadoh, D. and K. Ofori-Kwakye, Dosage forms of herbal medicinal products and their stability considerations-an overview. *Journal of Critical Reviews*, 2017. 4(4): p. 1–8.
19. Sahu, A.N. and D. Mohapatra, *Herbal Drug Formulation and Standardization*. 2021, New Delhi: Ane Books Pvt. Ltd. ISBN-10: 9390658365, ISBN-13: 978–9390658367.

4 Instrumental Methodology for Qualitative and Quantitative Analysis, Quality Control, and Standardization of Herbal Drugs and Formulations

ABBREVIATIONS

Abbreviation	Description
AFLP	Amplified Fragment Length Polymorphism
AKC	Ayush Kvatha Churna
APCI	Atmospheric Pressure Chemical Ionization
AYUSH	Ayurveda, Yoga and Naturopathy, Unani, Siddha and Homoeopathy
CV	Coefficient of Variation
DESI	Desorption Electrospray Ionization
DNA	Deoxyribonucleic Acid
dNTPs	Deoxynucleoside Triphosphates
ESI	Electrospray Ionization
FTIR	Fourier-Transform Infrared
GC	Gas Chromatography
GC-MS	Gas Chromatography-Mass Spectrometry
HPLC	High-Performance Liquid Chromatography
HPTLC	High-Performance Thin-Layer Chromatography
HPTLC-MS	High-Performance Thin-Layer Chromatography-Mass Spectrometry
ICH	International Council for Harmonisation
ISSR	Inter-Simple Sequence Repeats
ITS	Internal Transcribed Spacer
k′	Capacity Factor
LAMP	Loop-Mediated Isothermal Amplification
LC-MS	Liquid Chromatography-Mass Spectrometry
LOD	Limit of Detection
LOQ	Limit of Quantification
m/z	Mass-to-Charge Ratio
MALDI	Matrix-Assisted Laser Desorption/Ionization
matK	Maturase K
MS	Mass Spectrometry
NGS	Next-Generation Sequencing

DOI: 10.1201/9781003540823-4

NMR	Nuclear Magnetic Resonance
PCA	Principal Component Analysis
PCR	Polymerase Chain Reaction
RAPD	Random Amplified Polymorphic DNA
rbcL	Ribulose-1,5-Bisphosphate Carboxylase/Oxygenase Large Subunit
$\mathbf{R_f}$	Retention Factor
RFLP	Restriction Fragment Length Polymorphism
RNase	Ribonuclease
RSD	Relative Standard Deviation
SCAR	Sequence-Characterized Amplified Regions
S/N	Signal-to-Noise Ratio
SSR	Simple Sequence Repeats
$\mathbf{T_f}$	Tailing Factor
TLC	Thin-Layer Chromatography
TOF	Time-of-Flight
$\mathbf{R_t}$	Retention Time
UPLC	Ultra-Performance Liquid Chromatography
UV	Ultraviolet
WHO	World Health Organization

4.1 INTRODUCTION

Plants have been used as natural remedies to treat and alleviate various diseases since prehistoric times. As time progresses, this traditional knowledge has advanced, contributing significantly to modern drug discovery and development. The exploration of medicinal plants continues with the aim of developing enhanced drugs that offer higher efficacy, improved safety, and increased cost-effectiveness. However, with increasing demand for medicinal plants, attempts to meet the needs have resulted in increased adulteration. Simultaneously, plants are highly vulnerable to contamination at every stage of processing, from harvesting to final production. The chemical composition of herbs and herbal products can also vary significantly due to multiple factors like the stage of collection, the specific plant parts harvested, seasonal variations, geographical origin, drying methods, and other processing parameters [1]. Such variations can impact the quality, consistency, and efficacy of the final product. Substitutes of inferior quality and lower cost are often introduced, which may lack the necessary number of bioactive compounds required for effective therapeutic outcomes [2]. Herbal extracts are typically complex mixtures composed of numerous constituents. Identifying the active principles responsible for therapeutic effects is often challenging, as they may remain unknown in many cases. Additionally, the lack of commercially available selective analytical methods or reference standards further complicates the standardization and quality control of herbal formulations [1]. Therefore, proper identification of the plant is essential, as its misuse or incorrect utilization could lead to adverse interactions and even fatal outcomes.

With the increasing commercialization of herbal medicines, ensuring the safety, quality, and efficacy of medicinal plants and herbal formulations has become a critical concern. The major challenge in the herbal drug industry is the absence of stringent quality control guidelines for raw materials and formulations. Therefore, standardization, as defined by the American Herbal Product Association, involves implementing measures to ensure consistency in products. This process assesses the quality and purity of herbal drugs through various parameters, including morphological, microscopic, physical, chemical, and biological evaluations [3].

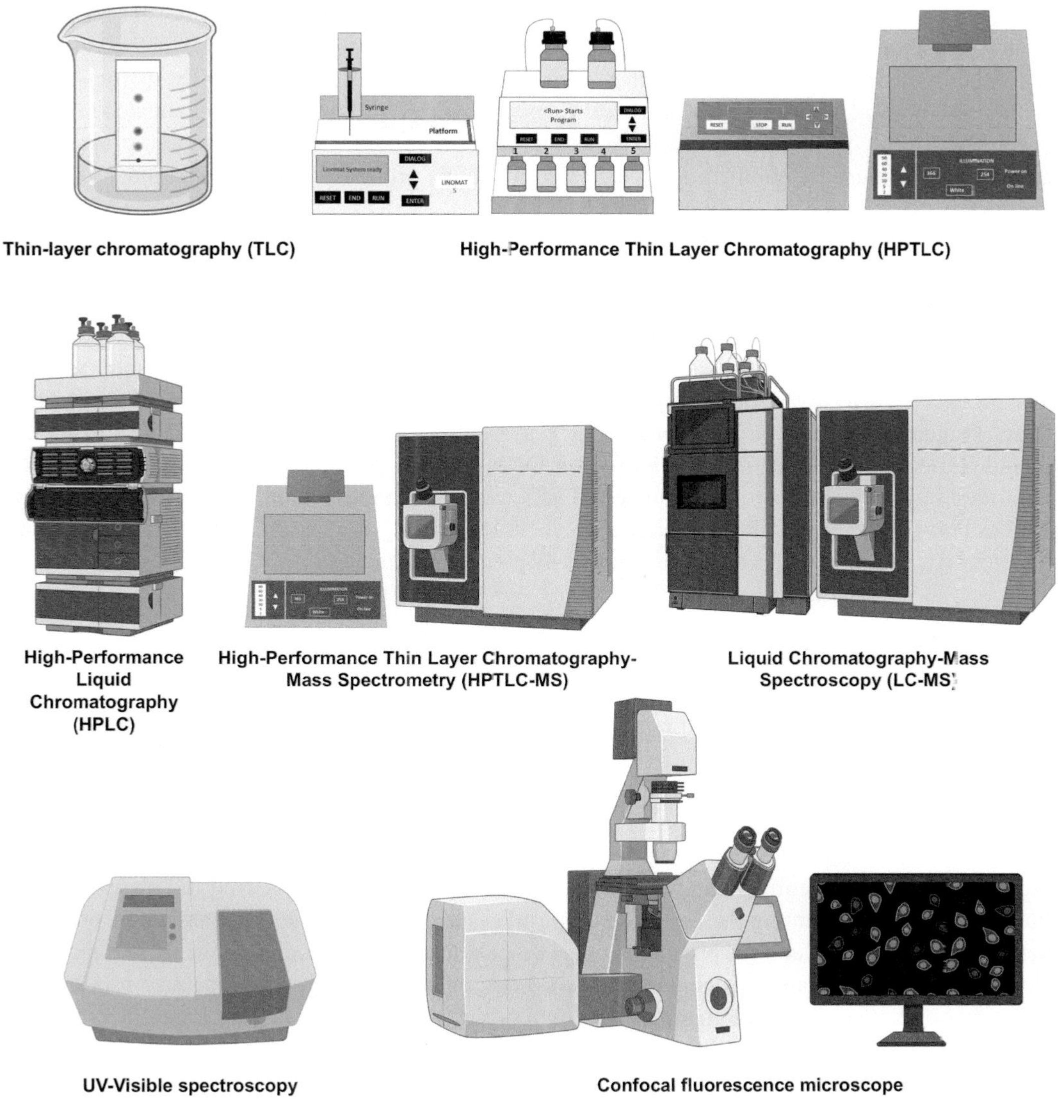

FIGURE 4.1 Instruments for structural elucidation and phytochemical integrity analysis of herbal drugs and formulations

This chapter focuses on instrumental methodologies essential for the authentication, quality control, and standardization of phytopharmaceuticals. It discusses molecular-based authentication techniques, followed by the application of various techniques (Figure 4.1), HPLC, HPTLC, and LC-MS UV-Visible spectroscopy and confocal fluorescence microscopy for bioactive compound analysis. The integration of these methods with chemometric tools is also addressed to improve batch-to-batch consistency and compliance with regulatory standards.

4.2 PLANT AUTHENTICATION

Plant authentication is the process of accurately verifying the identity of medicinal plant species to ensure the quality, safety, and efficacy of herbal products. It is a critical quality control measure that prevents adulteration, substitution, and the use of spurious materials, which can compromise therapeutic effectiveness and pose health risks. Authentication methods include morphological and anatomical assessments, chemical profiling, and molecular approaches such as DNA marker analysis. Among these, DNA-based techniques offer superior specificity and reliability, as they are not influenced by plant age, environmental conditions, or tissue type. The use of authenticated reference specimens from certified herbaria further strengthens the accuracy of identification, providing an essential benchmark for confirming species identity before processing and formulation.

4.2.1 Genetic Plant Fingerprinting

DNA fingerprinting, also referred to as DNA typing, genetic fingerprinting, or DNA profiling, is widely utilized in plants for purposes such as plant identification, species identification, biodiversity conservation, identifying trait-associated markers, and analyzing genetic diversity and variation. The process involves isolating DNA from plant cells, followed by quantification and quality assessment. Fingerprinting can be performed using polymerase chain reaction (PCR)-based techniques, including random amplification of polymorphic DNA (RAPD), inter-simple sequence repeat (ISSR), amplified fragment length polymorphism (AFLP), and DNA amplification fingerprinting (DAF), as well as non-PCR methods like restriction fragment length polymorphism (RFLP). Table 4.1 shows different types of molecular techniques for DNA plant authentication.

4.2.2 Current Techniques of Genetic Fingerprinting

4.2.2.1 DNA Fingerprinting Technique

DNA fingerprinting is an essential technique for standardizing and ensuring the quality of herbal drugs by distinguishing authentic products from substituted or adulterated ones. Unlike phytochemical profiles, which can vary due to environmental factors, plant parts, or physiological conditions, DNA remains consistent, making it a reliable marker for identification. This is especially useful when chemical composition alone cannot confirm the authenticity of a herb. DNA fingerprinting is effective in detecting adulterants, even in processed materials, ensuring the integrity of the final product and preventing the inclusion of harmful substances. Studies have shown a correlation between DNA markers and the phytochemical composition of plants, aiding in the differentiation of closely related species and ensuring that the correct species with desired therapeutic properties are used. Techniques such as random amplified polymorphic DNA RAPD and RFLP have been applied to detect interspecies variation in genera like *Glycyrrhiza*, *Echinacea*, *Curcuma*, and *Arabidopsis.* This allows for precise species identification within complex plant families. The integration of molecular techniques like DNA fingerprinting with analytical tools has created a comprehensive botanical characterization system, improving the quality, safety, and efficacy of large-scale herbal drug production by identifying authentic species and detecting adulterants [4].

4.2.2.2 DNA Barcoding Technique

DNA barcoding may be defined as the use of short nuclear or organelle DNA sequences for the identification of organisms. This technique identifies the unknown plant species by referring to the preexisting classification. Utilizing a particular region of the chloroplast/nuclear gene (rbcL, matK, trnH-psbA, trnL-F, rpl36-rps8, ITS, and 5S rRNA), barcoding analysis is carried out. However, in the case of the animal mitochondrial gene region (Cytochrome Oxidase 1), coding DNA could be applied to generate DNA barcodes in animals. DNA barcoding focuses on finding a universal DNA sequence that can differentiate organisms. To ensure the reliability of DNA barcoding, an approach of bi-directional sequencing of the selected genetic amplicon is necessary, which allows for the precise organism and taxonomic classification [4]. In a recent study, DNA-based molecular authentication was used for the authentication of *Piper longum* fruits (Figure 4.2). The BLAST analysis results of the rbcL sequence showed a percentage identity of 100% to *Piper longum* (Accession: ON720789.1), confirming the authenticity of the collected fruits [1].

TABLE 4.1
Different Types of Molecular Techniques for DNA Plant Authentication

Marker Type	Description	Techniques	Application	Limitation	Examples of Plants	References
RAPD (Random Amplified Polymorphic DNA)	Uses random primers for DNA amplification and provides random patterns for genetic diversity and plant authentication.	PCR with random primers. Amplified fragments are analyzed on agarose gels.	Genetic diversity Species authentication Population studies	Low reproducibility Sensitive to PCR conditions	Useful in differentiating the two varieties of *Silybum marianum* (var. album and var. purple), which are difficult to distinguish in dried form	3, 5
RFLP (Restriction Fragment Length Polymorphism)	Detects DNA polymorphisms by restriction enzyme digestion and hybridization.	DNA is digested with restriction enzymes, separated on gels, and hybridized with labelled probes.	Identifying genetic diversity and genetic mapping	Labor intensive; requires large amounts of high-quality DNA	*Boerhavia diffusa* L. was differentiated from *T. portulacastrum* and *T. monogyna* by PCR amplification *Angelica sinensis* was differentiated from seven different adulterant species of Angelica	3, 5
AFLP (Amplified Fragment Length Polymorphism)	Combines of RAPD and RFLP to create unique DNA patterns.	Restriction enzymes digest DNA. Adapters and selective primers are used for amplification.	High-resolution genetic analysis Species authentication	Expensive Labor intensive Requires high-quality DNA	Differentiate *E. purpurea* from *E. angustifolia* and *E. pallida* by generating specific fragments with different primer combinations	3, 5
SSR (Simple Sequence Repeats)	Targets short tandem repeats in DNA. Highly polymorphic and co-dominant.	PCR amplification with primers flanking SSR regions. Fragments are resolved on agarose/ polyacrylamide gels.	Genetic fingerprinting, cultivar identification, and marker-assisted breeding	Requires prior sequence information; limited to known SSR loci	Analyze both American and Oriental ginseng roots Study of Capsicum species uses microsatellites to map markers onto the tomato genome	3, 5
ISSR (Inter-Simple Sequence Repeats)	Amplifies regions between microsatellite repeats. Highly reproducible and dominant marker.	PCR amplification with primers complementary to microsatellite repeats.	Species identification, genetic diversity studies	Dominant marker (cannot distinguish homozygotes from heterozygotes); lower resolution compared to SSR	*Phyllanthus amarus* (authentication of anti-hepatotoxic herbal extracts), *Crocus sativus* (detection of saffron adulteration by related species)	5

(Continued)

TABLE 4.1 ***(Continued)***
Different Types of Molecular Techniques for DNA Plant Authentication

Marker Type	Description	Techniques	Application	Limitation	Examples of Plants	References
SCAR (Sequence-Characterized Amplified Regions)	Converts polymorphic DNA markers (RAPD, AFLP, ISSR) into sequence-based markers. Highly specific and reproducible.	Elution and sequencing of unique DNA fragments; primers designed for sequence amplification.	Species-specific identification, and authentication of herbal drugs	Requires prior sequence data; costly and time consuming for marker development	*Bacopa monnieri* (differentiation of species with overlapping morphology), *Cyperus rotundus* (distinguishing medicinal roots from adulterants), *Aconitum heterophyllum* (species-level discrimination)	3
DNA Barcoding	Uses short nuclear or organelle DNA sequences for species identification. Combines conserved and variable regions.	Amplification and sequencing of barcoding regions (e.g., *rbcL*, *matK*, ITS).	Species-level identification and detection of adulterants	May fail to distinguish closely related species; degraded DNA amplification can be challenging	*Crocus sativus* (authentication of saffron and detection of adulterants like *Calendula officinalis*), *Schisandra chinensis* (population-level differentiation), *Astragalus spp.* (species discrimination)	3
LAMP (Loop-Mediated Isothermal Amplification)	Amplifies DNA at a constant temperature using four primers. Detects species-specific differences with high sensitivity and specificity.	Amplification occurs at 60–66°C; detection is via turbidity or white precipitate (magnesium pyrophosphate).	Rapid identification of medicinal plants and adulterants	Complex primer design; limited studies in medicinal plants	*Nigella sativa* (development of molecular identification tools), *Taraxacum formosanum* (discrimination from adulterants like *Taraxacum officinale*), *Zingiber officinale* (validation of authentic ginger)	3
NGS (Next-Generation Sequencing)	High-throughput sequencing generates vast sequence data. Enables whole-genome and transcriptome analysis.	DNA is fragmented, amplified, and sequenced on platforms (e.g., Illumina, Roche/454). Data was analyzed using bioinformatics tools.	Species identification, transcriptome profiling, pathway analysis	Expensive; requires bioinformatics expertise; limited read lengths for some platforms	*Costus pictus* (identification of anti-diabetic properties through transcriptome analysis), *Dendrobium officinale* (analysis of biosynthetic pathways for medicinal compounds), *Huperzia serrata* (gene discovery)	3

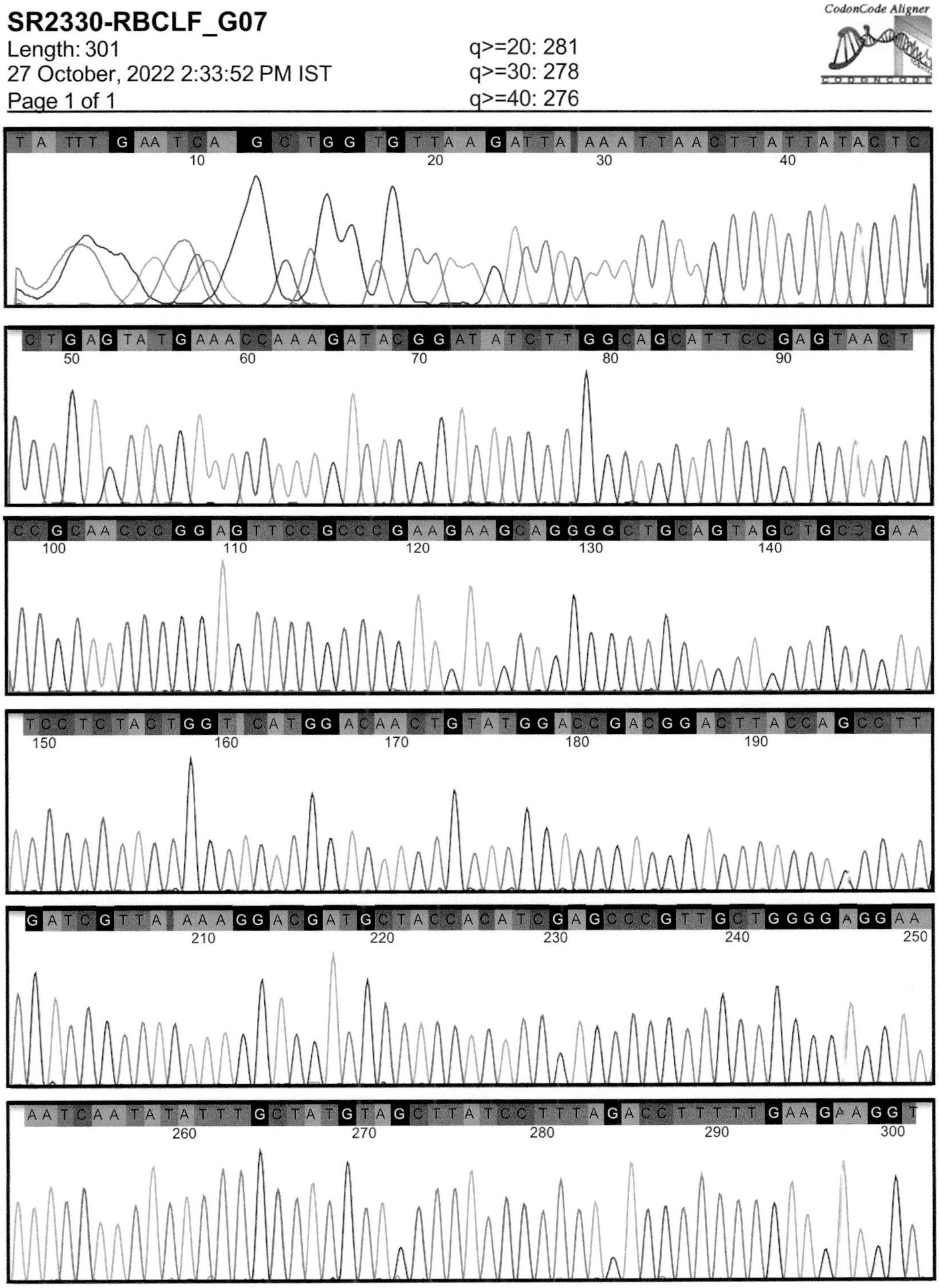

FIGURE 4.2 DNA sequencing chromatogram of rbcL gene of *Piper longum*

4.2.2.3 The ISSR Technique

ISSR is a PCR-based technique widely used in DNA fingerprinting due to its cost-effectiveness and versatility. It is applied in various areas such as genetic fingerprinting, gene tagging, detecting clonal variations, phylogenetic analysis, assessing genomic instability, and evaluating hybridization. For example, ISSR markers have been used to differentiate *Cannabis sativa* and *Arabidopsis thaliana* L. Heyne from their adulterated species. Another DNA fingerprinting tool, sequence-characterized amplified region (SCAR) markers, offers reliable authentication of herbs and effective discrimination from adulterants. SCAR markers are also useful for differentiating morphologically similar plant species. DNA-based molecular markers have proven valuable in distinguishing different accessions of plants collected from various geographical regions. Examples include *Taxus wallichiana*, *Azadirachta indica*, *Juniperus communis L.*, *Codonopsis pilosula*, *Allium schoenoprasum* L., and *Andrographis paniculata*. These markers play a crucial role in ensuring the authenticity of medicinal plants in herbal formulations [5].

4.2.2.4 Microsatellite Techniques

Microsatellites, also known as simple sequence repeat (SSR) markers, are DNA sequences comprised of 2–5 base pairs that repeat several times at a specific site or location in the genome. These markers are fully codominant and are commonly used for applications such as marker-assisted selection, determining lineage, and studying population structure. Microsatellite markers were found to be widely distributed throughout the genome and locally replicated in tandem, so it is necessary to amplify specific markers using primers designed for each unique location. The process of amplifying includes first identifying the exact sequence of the microsatellite through sequence analysis and comparing it with the available DNA library to help find and study these markers. The identification process may include cloning, sequencing, or using labelled probes to highlight the repeat regions [2].

4.2.3 Process of DNA-Based Molecular Authentication

The process of DNA-based molecular authentication of plants is shown in Figure 4.3. It starts by isolating the DNA through the homogenization of the plant tissue and extracting DNA using kits like the NucleoSpin® Plant II Kit, followed by RNase treatment and filtration to remove RNA and debris. The DNA is further purified through spin-column washing, eluted, and stored. DNA quality is checked via agarose gel electrophoresis. PCR is performed using rbcL primers and a master mix in a thermal cycler, and the amplified products are confirmed by electrophoresis. Post-PCR, ExoSAP-IT is used to remove residual primers and dNTPs. Sequencing is conducted with the BigDye Terminator v3.1 Kit, and the products are cleaned via ethanol precipitation. Sequenced data is analyzed using Sequence Scanner and Geneious Pro for alignment and editing, followed by BLAST to compare sequences with databases. BLAST results are interpreted using bit scores, which reflect alignment quality, and E-values, which indicate the statistical significance of the match [2, 3].

4.2.4 Advantages of Genetic Fingerprinting

- A very small amount of sample is required; the whole plant is not required for authentication, as in the case of taxonomy.
- Genetic fingerprinting provides a unique genetic profile for individuals or species, ensuring precise identification.
- It minimizes errors compared to traditional morphological, microscopical, and biochemical methods.
- It helps differentiate between closely related plant species.
- It detects adulteration and ensures the purity of medicinal plant extracts.
- It aids in tracking genetic diversity among populations and assists in species conservation and breeding programs.

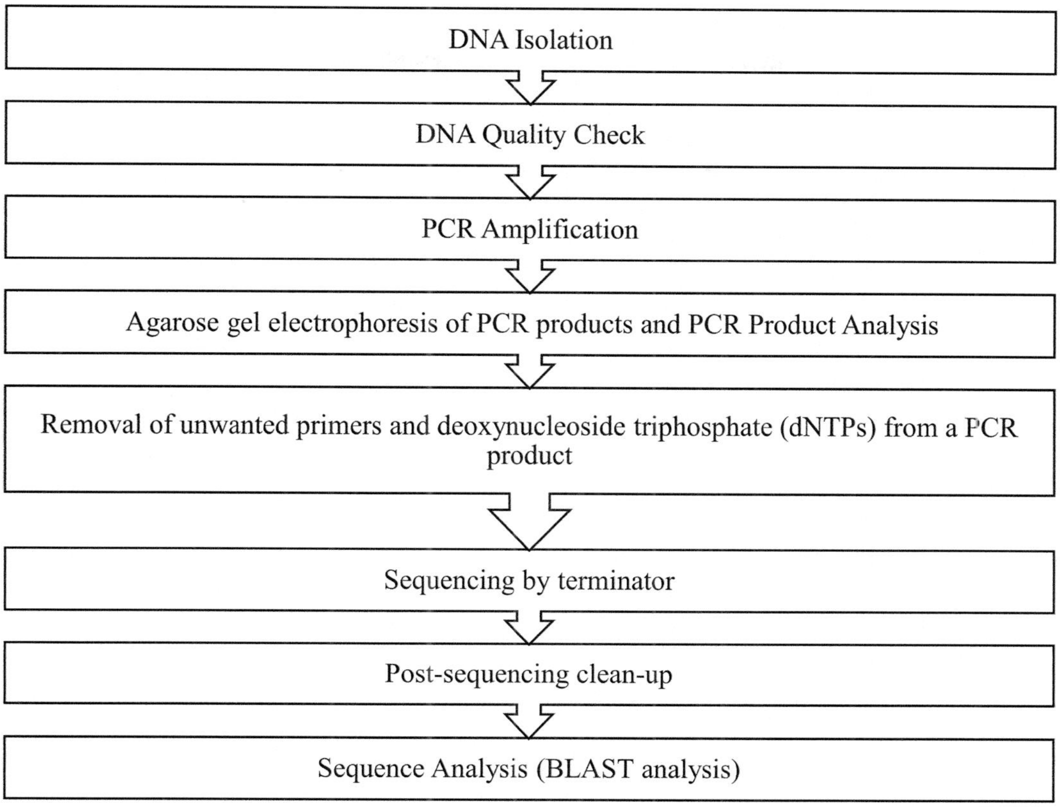

FIGURE 4.3 Process of DNA-based molecular authentication of plants

4.2.5 Challenges with the Genetic Fingerprinting

- Genetic fingerprinting is a precise tool for plant identification, but several challenges limit its practical use in herbal product standardization.
- DNA degradation during the processing (drying, extraction) often results in poor-quality DNA, especially in powdered or aged samples.
- A lack of standard protocols leads to variations in DNA extraction, primer design, and PCR conditions, which affect reproducibility.
- Complex polyherbal formulations contain multiple species, making it difficult to detect low-abundance components.
- Intraspecific genetic variability within the same species can result in misidentification.
- Public DNA databases are often incomplete or contain misannotated reference sequences, limiting accurate species identification.
- The technique is resource intensive, requiring expensive instruments, reagents, and skilled personnel.
- Legal and regulatory constraints, such as those under the Nagoya Protocol, may limit access to genetic resources and reference data.
- Plant DNA fingerprinting identifies genetically diverse populations, unlike human fingerprinting, which targets individuals. Its complexity and limited advancement stem from multiple species and diverse mating systems [4].

4.2.6 Chemical Fingerprinting

Chemical fingerprinting involves identifying and quantifying specific chemical substances or components in a plant using chemical markers. It is also referred to as chromatographic fingerprinting, as it generates a chromatographic profile of a plant or plant product, representing the extract's chemically significant components. The chemical profile is characterized based on the integrity, similarity, and reproducibility, providing a chemical representation of the plant under investigation. Chromatographic fingerprinting is performed using techniques such as thin-layer chromatography (TLC), high-performance thin-layer chromatography (HPTLC), high-performance liquid chromatography (HPLC), gas chromatography (GC), and various hyphenated methods like LC-MS and GC-MS [3].

4.2.6.1 Thin Layer Chromatography

4.2.6.1.1 Principle

TLC is an analytical technique based on the principle of differential adsorption. In this method, a sample is applied near the base of a plate coated with an adsorbent material such as silica gel or alumina, which serves as the stationary phase. When the plate is placed in a developing chamber containing an appropriate solvent (the mobile phase), the solvent rises through the plate by capillary action. As it ascends, the components of the sample separate based on their relative affinities for the stationary and mobile phases. Compounds with greater solubility in the mobile phase travel farther, while those with stronger interactions with the stationary phase move more slowly. This results in the formation of distinct spots on the plate, which can be visualized and identified using suitable detection techniques (Figure 4.4).

4.2.6.1.2 Advantages

- Simple, cost effective, and time efficient with a short development time
- Requires only a small amount of sample
- No need for complex instrumentation
- Allows simultaneous analysis of multiple samples
- Suitable for both qualitative and semi-quantitative analysis

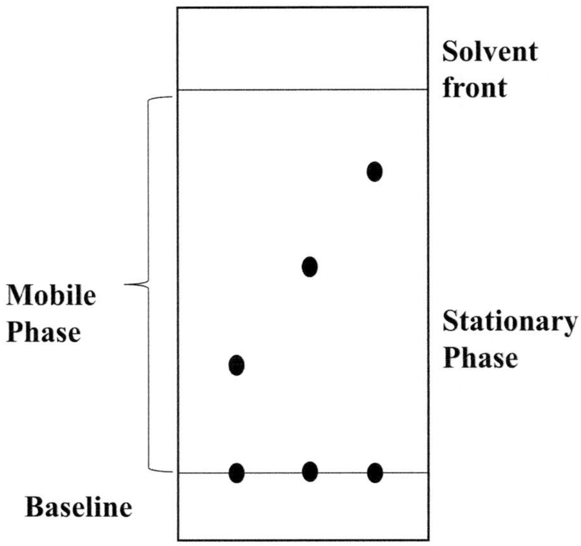

FIGURE 4.4 Principle of TLC

- Enables visual detection using UV light or specific reagents
- Facilitates easy identification and purity assessment of compounds
- Provides fast separation with high selectivity, even for closely related compounds
- Supports effective isolation of most components
- An economical chromatographic technique ideal for rapid and reliable analysis

4.2.6.1.3 Instrumentation

The components of TLC instrumentation are illustrated in Figure 4.5 and are briefly described here.

- **TLC plate:** A flat, inert support (usually glass, aluminum, or plastic) coated with a thin layer of stationary phase (commonly silica gel, alumina, or cellulose).
- **Sample applicator (spotting device):** A capillary tube or micropipette used to apply small, precise volumes of the sample solution onto the TLC plate.
- **Developing chamber (TLC chamber):** A closed container (e.g., glass jar or tank) that holds the mobile phase (solvent system). It maintains a saturated atmosphere to promote even solvent rise and consistent development.
- **Mobile phase (solvent system):** A suitable solvent or mixture of solvents that moves through the stationary phase via capillary action, enabling separation based on compound polarity.
- **Saturation filter paper (optional):** A filter paper lining the inner wall of the developing chamber to help saturate the atmosphere with solvent vapors, improving reproducibility and resolution.
- **Drying Station:** An area or device (often a hot air blower or drying oven) used to dry the developed TLC plate before visualization.
- **Visualization chamber or UV cabinet:** A UV light box (typically at 254 nm or 365 nm) used to observe UV-active compounds. Visualization can also be done using chemical reagents like iodine vapor, ninhydrin, or anisaldehyde.
- **TLC scanner**: Used for measuring Rf values. For quantitative analysis, a TLC densitometer or scanner may be used to detect and quantify compound spots.

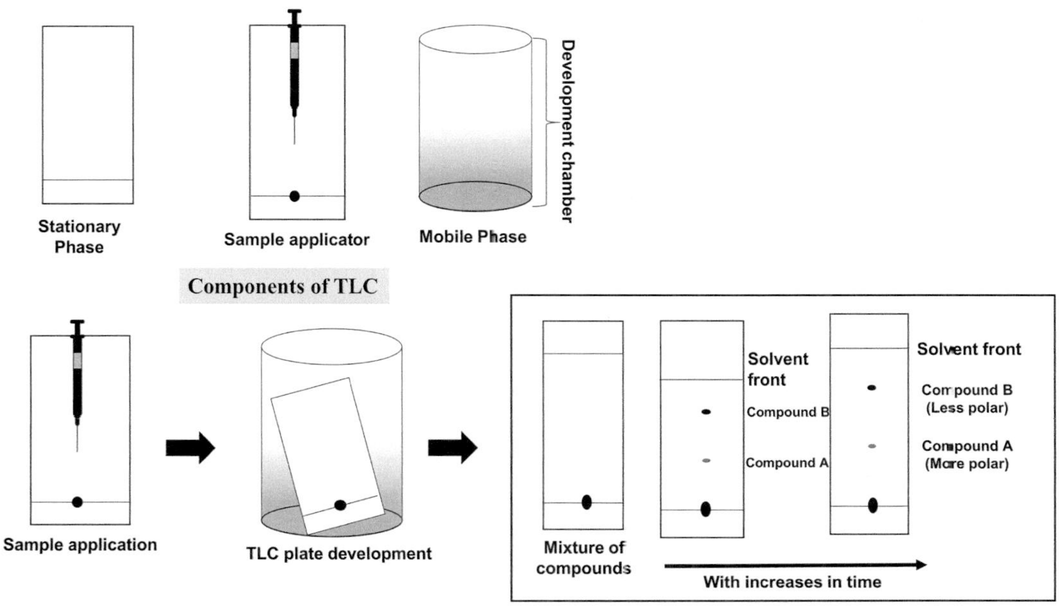

FIGURE 4.5 Components of TLC setup

4.2.6.1.4 *Procedure*

The procedure of TLC is shown in Figure 4.6.

4.2.6.1.5 *Applications to Herbal Drugs*

- Identification of marker compounds in plant extracts
- Detection of adulterants or substitutes in herbal raw materials
- Fingerprint profiling for standardization and authentication
- Monitoring stability and degradation of herbal formulations
- Biological detection TLC is used to locate antimicrobial, antioxidant, or enzyme-inhibitory constituents
- Comparative phytochemical screening across different species or batches

4.2.6.1.6 *Limitations of TLC*

- Uneven or overloaded spots can cause streaking, poor separation, or overlapping of compounds.
- Using a non-optimized mobile phase can lead to poor resolution, non-separation, or all compounds moving to the solvent front or remaining at the baseline.
- Touching the plate or exposure to dust, oils, or moisture can interfere with the migration and separation of analytes.
- If the developing chamber is not properly saturated with solvent vapors, the solvent front may rise unevenly, affecting reproducibility and resolution.

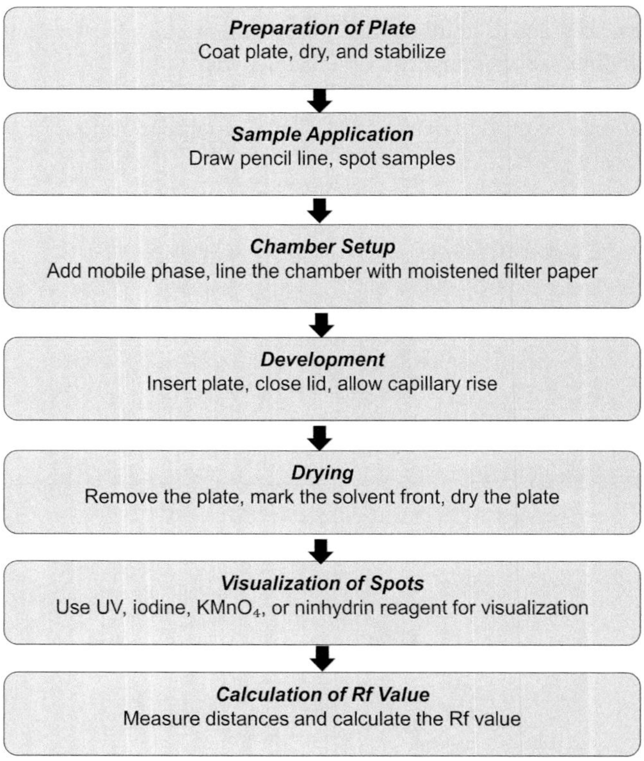

FIGURE 4.6 Procedure of TLC

- Uneven plate coating can result in irregular migration of compounds.
- Allowing the solvent front to run too far can cause diffusion of the spots, making them hard to distinguish or interpret.
- Insufficient drying of the plate before visualization can affect the interaction with detection reagents or UV light.
- Non-UV-active or colorless compounds may be hard to detect without appropriate visualization reagents.
- Small variations in environmental conditions, solvent composition, or plate handling can affect the reproducibility of results.

4.2.6.2 High-Performance Thin Layer Chromatography

HPTLC is a planar chromatographic technique that works on principles similar to TLC [5]. In particular, design for the analysis of the natural product offers a way to visually represent the complex mixture of substances found in a plant matrix which can be visualized using multiple methods; individual spots can be quantitatively analyzed by densitometry along a specific track, known as a fingerprint, which is characterized by the arrangement of zones in the sample's chromatogram, which acts as the primary visual tool for verifying identity and purity. This analysis considers the number, R_f values, and color or fluorescence of the zones under different detection modes at different wavelengths [6, 7]. The versatility of HPTLC lies in its multiple detection capabilities, allowing for simultaneous detection of chemicals from various phytochemical groups in a single analysis [7]. HPTLC is an advanced version of TLC with significant advantages, including higher efficiency, quicker detection, improved mass transfer, and faster analysis [5].

HPTLC has long been utilized to authenticate herbal drugs and extracts while detecting adulteration or falsification, serving as a critical tool for ensuring their purity and integrity [7]. HPTLC fingerprinting tool identifies and standardizes medicinal plants by analyzing molecular markers and chemical profiles. It allows the quantitative analysis of standard marker compounds, ensuring quality in herbal formulations. Standardizing the extraction process and using well-characterized chemical standards are essential for accurate monitoring and quantification [5].

4.2.6.2.1 Principle

HPTLC works on the fundamental principles of adsorption chromatography, separating sample components based on their interactions with the stationary and mobile phases. Capillary action facilitates the movement of the mobile phase, allowing analytes to migrate through the stationary phase according to their respective affinities. Components with a stronger affinity for the stationary phase move more slowly, while those with a weaker affinity travel faster. As a result, the components become separated on the chromatographic plate [7].

4.2.6.2.2 Instrumentation

Figure 4.7 depicts the key components of HPTLC and their arrangement for high-resolution chromatographic analysis. The setup includes a semi-automatic applicator, such as the Linomat 5, fitted with 100 μL precision syringes and connected to a nitrogen gas tank for controlled sample application; a twin-trough glass chamber for chromatogram development, ensuring saturation and uniform migration of the mobile phase; and a specialized plate heater for controlled heating of chromatographic plates post-development to activate or stabilize analytes. A TLC visualizer chamber is used for documentation and visualization of chromatographic plates under UV light at wavelengths of 254 nm and 366 nm, while densitometric scanning is performed using a TLC scanner interfaced with analytical software such as WinCATS or the updated version visionCATS, for data integration, quantification, and documentation. Pre-coated silica gel 60 F_{254} plates (dimensions 10 × 10 cm or 20 × 10 cm) on aluminum backing are used as the stationary phase for optimal resolution and sensitivity [6, 8].

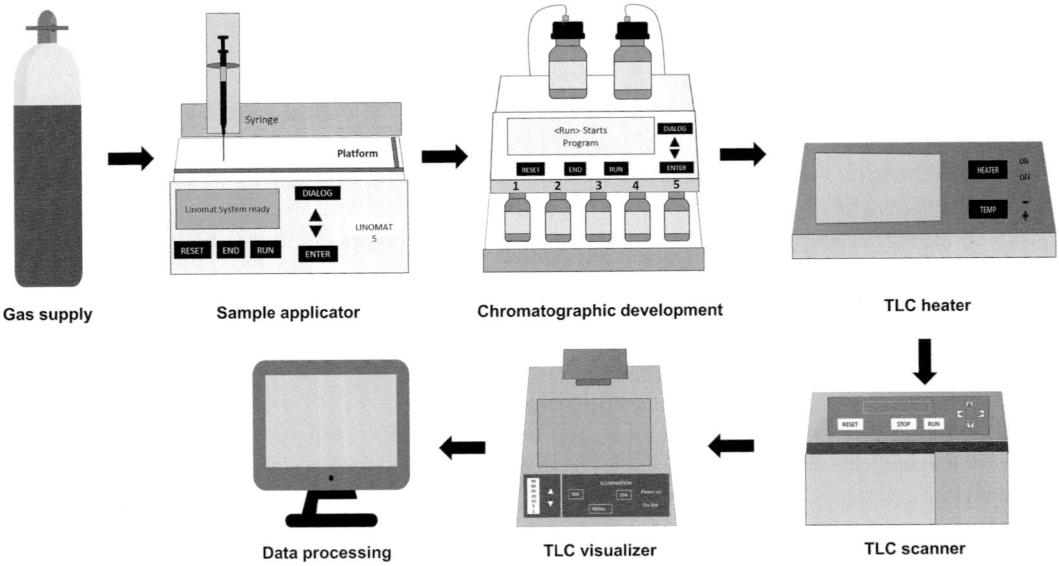

FIGURE 4.7 HPTLC instrumentation

4.2.6.2.3 Procedure

(A) **HPTLC Method Development**

- **Standard and extract sample preparation:** For chromatographic analysis, solvents must be HPLC or high-purity grade, selected based on the solubility of the standard and extract. Stock solutions are prepared with suitable solvents and sonicated; extracts are also filtered through a 0.22 µm membrane. Proper solvent choice is crucial to avoid solubility issues and ensure accurate results. In HPTLC, volatile non-polar solvents are preferred, while polar solvents are used in reversed-phase chromatography. Common solvents include methanol, ethanol, water, acetonitrile, ethyl acetate, and hexane [7].
- **Selection of the stationary phase:** HPTLC plates are pre-coated with uniformly distributed fine particles, commonly silica gel, to ensure consistent and smooth layers. Other adsorbents include alumina, cellulose, Kieselguhr G, polyamide powder, Fuller's earth, Magnesol, and Sephadex, selected based on the sample type [7]. Plates are available in standard sizes (e.g., 20 cm × 10 cm, 20 cm × 20 cm, 10 cm × 10 cm), allowing multiple samples to be run simultaneously [9]. Silica-gel-based normal-phase TLC using less polar solvents like chloroform–methanol is widely used in pharmaceutical analysis [10]. Reversed-phase TLC employs modified silica gels (e.g., C-18, C-8) with polar mobile phases like methanol–water [10]. Silica gel with 60 Å pores and a 500 m^2/g surface area remains the most common stationary phase, used in about 90% of applications [9]. The HPTLC plates are advised to be stored properly, inspected under white and UV light for defects, and prewashed to improve reproducibility and robustness.
- **Selection of mobile phase:** The mobile phase is selected based on the stationary phase and analyte properties, using precise solvent mixtures. It may include polar and non-polar solvents, with or without modifiers, to improve resolution. In normal-phase TLC, less polar solvents are used, while reversed-phase TLC uses polar mixtures like methanol–water. Solvent polarity is adjusted to optimize R_f and separation.
- **Preconditioning of the development chamber:** Saturation is essential for highly polar mobile phases, but not required for low-polarity ones. An unsaturated chamber

can lead to edge effects, high R_f values, and band distortion. Pre-saturating the chamber by lining it with filter paper for 20 minutes before development ensures uniform solvent vapor distribution, improving separation and reducing R_f values.

- **Sample application:** Samples are loaded into a Hamilton syringe (up to 100 μL) with care to avoid air bubbles. Using the Linomat 5 applicator, connected to a nitrogen chamber at 60 to 90 psi, samples are applied to plates sized as per sample number and dried in a desiccator.
- **Plate development:** HPTLC plates are developed in an automated chamber saturated with the mobile phase for 20 minutes at room temperature. Solvent types and ratios are optimized, with a run length of 80 mm. If unavailable, a manual twin trough chamber is used.
- **Chromatogram derivatization and evaluation:** The developed layers were dried on the plate heater for 5 minutes and then derivatized using a selected solution. Plates were examined under UV light at 254 nm, 366 nm, or white light before and after derivatization. Fluorescent compounds were detected directly, while non-fluorescent ones required treatment with fluorescent materials. Derivatization with sulfuric acid or vanillin-sulfuric acid induced fluorescence or charring, revealing compound spots [6]. Various derivatizing agents for HPTLC are listed in Table 4.2 [11].
- **Scanning:** Quantification is done at the compound's λ_{max} (200–400 nm) using a scanner, which detects each band *in situ*, records peaks, and compares them to standards for accurate HPTLC analysis

(B) **Validation of HPTLC**

TABLE 4.2
Various Derivatizing Agents

Serial Number	Derivatizing Agent	Chemical Compounds Detected	Color of Identification
1	Dragendorff's reagent	Alkaloids	Red-brown zone (visible)
2	Polyethylene glycol reagent (diphenylboric acid-2-aminoethyl ester)	Flavonoids	Intense yellow, orange, and green fluorescent zones (UV 366 nm)
3	Vanillin-sulfuric acid/ anisaldehyde-sulfuric acid	Bitter principles	Red-brown, yellow-brown, and dark green zones (visible)
		Saponins	Colored zones (visible)
		Essential oils	Blue, brown, or red zones (visible)
4	Concentrated sulfuric acid	Flavonoids, alkaloids, terpenoids	Brown, yellow, or black zones (visible)
5	Vanillin	Phenolic compounds, flavonoids, lignans, tannins	Pink, violet, red, or blue zones (visible)
6	Anisaldehyde	Terpenoids, alkaloids, flavonoids, steroids	Purple, blue, green, or red zones (visible)
7	Ehrlich's reagent	Indole alkaloids, quinolines	Violet, purple, or blue zones (visible)
8	Tungstophosphoric acid	Flavonoids, terpenoids, saponins	Yellow, brown, or blue zones (visible)
9	P-anisidine	Phenolic compounds, flavonoids	Yellow to orange zones (visible)
10	Iodine vapors	Terpenoids, essential oils	Brown or yellow-brown zones (visible)

Validation and method development are interconnected processes. It is carried out as per the ICH guideline on analytical method validation. It begins with defining an analytical goal, followed by method selection, optimization, and pre-validation, leading to a validation protocol. Data from validation experiments are assessed against acceptance criteria, and if met, the method is considered valid. Some validation data may come from prior method development. Qualitative HPTLC validation follows this approach, but may require modifications based on conditions like humidity or precision. The validation protocol ensures consistency, compliance, and reliability within the quality management system [12].

- **Selectivity:** The ability of the method to distinguish the target analyte from other components, like impurities or excipients. It is typically evaluated by resolution (Rs), retention factor, tailing factor, and system suitability studies, often using known impurities spiked into the sample [13].
- **Sensitivity:** The method's capability to detect small changes in analyte concentration, ensuring results are directly proportional to concentration over a defined range. It is usually assessed using the slope of the calibration curve or signal-to-noise ratios [13].
- **Precision:** The degree of reproducibility under normal operating conditions, expressed as relative standard deviation (RSD) or coefficient of variation (CV).
- **Repeatability**: The same analyst and equipment are used; acceptable RSD: ≤1% (peak area), ≤2% (peak position), ≤3% (sample application).
- **Intermediate precision**: Different analysts, days, or equipment batches within the same lab are used; acceptable RSD: ≤10% [13].
- **Accuracy:** The closeness of the measured value to the true value, often evaluated through recovery studies by spiking known amounts of analyte into a blank matrix. Accuracy should align with the method's RSD limits [13].
- **Robustness and ruggedness:** The method's ability to remain unaffected by small, deliberate variations in method parameters (e.g., solvent composition, detection wavelength, application technique). Robustness ensures method reliability under routine and slightly altered conditions, particularly crucial for techniques like HPTLC [12].
- **Limit of detection (LOD):** It is the lowest amount of an analyte that can be detected but not necessarily quantified under the stated experimental conditions. It is typically validated using the signal-to-noise ratio method (where LOD corresponds to an *S/N* ratio of about 3:1) or by statistical calculation. Visual evaluation may also be used for non-instrumental methods [13].

$$\mathrm{LOD} = \frac{3.3 \times \sigma}{S} \quad (4.1)$$

where,
σ = standard deviation of the response,
S = slope of the calibration curve.

- **Limit of quantification (LOQ):** It is the lowest concentration of an analyte that can be quantified with acceptable precision and accuracy. LOQ is validated similarly, where an *S/N* ratio of about 10:1 is considered acceptable. After determining LOQ, validation includes assessing accuracy (through recovery studies) and precision (by evaluating %RSD) [13].

$$\mathrm{LOQ} = \frac{10 \times \sigma}{S} A = \varepsilon c l \ . \quad (4.2)$$

where,
σ = standard deviation of the response,

S = slope of the calibration curve.

- **System suitability testing:** System suitability tests are a set of evaluations performed to confirm that the entire analytical system, surrounding the instrument, method, and sample preparation, is functioning correctly and capable of generating reliable and reproducible results.

4.2.6.2.3.1 *Application to Herbal Drugs*

Identification of similar plants: Through quantitative analysis, species of plants highly similar to one another with chemical variations can be identified [6]. HPTLC fingerprinting serves as a valuable tool for identification by comparing species within the same genus or related taxa, analyzing molecular variations across different habitats and seasons, verifying morphological and physiological changes within the same taxon, and distinguishing plant parts used in analysis [6].

Quantification of bioactive compounds in an extract: A validated HPTLC method was developed for the estimation of β-carotene in *Gymnosporia senegalensis* (Lam.) Loes. HPTLC separation was performed on silica gel 60 F254 plates using n-hexane: ethyl acetate (6:4, v/v) as the mobile phase. Densitometric analysis at 254 nm showed good linearity ($r^2 = 0.9645$, $R_f = 0.27$) with LOD and LOQ of 10.15 and 30.76 ng/band, respectively. The method demonstrated high precision, accuracy, and robustness, making it suitable for routine quality control. Additionally, terpenoids isolated via TLC exhibited notable antioxidant activity, with the stem showing the highest total terpenoid content (51.6 ± 0.06). This eco-friendly HPTLC approach is both effective and sustainable for β-carotene analysis in herbal formulations [14]. In a study, an HPTLC method was developed and validated for the simultaneous estimation of quercetin, curcumin, and ascorbic acid in polyherbal formulations. The method proved to be simple, rapid, accurate, and sensitive, demonstrating its suitability for routine quality control of antioxidant-rich herbal products [15].

Standardization and validation of multicomponent ayurvedic formulations: HPTLC was successfully applied for the standardization of Ayush Kvatha Churna (AKC), a classical Ayurvedic formulation widely used for immunity boosting, especially during the COVID-19 pandemic. A simple, accurate, and sensitive HPTLC method was developed for the identification and quantification of key marker compounds – rosmarinic acid, trans-cinnamaldehyde, and piperine. The analysis was performed using a solvent system of Toluene: Ethyl acetate: Ethyl alcohol: Formic acid (5.6:2.4:2:0.3 v/v/v/v), yielding R_f values of 0.33, 0.77, and 0.69, respectively. The method was validated and proved effective for the standardization and quality control of AKC, demonstrating HPTLC as a powerful tool in the authentication and routine analysis of herbal formulations [6].

Quality control of herbal medicines: In a recent study, HPTLC was employed as a key tool for the quality control and standardization of the Ayurvedic formulation Anu Taila. The HPTLC fingerprinting method successfully identified major bioactive compounds such as berberine chloride, marmelosin, negundoside, glycyrrhizin, PHBA, lupeol, embellin, and solasodine. This chromatographic profiling served as a preliminary authentication method for ensuring the consistency and integrity of the formulation. The use of HPTLC establishes a robust quality control protocol for complex polyherbal formulations like Anu Taila [7].

Drug-excipient compatibility study: HPTLC is also used for investigating drug-excipient compatibility in herbal formulations. Typical HPTLC fingerprint and three-dimensional chromatogram of *Piper longum* fruits ethanolic extract (PLFEE) with solid dispersion (SD) formulation components are shown in Figure 4.8. The fingerprints (under 245 nm) reflected the presence of chromatographic bands for pure piperine (PIP), piperlonguminine (PLGN) in extract, physical mixture of extract and polymer (PM), and solid dispersion (SD), representing the retention of their chemical integrity. All the chromatographic bands of the extract are shown in physical mixture and formulation, representing the compatibility of the extract with the used excipients of the SD [1].

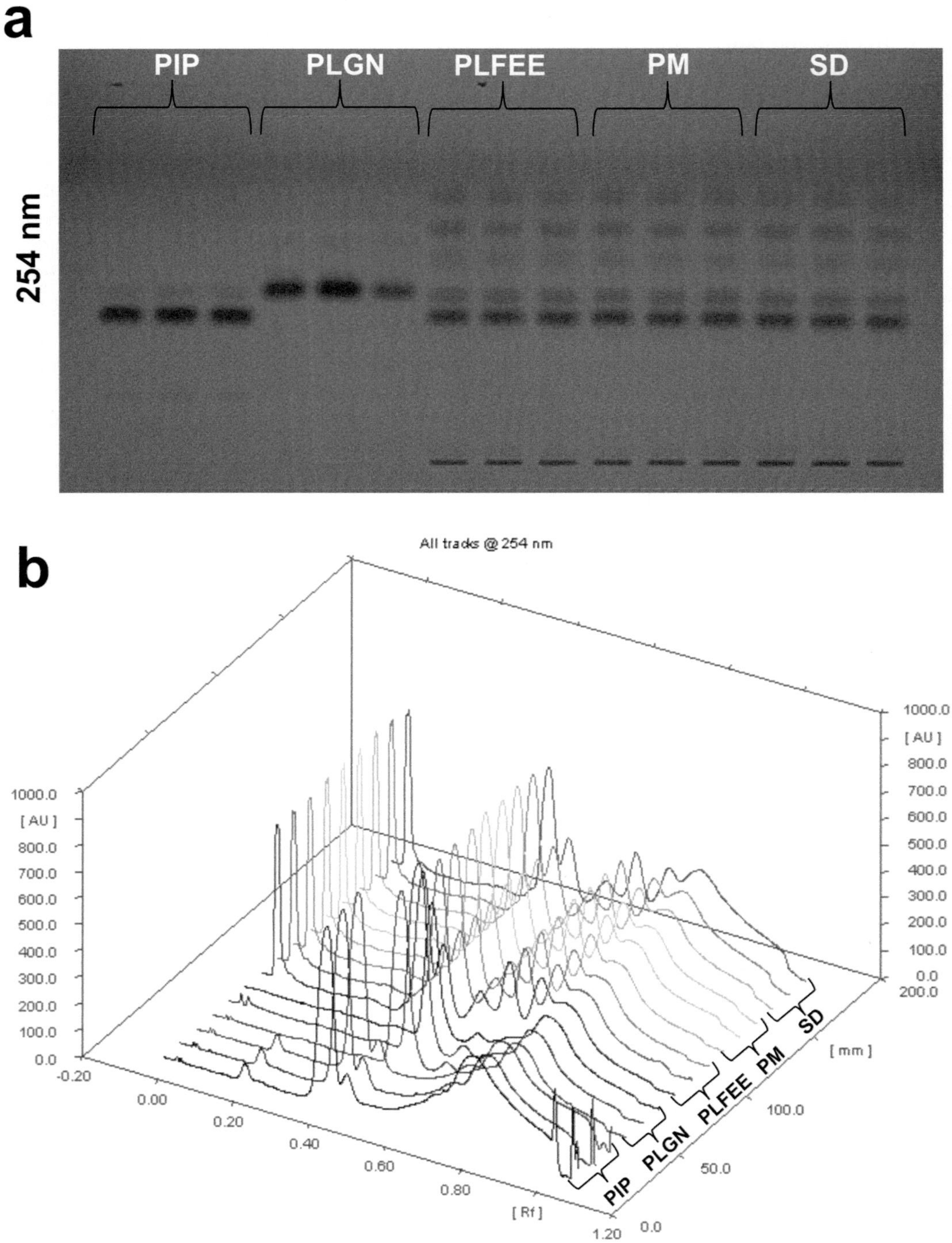

FIGURE 4.8 HPTLC fingerprints and chromatogram: (a) fingerprint of piperine (PIP), piper longuminine (PLGN), *Piper longum* fruits ethanolic extract (PLFEE), physical mixture of extract and polymer (PM), and solid dispersion (SD); (b) three-dimensional chromatogram of PIP, PLGN, PLFEE, PM, and SD.

4.2.6.2.3.2 Challenges in Performing HPTLC

- Tailing, merged zones, or distorted spots may occur due to overloading of the sample or poor pipetting technique.
- Plate contamination from dust, fingerprints, or previous use can introduce artifacts or interfere with analyte migration.
- Improper plate activation (e.g., insufficient drying or heating) affects adsorption and solvent migration behavior.
- Variability in plate batches may lead to inconsistent performance and affect reproducibility.
- Irregular solvent front development can result from poor chamber saturation, improper plate placement, or uneven pre-equilibration.
- Solvent volatility or composition changes during development may reduce separation quality and reproducibility.
- Environmental factors such as fluctuations in humidity or temperature can alter chromatographic performance.
- Traditional UV or fluorescence detection may fail to detect weakly absorbing or trace-level compounds.
- Quantitative accuracy is limited by densitometric challenges, particularly when spots overlap or analytes have poor UV absorbance.
- Baseline drift, noise, or improper scanning conditions may compromise densitogram quality and quantification.
- Solvent system selection often requires trial and error and may not always achieve the desired separation or be compatible with the detection method.
- Lack of standardized protocols across laboratories leads to inconsistent parameters (e.g., sample volume, solvent ratios, development time), reducing result comparability.
- Manual intervention at multiple stages (spotting, development, detection) increases the likelihood of operator error and reduces throughput.
- Reproducibility is operator dependent, requiring technical skill to maintain consistency and accuracy.
- Limited automation and documentation systems in some setups reduce repeatability and make result recording subjective.
- Training and experience are essential for correct technique, optimization, and troubleshooting; a lack thereof can compromise overall data quality.

4.2.6.2.3.3 Advances with HPTLC HPTLC is an advanced form of classic TLC, offering enhanced resolution, sensitivity, and reproducibility. In recent years, several technological and methodological advancements have expanded its applications in pharmaceutical, herbal, forensic, and environmental analysis. Modern HPTLC systems feature advanced automation, including piezoelectric or syringe-based sample applicators, precision-controlled development chambers, and densitometric scanners with multi-wavelength and 3D imaging capabilities. These innovations enhance throughput, ensure reproducibility, and reduce analytical errors. Additionally, HPTLC can be coupled with techniques such as mass spectrometry (HPTLC-MS), FTIR (HPTLC-FTIR), and NMR (HPTLC-NMR), allowing for direct identification and characterization of compounds from the chromatographic plate, which is especially valuable in complex sample environments. HPTLC is also widely used in the fingerprinting and quality control of herbal formulations, enabling the simultaneous quantification of multiple phytoconstituents, and is endorsed by regulatory bodies like WHO, ICH, and AYUSH for herbal drug standardization. Moreover, the adoption of green solvents and miniaturized plate formats contributes to sustainable analytical practices by reducing solvent consumption and minimizing environmental impact. Advanced software solutions further support HPTLC analysis by providing digital image acquisition, peak integration, and data interpretation, with chemometric techniques like principal component analysis (PCA) aiding in the classification, pattern recognition, and authentication of complex samples.

4.2.6.3 HPTLC-MS: High-Performance Thin Layer Chromatography-Mass Spectrometry

HPTLC-MS is a powerful hyphenated analytical technique that combines the separation efficiency and versatility of high-performance thin layer chromatography (HPTLC) with the structural elucidation capabilities of mass spectrometry (MS). This integration allows for the rapid identification, quantification, and confirmation of analytes in complex mixtures such as herbal formulations, pharmaceuticals, foods, and biological samples. HPTLC enhances resolution and reproducibility over conventional TLC, while MS provides sensitive and specific molecular characterization, making the combined technique ideal for quality control, bioactivity-guided fractionation, and forensic analysis [8].

4.2.6.3.1 Principle

The principle of HPTLC lies in the planar chromatographic separation of analytes based on their differential affinities to a stationary phase (usually silica gel) and a mobile phase (solvent system). Compounds migrate at different rates depending on their polarity, size, and interaction with the stationary phase. Once separated, the components are visualized, and the targeted spots or zones are eluted or desorbed and introduced into a mass spectrometer, where they are ionized. The MS component separates and detects ions based on their mass-to-charge (m/z) ratio, allowing for the identification of compounds through their mass spectra and fragmentation patterns. Thus, HPTLC-MS synergistically combines the high-throughput separation of planar chromatography with the molecular specificity of mass spectrometry [8].

4.2.6.3.2 Instrumentation

HPTLC-MS involves precise and automated sample application using devices like CAMAG Linomat or ATS-4, ensuring accurate band formation. Pre-coated TLC plates, typically silica gel 60 F_{254} or RP-18, with particle sizes around 5 to 7 µm, provide high-resolution separation. Chromatographic development occurs in manual or automated chambers under controlled conditions. Post-separation, derivatization units are used to visualize analytes either chemically or via UV detection. Documentation is done using densitometric or video-based scanners. The MS interface can be elution based (e.g., CAMAG TLC-MS Interface) or desorption based (e.g., MALDI-MS, DESI-MS), enabling direct transfer or ionization of analytes. The mass spectrometer may use ESI, APCI, or MALDI ionization and can be equipped with quadrupole, TOF, or Orbitrap analyzers for compound identification [8].

4.2.6.3.3 Procedure

- **Sample preparation:** Extracts or samples are prepared in a suitable solvent such as methanol or chloroform to ensure solubility and minimize matrix effects. Filtration is done to remove particulates.
- **Plate preparation:** HPTLC plates are pre-washed (usually with methanol), dried, and activated at 110–120°C for 15–20 minutes.
- **Sample application:** Samples are applied as narrow bands using an automated applicator to ensure reproducibility and reduce spot overlapping.
- **Development:** The plate is placed in a saturated chamber with the mobile phase (solvent system selected based on analyte polarity) for chromatographic separation.
- **Visualization:** After development, analyte bands are visualized under UV light or post-derivatization.

Extraction or Desorption for MS:

1. For elution-based HPTLC-MS, the targeted spot is eluted using a solvent (e.g., methanol/water) through a TLC-MS interface and directed to the MS inlet via syringe or flow cell.

2. For desorption-based coupling, the plate is directly placed in the ionization source (e.g., MALDI or DESI), and MS data are collected from the surface.
3. Mass spectrometry analysis: Ionized molecules are separated by their m/z ratio and detected, yielding spectra that allow for identification and structural interpretation.

4.2.6.3.4 Applications to Herbal Drugs

HPTLC-MS finds wide applications across pharmaceutical, herbal, forensic, food, and environmental sciences.

Pharmaceuticals: Detection of counterfeit drugs, impurity profiling, and content uniformity testing. In a recent study, HPTLC-MS was used to identify the flavonoid glycosides in Caigua leaves, fruits, and supplements. Particularly, compounds such as 6-acetyl-β-D-glucopyranoside (m/z 473), apigenin 200-O-rhamnose-C-hexoside (m/z 577), and chrysin 6-C-fucopyranoside (m/z 415) were detected, with fragmentation patterns confirming their identity. Additionally, chrysin derivatives like O-hexoside (m/z 457) and other C-glycosides were also identified. Luteolin 8-C-glucoside (orientin, m/z 477) was reported for the first time in Caigua and its supplements, which showed the least diversity in glycosylated flavones per HPTLC data [9].

Herbal and natural products: In case of herbal drugs, HPTLC-MS is used for the authentication of botanical species, standardization of plant extracts, and identification of bioactive compounds. In a recent study, HPTLC-MS/MS was used for the analysis of flavonoids in *Cyclanthera pedata* leaves, fruits, and dietary supplements. The HPTLC-MS/MS analyses enabled tentative identification of several compounds in crude extracts from leaves and fruits. Further, HPTLC-DPPH• assay combined with image analyses and HPTLC-MS/MS analyses was used for direct screening of antioxidant activity of separated compounds (chromatographic zones) on HPTLC silica gel plates and tentative identification of active compounds [10].

In another study, HPTLC-MS was used for rapid screening of undeclared hypoglycemics in counterfeit herbal antidiabetic products. A precise, sensitive, and simple HPTLC-MS method was reported for the simultaneous determination of synthetic antidiabetics (pioglitazone, metformin, glipizide, and glimepiride) in counterfeit herbal products as adulterants. The study revealed the presence of metformin in two of eight market products [12].

4.2.6.4 High-Performance Liquid Chromatography

HPLC is an advanced chromatographic technique used for the separation, identification, and quantification of compounds in a mixture. It operates by passing a liquid mobile phase through a column packed with a stationary phase, where different molecules interact based on their affinities, resulting in their separation [2]. HPLC is an advanced form of column chromatography widely used in pharmaceutical analysis. It plays a crucial role in drug discovery, development, and quality control by separating, identifying, and quantifying compounds in a liquid sample. HPLC injects a sample into a column containing a stationary phase, while a high-pressure mobile phase facilitates separation based on solute affinity [16].

4.2.6.4.1 Principle of HPLC

HPLC is based on the separation of components in a mixture through their differential interactions with a stationary phase and a mobile phase. The stationary phase is usually silica based and modified with hydrophobic alkyl chains, such as C18, which makes it suitable for separating natural compounds. The mobile phase typically consists of solvent mixtures such as alcohol – water or acetonitrile – acidified water, prepared in varying ratios to optimize resolution. As the analyte moves with the mobile phase across the stationary phase, each compound interacts differently with the stationary phase, resulting in distinct retention times and enabling separation.

The degree of interaction, and therefore the efficiency of separation, is influenced by specific physicochemical properties of the compounds. These include polarity, charge, molecular weight, and the presence of functional groups. Compounds with stronger interactions with the stationary phase elute later, while those with weaker interactions elute earlier, allowing for accurate qualitative and quantitative analysis [2].

4.2.6.4.2 Instrumentation

The different components of the HPLC systems are described in the following and shown in Figure 4.9.

- **Solvent reservoir:** The solvent reservoir stores the mobile phase, which is typically a mixture of organic solvents and water. The mobile phase is critical for transporting the sample through the chromatographic system. A consistent and steady flow of the mobile phase is essential for accurate analysis [16].
- **Pump:** The pump delivers the mobile phase through the system at a controlled flow rate, typically ranging from 0.1 to 10 mL/min. It ensures precise flow and pressure, maintaining the proper solvent delivery to the column. Pumps may be single or dual to allow for gradient elution, where the composition of the mobile phase changes during the analysis [16].
- **Injector:** The injector is responsible for introducing the sample into the mobile phase. A small, precise volume of the sample (usually in microliters) is injected into the system. Automatic or manual injectors are used to ensure consistent sample introduction for reproducible results. Proper mixing with the mobile phase ensures efficient separation in the column [16].
- **Column:** The column is the core of the HPLC system, where separation occurs. It is packed with a stationary phase material, typically silica or polymer-based particles. The stationary phase interacts with the sample components, causing them to separate based on their affinity for the stationary phase, resulting in different retention times. The choice of column depends on the type of chromatography [16]:
  - Normal phase: polar stationary phase
  - Reverse phase: non-polar stationary phase
- **Detector:** The detector monitors the separated components as they elute from the column. The detector converts the signal into a chromatogram, which provides information on the concentration and identity of the analytes. Common detector types include the following [16]:
  - UV/Vis detector: Measures absorbance at specific wavelengths.
  - Photodiode array detector (PDA) detector: It detects compounds by measuring their UV-visible absorbance across a wide wavelength range. Compared to a single-wavelength UV detector, it provides the entire absorbance spectrum of analytes, helping in compound identification, purity testing, and peak characterization.
  - Fluorescence detector: Detects compounds that emit fluorescence upon light excitation.
  - Refractive index detector: Measures changes in the refractive index of the eluting compounds.
- **Data system:** The data system collects and analyzes the signals from the detector, displaying them as chromatograms (graphs of signal versus time). The software helps in identifying retention times and calculating peak areas for quantification. It also stores and processes data, enabling the optimization of methods and facilitating data reporting [16].

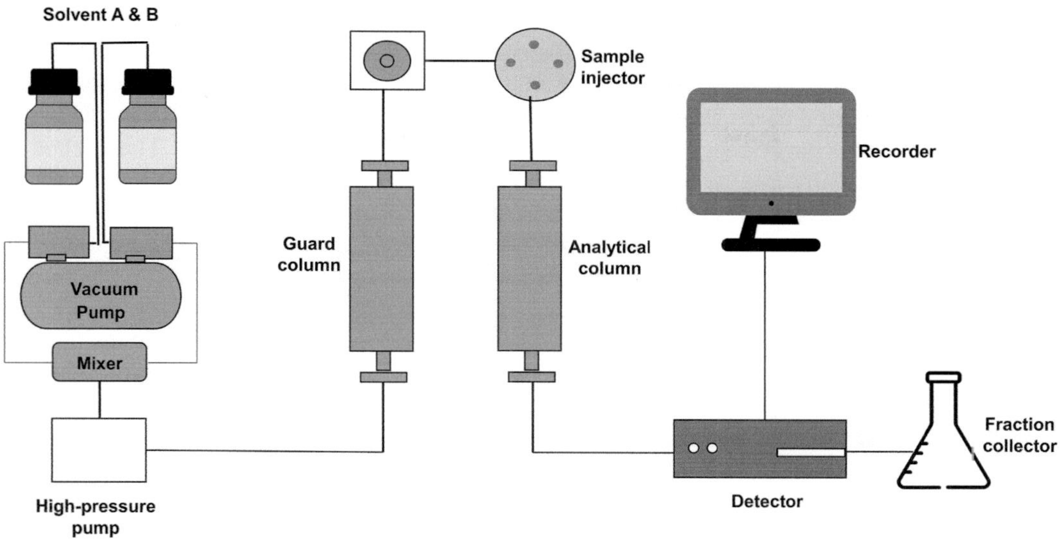

FIGURE 4.9 Instrumentation of HPLC

4.2.6.4.3 Procedure

(A) **HPLC Method Development**

- **Selection of the stationary phase or columns:** The HPLC column is crucial for effective separation and accurate analysis. Proper selection depends on stationary phase chemistry, retention capacity, particle size, and dimensions, while poor use can lead to unreliable results. HPLC columns consist of hardware, a matrix, and a stationary phase. Silica is the most common matrix due to its robustness, chemical stability, and consistent particle size, though it dissolves in solutions above pH 7. The stationary phase determines whether a column is suited for normal-phase (polar stationary phase) or reverse-phase chromatography (non-polar stationary phase). Reverse-phase columns like C3–C5 (for peptides and ion-pairing), C8 (for pharmaceuticals and steroids), and C18 (high retention) are widely used. A stable, high-performance column is essential for reproducible method development, as separation selectivity varies across manufacturers and batches [16]. The commonly used column in the case of HPLC is reverse phase chromatography, which is the reverse phase column that contains silica bonded with hydrophobic alkyl chains, such as C18, providing efficient separation of natural molecules [2].
- **Mobile phase:** The mobile phase typically consists of a solvent mixture like alcohol–water or acetonitrile with acidified water, with varying ratios throughout the analysis [2]. For highly ionizable compounds with a low pKa, adjusting the mobile phase pH is crucial to control ionization, retention time, and selectivity, improving separation. Small pH changes can significantly affect the ionization state, retention, and sensitivity. Buffers such as potassium phosphate, sodium phosphate, and acetate are evaluated for system suitability and chromatographic performance [16]. In reverse-phase HPLC, acetonitrile and methanol are common modifiers. Gradient elution is preferred for complex samples, as it offers better control over solvent strength and ensures optimal retention and separation of components with varying affinities for the stationary phase [16].
- **Sample preparation:** Sample preparation is crucial in method development to ensure accuracy. The sample solution should be filtered with a 0.22 or 0.45 μm pore-size filter

for the removal of particulates. Filtration is a preventive maintenance tool for HPLC analysis. The efficiency of syringe filters depends on their ability to remove contaminants and insoluble particles without introducing unwanted substances into the filtrate. If unexpected peaks appear in the filtered samples, the diluent should also be filtered to check if the syringe filter or its housing is leaching any components. The method should specify filter type, manufacturer, and pore size, ensuring the prepared sample is interference free, is HPLC compatible, and does not harm the column [16].

- **Method development and optimization:** Method development focuses on optimizing mobile phase composition, flow rate, and injection volume for high peak accuracy and efficiency. Mobile phase optimization is prioritized over stationary phase changes. Important factors include solvent acidity, solvent selection, gradient conditions, flow rate, temperature, sample concentration, injection volume, and diluent type. Once selectivity is optimized, additional adjustments to column dimensions, particle size, and flow rate can be made without affecting capacity factor or selectivity. Only parameters impacting selectivity should be evaluated to minimize trial chromatograms [16].
- **Detectors:** The choice of detector depends on factors like sample needs, analyte properties, potential interferences, detection limits, availability, and cost, with common detectors including UV, fluorescence, electrochemical, RI, and MS. The UV-visible detector is versatile, used for UV-based applications, impurity identification, and quantitative analysis, with a diode array detector (DAD)/PDA for simultaneous multi-wavelength detection. For less UV-active compounds, an evaporative light scattering (ELS) detector is used, which scatters light in proportion to non-volatile components [2]. The PDA detector offers high chromatographic and spectral sensitivity, suitable for HPLC and LC/MS. The refractive index (RI) detector is ideal for compounds with low or no UV absorption. The multi-wavelength fluorescence detector is highly sensitive and selective, useful for quantifying low concentrations of target compounds. For multicomponent analysis, absorption spectra may shift to different wavelengths [16].

(B) **Validation of HPLC**

Validation ensures that an analytical method produces reliable, consistent, and reproducible results. It confirms accuracy, precision, and robustness across different conditions, guaranteeing data quality and regulatory compliance. HPLC method validation is crucial for meeting regulatory standards (FDA, ICH, USP) and ensuring reproducibility across analysts and conditions. It helps detect impurities, ensures drug stability, and assesses robustness against variations. Validated methods reduce errors, costs, and support GLP/GMP compliance, ensuring high-quality standards in analytical labs and pharmaceutical manufacturing. Validation is required before routine use and when conditions or methods change [16].

Parameters for validation are as follows:

- **Specificity:** The ability of a method to accurately measure the analyte in the presence of potential interferences, such as excipients, synthetic precursors, enantiomers, and degradation products [16].
- **Linearity and range:** Assesses whether test results are directly proportional to analyte concentration within a specified range. Linearity is evaluated across at least six concentrations, and results are expressed as a regression line with confidence limits. The range defines the upper and lower concentration limits with demonstrated accuracy, precision, and linearity [16].
- **Precision:** Measures the consistency of results and is evaluated at three levels:
  - Repeatability (method precision): assesses variation under the same conditions over a short period.
  - Intermediate precision (reproducibility/ruggedness): evaluates variations within the same laboratory (different days, analysts, or instruments) [16].

- Reproducibility: assesses precision between different laboratories. Precision is typically expressed as standard deviation (SD) or relative standard deviation (RSD) [16].
- Accuracy (recovery): determines how close the measured value is to the true value by analyzing samples spiked with known amounts of analyte. Results are calculated as percentage recovery to ensure no interference affects quantification [16].

- **Solution stability:** Evaluates the stability of samples and standards under normal storage and analytical conditions, ensuring no degradation occurs during analysis. Special conditions such as refrigeration or light protection may be required [16].
- **Limit of detection (LOD):** The lowest analyte concentration that can be detected but not quantified with certainty. It is often determined based on a signal-to-noise (*S/N*) ratio of ~3:1.
- **Limit of quantification (LOQ):** The lowest analyte concentration that can be quantified with acceptable accuracy and precision, typically determined by an *S/N* ratio of ~10:1 and verified by injecting standards [16].
- **Robustness:** Evaluates the method's reliability under small but deliberate variations in conditions, such as changes in pH, mobile phase composition, temperature, and instrument settings. A robust method ensures consistent performance under slightly altered conditions [16].
- **System suitability:** A set of tests ensuring the chromatographic system functions properly before analysis. Parameters such as peak resolution, theoretical plates, peak tailing, and capacity factor are measured to confirm system performance [16].

4.2.6.4.4 *Application to Herbal Drugs*

Authentication of herbal drugs: HPLC is widely used for phytochemical authentication, ensuring the reliability of herbal extracts by detecting specific marker compounds and preventing adulteration. In this study, HPLC was effectively utilized to generate chromatographic fingerprints for the identification and differentiation of four taxonomically distinct medicinal plants. Detection at 270 nm provided high resolution, and data alignment significantly enhanced fingerprint similarity. Combined with multivariate analyses such as PCA and HCA, HPLC enabled clear clustering of plant species, reflecting their evolutionary relationships. This demonstrates HPLC's strong potential as a reliable tool for distinguishing and authenticating medicinal plants, particularly in the context of herbal standardization and trade [17].

Phytochemical profiling: HPLC is used to identify and quantify bioactive constituents such as alkaloids, flavonoids, glycosides, tannins, and phenolic compounds in herbal extracts. In a study, HPLC was effectively applied for phytochemical profiling of the most active fraction of *Cordia dichotoma* bark, which showed cytotoxic activity against A-549 human lung cancer cells. HPLC analysis revealed the presence of key bioactive phenolic compounds, including gallic acid, caffeic acid, ferulic acid, and vanillic acid, contributing to the plant's anticancer potential. This application of HPLC played a crucial role in identifying and quantifying phytoconstituents responsible for the observed biological activity, supporting its value in bioactivity-guided fractionation and chemoprofiling of medicinal plants [18].

Quality control and standardization: HPLC ensures batch-to-batch consistency by monitoring marker compounds. It is used to develop fingerprint chromatograms for herbal raw materials and formulations to detect adulteration or substitution. HPLC-PDA fingerprinting, combined with multivariate analysis, was effectively used for the quality control and authentication of *Picrorhiza kurroa* (Kutki) rhizomes. The method generated distinct chromatographic profiles with 28 peaks, of which 9 were key markers, and enabled the classification of 11 samples using PCA and hierarchical cluster analysis (HCA). This approach demonstrated high precision and reliability in differentiating authentic and market samples, making it a robust tool for standardization of Kutki [19].

Purity and stability testing: HPLC helps assess the purity of isolated compounds and detect degradation products over time or under various storage conditions. It was used for evaluating the stability of hyperforin, enabling precise monitoring of degradation under various storage conditions. The study demonstrated that hyperforin remains most stable when stored under nitrogen at –20°C or –30°C, with HPLC confirming no significant degradation compared to samples stored at –196°C in liquid nitrogen [20].

Pharmacokinetic studies: Used to measure the bioavailability of phytochemicals in biological samples (e.g., blood or plasma) during clinical or preclinical trials. A validated reverse-phase HPLC (RP-HPLC/UV) method was used for the bioanalysis of glipizide in rat plasma, supporting the evaluation of herb–drug interactions with *Andrographis paniculata* extract and andrographolide. The method demonstrated high sensitivity, selectivity, and linearity (R^2 = 0.998) across a concentration range of 25–1500 ng/mL. This HPLC-based approach enabled reliable pharmacokinetic profiling, showing that APE significantly increased GLZ bioavailability, while andrographolide reduced it in diabetic rats [21].

Identification of active constituents: Enables separation and identification of individual active compounds from complex herbal matrices. In a study, the presence of bioactive piperine in *Piper longum* extract was investigated by a validated HPLC method. Typical HPLC chromatogram and PDA spectrum of piperine and ethanolic extract are shown in Figure 4.10. The appearance of piperine peak in the extract at the retention time (7.087 min) of pure piperine and similar absorption maxima (342 nm) in both cases confirmed the presence of piperine in standardized extract [1].

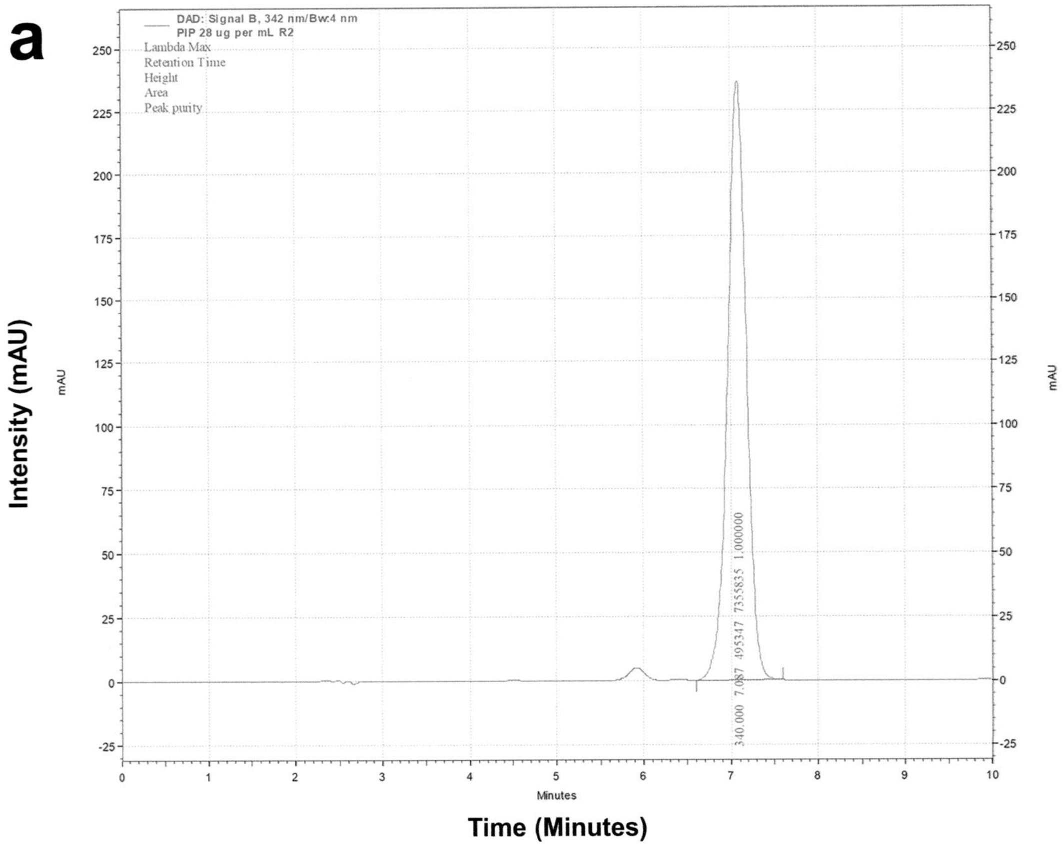

FIGURE 4.10 HPLC chromatograms and absorption spectra of piperine in pure piperine and ethanolic extract of *Piper longum*. Chromatogram (a, b) of pure piperine and extract (c, d) spectrum of piperine and extract showing the absorption maxima at 342 nm.

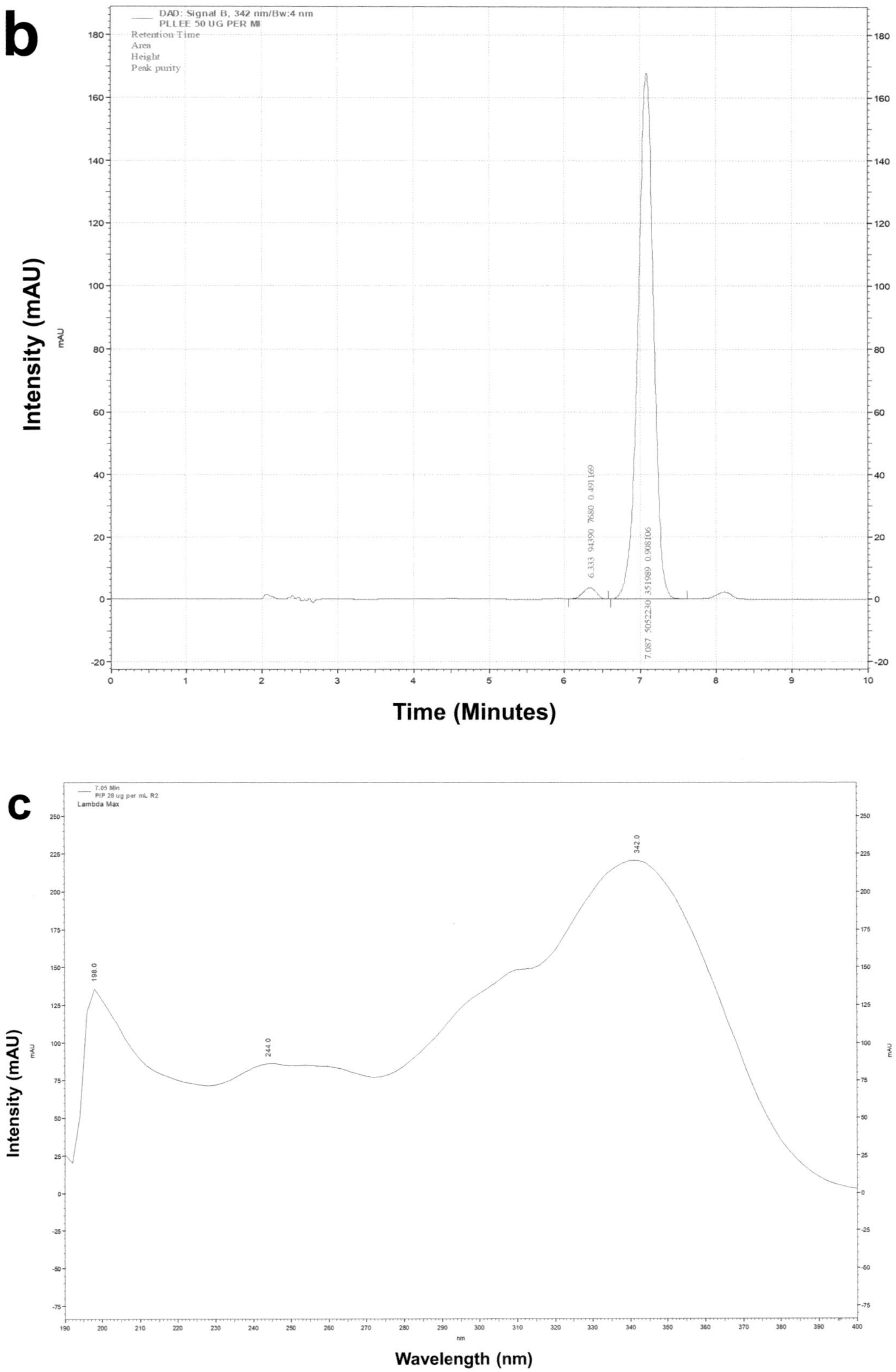

FIGURE 4.10 (Continued)

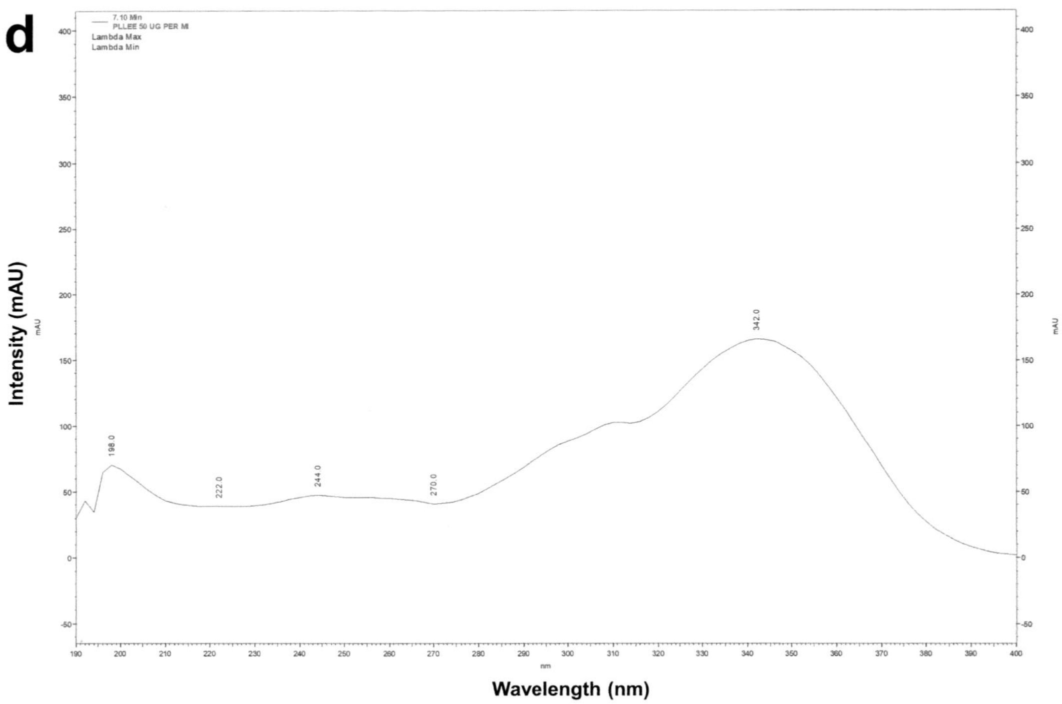

FIGURE 4.10 (Continued)

Detection of contaminants: HPLC is employed to detect pesticide residues, heavy metals, or synthetic drugs illegally added to herbal preparations. A validated LC-MS/MS method, based on liquid chromatography, was employed to detect and quantify 14 mycotoxins, including emerging ones like enniatins, beauvericin, and sterigmatocystin, in herbs and herbal infusions. This HPLC-based approach enabled accurate contaminant profiling, revealing that 72% of herb samples were contaminated, with BEA being the most prevalent. Despite reduced levels in infusions, some compounds like HT-2 toxin still posed potential health concerns, demonstrating the critical role of HPLC in food safety and contaminant exposure assessment [22].

Formulation development: Used to evaluate drug-excipient interactions, release profiles, and stability in herbal formulations.

4.2.6.4.5 HPLC: Challenges and Limitations

- Achieving analyte solubility: Ensuring the target compound dissolves adequately in a solvent compatible with the HPLC system is often the first hurdle, especially with complex samples.
- Dealing with matrix effects: The presence of other components in the sample (the matrix) can interfere with the separation or detection of the analyte, necessitating cleanup steps.
- Separating interferences: Removing compounds that might co-elute with the analyte on the column is vital for accurate quantification.
- Preventing analyte degradation or loss: Sensitive analytes can degrade during preparation or be lost through adsorption or incomplete extraction, requiring careful handling and optimized procedures.
- Avoiding contamination: Introducing impurities from reagents or equipment can lead to false results, emphasizing the need for clean practices.

- Removing particulate matter: Solid particles can damage the HPLC column, so filtration is a mandatory step.
- Developing efficient methods: Creating a sample preparation workflow that is both effective and not excessively time consuming can be a significant undertaking.
- Ensuring reproducibility: Consistent sample preparation is essential for reliable results across different analyses and analysts.
- While carrying out the analysis, there are chances of column clogging, contamination, or degradation over time, affecting separation efficiency.
- Selecting and preparing the correct mobile phase composition, ensuring proper degassing and pH, is critical for achieving good separation and detector compatibility.
- The detector must be sensitive enough to measure the analyte accurately, and baseline noise or drift can hinder this.
- Poor resolution between peaks, asymmetrical peak shapes (tailing or fronting), and inconsistent retention times can complicate analysis.
- There may be issues with the HPLC instrument itself, such as pump malfunctions, leaks, or injector problems, which can compromise the reliability of the results.
- Correctly identifying and integrating peaks and using appropriate calibration methods are essential for accurate quantitative analysis.

4.2.6.4.6 Advances of HPLC

Ultra-performance liquid chromatography: UPLC offers improved speed and efficiency, reducing analysis time to around 15 minutes while maintaining high resolution and sensitivity. UHPLC is quickly becoming the preferred HPLC system in research and development, though its use in quality control labs is growing more slowly. Advances in column technology enable faster and more efficient analysis of complex samples, including chiral molecules and biomolecules [17]. A comparison between HPLC and UHPLC techniques is shown in Table 4.3 [17].

TABLE 4.3
Comparison Between HPLC and UHPLC Techniques Based on Operational and Performance Parameters

Feature	HPLC (High-Performance Liquid Chromatography)	UHPLC (Ultra-High-Performance Liquid Chromatography)
Pressure	Operates at lower pressure (≤400 bar)	Operates at much higher pressure (≥1000 bar, up to 1500 bar)
Column Particle Size	Uses larger particles (3–10 μm)	Uses smaller particles (≤2 μm)
Separation Efficiency	Lower efficiency and resolution	Higher efficiency and better resolution due to smaller particle size
Analysis Time	Longer run times	Shorter run times (up to 10× faster)
Solvent Consumption	Higher solvent usage	Lower solvent usage (due to reduced column size and smaller particles)
Sensitivity	Moderate sensitivity	Higher sensitivity due to better peak resolution
Instrument Cost	Generally lower	Higher cost due to high-pressure system requirements
Maintenance	Lower maintenance requirements	Requires more frequent maintenance due to high pressure and smaller particles
Sample Throughput	Slower sample analysis	Faster sample analysis, making it ideal for high-throughput applications
Application	Suitable for routine analysis	Preferred for advanced research, high-resolution separations, and rapid analysis

4.2.6.5 Liquid Chromatography-Mass Spectroscopy

LC-MS has become the preferred method in various stages of drug development. The chemical standardization of an aqueous extract from a mixture of 20 herbs identified 20 chemical compounds as reference markers using LC-MS [6]. Two HPLC methods have been reported for analyzing aristolochic acid I and II in herbal medicines – one utilizing a photodiode array detector (LC/UV) and the other employing mass spectrometry (LC/MS). The LC/UV method was conducted using a Cosmosil 5C18-MS column with a gradient solvent system of phosphate buffer–acetonitrile and a UV detector set at 390 nm. Meanwhile, the LC/MS method employed an acetate buffer–acetonitrile solvent system and positive-ion electrospray ionization MS. The characteristic fragment ions for aristolochic acid I were detected at m/z 359, m/z 324, m/z 298, and m/z 296, while those for aristolochic acid II were observed at m/z 329, m/z 294, and m/z 268 [6].

4.2.6.6 Spectroscopic Techniques

Spectroscopic techniques play a pivotal role in qualitative and quantitative analyses, quality control, and standardization of phytopharmaceuticals. These non-destructive methods enable rapid and accurate identification of bioactive compounds and their structural features. UV-visible spectrophotometry is commonly employed to estimate total phenolics, flavonoids, and other chromophoric compounds. Fourier-transform infrared spectroscopy (FTIR) facilitates the identification of functional groups and detection of adulterants through unique spectral fingerprints. Nuclear magnetic resonance (NMR) spectroscopy provides detailed structural information, aiding in the elucidation of complex phytoconstituents. Mass spectrometry (MS), often coupled with chromatographic methods, allows for high-sensitivity molecular characterization. These techniques, individually or in combination, ensure the integrity, efficacy, and safety of herbal formulations, aligning with regulatory standards for phytopharmaceuticals. UV-visible (UV-Vis) spectrophotometry plays a key role in herbal standardization by enabling the quantitative estimation of major phytochemicals like flavonoids, phenolics, and tannins. It supports quality control, authentication through spectral fingerprinting, detection of adulterants, and assessment of potency. Its simplicity, speed, and cost-effectiveness make it ideal for routine standardization, though it is less selective in complex herbal mixtures. FTIR spectroscopy contributes to herbal standardization by identifying functional groups and detecting compositional changes or adulteration. It provides unique infrared fingerprints for authentication and differentiation of herbal materials. The method is rapid and non-destructive, but spectral complexity and overlapping peaks may limit its specificity for minor components. NMR spectroscopy is a powerful tool for herbal standardization, offering detailed structural information and both qualitative and quantitative analysis of active constituents. It ensures batch-to-batch consistency, identifies structural markers, and distinguishes isomers. However, it requires pure samples, is expensive, and demands expert handling. MS is widely used in herbal standardization for identifying and quantifying bioactive compounds, detecting trace constituents, and profiling complex formulations. It provides high sensitivity and specificity and is effective when coupled with chromatography. Despite its advantages, it requires advanced instrumentation, is costly, and can be affected by matrix interferences.

4.2.6.7 UV-Visible Spectroscopy

4.2.6.7.1 Principle

UV-visible spectroscopy is based on the absorption of ultraviolet (200–400 nm) and visible (400–800 nm) radiation by molecules. When energy is absorbed, valence electrons are promoted from the ground state to excited state orbitals, producing a characteristic absorption spectrum that depends on the electronic structure of the compound. Colorless compounds are usually absorbed in the UV region, whereas colored compounds absorb in the visible region. Absorption spectra arise from electronic excitation, while emission spectra occur when excited electrons release energy and return to the ground state.

The main types of electronic transitions, as shown in Figure 4.11, involve excitation from σ (bonding), π (bonding), or n (non-bonding) orbitals to the higher-energy antibonding orbitals σ^* and

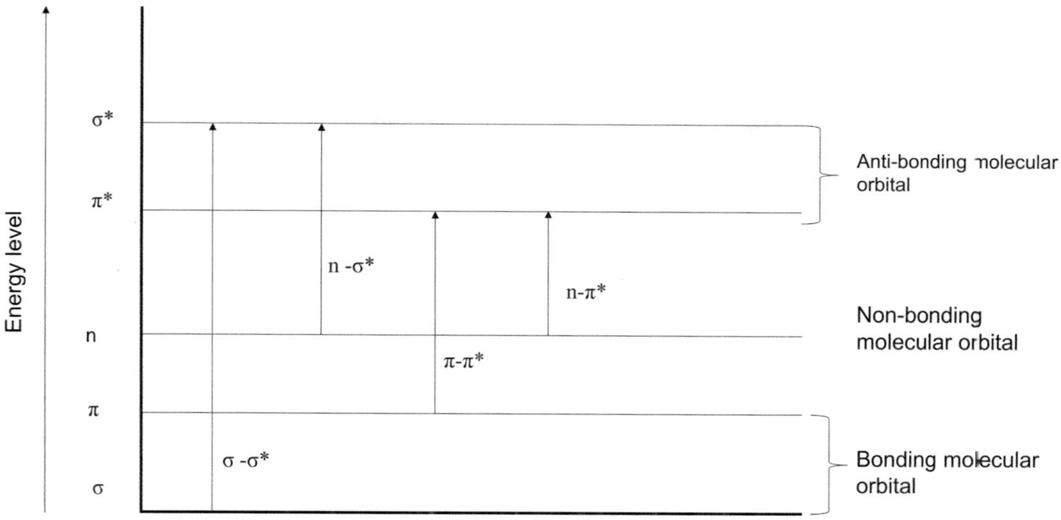

FIGURE 4.11 Energy levels of electronic transitions

π*. The possible transitions are σ → σ*, n → σ*, n → π*, and π → π*, and these are characteristic of functional groups such as alkanes, alkenes, aromatics, carbonyls, and heteroatom-containing compounds.

Chromophores are the functional groups directly responsible for absorption, while auxochromes such as —OH, $—NH_2$, and $—NO_2$ influence the absorption by shifting the absorption maximum (λ_{max}) or intensifying the spectrum, thereby affecting the color and absorption characteristics of molecules. The absorption spectrum is also influenced by the solvent: polar solvents generally cause shifts in λ_{max} and broadening of absorption bands, while non-polar solvents typically cause little change.

Quantitative analysis in UV-visible spectroscopy is explained by the Beer-Lambert law, which states that absorbance (A) is directly proportional to the concentration (c) of the absorbing substance and the path length (l) of the sample, with ε representing the molar extinction coefficient:

$$A = \varepsilon c l \tag{4.3}$$

where A is the absorbance, ε is the molar extinction coefficient, c is the concentration of analyte, and l is the path length.

The plot of UV spectrophotometer analysis is usually represented as an absorbance versus wavelength plot, which is used to determine λ_{max} and to estimate analyte concentrations [23].

4.2.6.7.2 *Instrumentation*

Components of the UV-vis spectrophotometer are described here, as shown in Figure 4.12 [23].

- **Light source:** For UV-vis spectroscopy, an ideal light source should be stable, intense, economical, easy to operate, and capable of emitting a continuous spectrum. The commonly used sources include hydrogen and deuterium lamps, which provide radiation in the UV region (185–380 nm), with deuterium favored for its stronger and more stable output. Xenon arc lamps generate highly intense radiation over a broad range (200–1000 nm) but require cooling due to high levels of heat production. Tungsten halogen lamps, containing iodine in a quartz envelope, operate in the 350–3000 nm range and are mainly applied in the visible to near-IR region, often with filters to minimize IR interference.

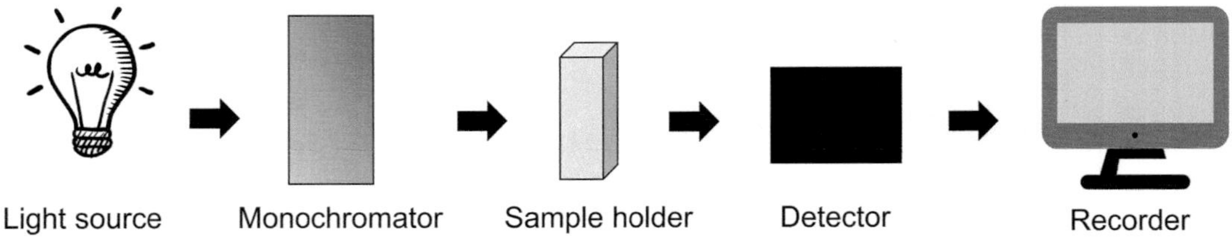

FIGURE 4.12 Instrumentation of UV-visible spectroscopy

- **Monochromators**: These are devices that convert polychromatic light into monochromatic light. They are mainly of two types:
  - Prism monochromators: Made of glass, quartz, or fused silica, and can be refractive or reflective in design.
  - Grating monochromators*:* Constructed from materials like glass, quartz, or alkyl halides (KBr, NaBr) with an aluminum-coated surface for reflection. These are of two kinds:
    - Diffraction gratings: Work on reinforcement of incident and reflected rays, with wavelength given by $\lambda = d\ (\sin i \pm \sin r)/n$.
    - Transmission gratings: Based on reinforcement of transmitted and refracted rays, expressed by $\lambda = d\ (\sin \theta)/n$.
- **Filters:** A filter is a device that transmits light of the desired wavelength while blocking unwanted radiation; the main types are absorptive filters and interference filters.
- **Sample cell:** Sample cells, or cuvettes, are used to hold solutions during UV-vis analysis and are made from different materials depending on the wavelength range required: optical glass covers 335–2500 nm, special optical glass covers 320–2500 nm, quartz for infrared covers 220–3800 nm, and quartz for far-UV covers 170–2700 nm.
- **Detectors**: These are devices that convert transmitted light into electrical signals for measurement. After the sample absorbs part of the radiation, the remaining light reaches the detector to measure its intensity. Common types include the following:
  - Photomultiplier tubes: Amplify photoelectrons through secondary emission.
  - Photovoltaic cells: Light generates electrons in a selenium layer to produce current measured by a galvanometer.
  - Phototubes: Light releases electrons from a photocathode that move to an anode, producing a current proportional to light intensity.

4.2.6.7.3 Advantages

- It is a rapid and simple measurement.
- It is quite less expensive compared to HPLC or LC-MS.
- The sample can often be recovered after analysis.
- It is suitable for routine screening, particularly for estimating total classes of compounds (e.g., total flavonoids, phenolics).
- It requires a minimal amount of sample.
- It is easy to operate, without need for professionals.

4.2.6.7.4 Procedure

- The instrument is switched on and stabilized for at least 15 minutes for accurate readings.
- The samples are standard solutions prepared along with the blank solution.
- Then, the cuvette is cleaned to remove dust, fingerprints, or water droplets.

- The reading of the blank absorbance is recorded with only the solvent used to dissolve the sample (no solute) by placing the cuvette in the spectrophotometer correctly; this will be subtracted from the sample readings.
- Then, the cuvette is rinsed twice and the sample is filled. Then the absorbance spectrum is recorded within 200–800 nm.
- The wavelength of maximum absorbance is identified.
- Absorbance (y-axis) versus wavelength (x-axis) is plotted to generate the absorbance spectrum, which helps identify compounds and estimate their proportions.

4.2.6.7.5 Challenges

- Herbal drugs contain multiple compounds in a single extract, many of which absorb UV light at similar wavelengths, making it difficult to resolve individual components.
- UV-vis detects chromophores, but different compounds can produce overlapping peaks, leading to ambiguous identification of specific phytochemicals.
- There might be an interference from impurities that contribute to the absorbance value, which might give an incorrect measurement.
- Compounds present in trace amounts may fall below the detection limit, preventing accurate detection of important bioactive molecules.
- In complex herbal extracts, the absorbance measured at a specific wavelength represents the sum of all compounds absorbing there, complicating quantification of a single constituent.
- For accurate quantification, pure standard compounds calibrations are required, which are difficult to attain commercially.
- Very low concentrations may not be detectable, and very high concentrations can exceed the linear range, necessitating careful sample preparation and dilution.
- Some herbal constituents degrade upon exposure to light during measurement, leading to inaccurate readings.
- It is quite a time-consuming process, in which extracts often need to be filtered and diluted.

4.2.6.7.6 Application to Herbal Drugs

UV-vis spectrophotometer in authentication of herbal drugs: UV-vis spectroscopy (200–900 nm) has a distinct spectral fingerprint for herbal extracts, helping in authentication, adulteration detection, and evaluation of batch variations. Recording extract spectra generates unique profiles for each plant, which can be compiled into a spectral library of authentic samples. Such libraries provide a cost-effective alternative to marker-based methods. For instance, Pippali (*Piper longum*) shows λ_{max} at 253–258 nm and 338–341 nm, Kankola (*Piper cubeba*) exhibits peaks at 295–296.5 nm and 352–359 nm, and Tea (*Camellia chinensis*) displays a strong λ_{max} at 272.5 nm [24].

Quantitative analysis of phytoconstituents: UV spectrophotometry is commonly used for estimation of secondary metabolites (like alkaloids, flavonoids, phenolics, and tannins), by preparation of calibration curves with their respective standards, and also in quality control and standardization of herbal formulations.

Estimation of total alkaloid: UV spectrophotometry is used for the estimation of total alkaloids in herbal extracts, utilizing bromocresol green (BCG) as a reagent. When the alkaloidal compounds react with BCG, they form a complex that produces a measurable yellow complex at 470 nm. An estimation of total alkaloids on *Chitrakadivati*, a classical polyherbal Ayurvedic formulation, was first done by confirming alkaloid presence qualitatively using Dragendorff's test and then quantifying them via UV spectrophotometry. Out of its 16 ingredients (plants), 9 were found to contain alkaloids, with the formulation showing a total alkaloid content of 0.16%, significantly higher than its individual components [25].

Estimation of flavonoid compounds: It is used for the estimation of flavonoid bioactive compounds in plant extracts, Ayurvedic formulations, and NDDS's. A validated method, that is, the aluminum chloride method, was developed for the quantification of hesperidin, detecting maximum absorbance at 284 nm. The method was applied to orange peel, bitter lime peel, the Ayurvedic formulation *Madhiphala Rasayana*, and a hesperidin-based topical nanoformulation, showing high recovery (97–99%) with good precision, linearity, and robustness [26].

Estimation of total phenolics compounds: The total phenolic content was quantified using the Folin-Ciocalteu method, where a calibration curve of a standard phenolic compound was employed to calculate the concentration of phenolics in the extracts and herbal formulation. A polyherbal formulation prepared from *Lantana camara*, *Magnolia obovata*, *Albizzia lebbeck*, *Clitoria ternatea*, *Cinnamomum cassia*, and *Coriandrum sativum* was analyzed using UV-visible spectrophotometry. The assay, performed at 765 nm, revealed a total phenolic content of 2.37 mg/g gallic acid equivalent (GAE) [27].

***In vitro* anti-oxidant activity:** Evaluation of the anti-oxidant potential of selected plant extracts using a UV-visible spectrophotometer is carried out through free-radical scavenging assays. Different assay methods, like the DPPH assay, ABTS•+ assay, and FRAP, were performed at different wavelengths. A study was carried out to evaluate the antioxidant activity of *Boswellia serrata*, *Commiphora wightii*, *Hemidesmus indicus*, *Aloe barbadensis*, *Withania somnifera*, *Zingiber officinale*, *Berberis aristata*, *Curcuma longa*, and their polyherbal formulation (vati). Antioxidant activities, including superoxide and hydrogen peroxide scavenging, were assessed, and quantitatively free-radical scavenging was assessed through standard assays such as DPPH at 517 nm, ABTS at 734 nm, superoxide radical scavenging at 560 nm, and nitric oxide scavenging at 540 nm using the Griess reaction [28].

***In vitro* anti-inflammatory activity:** *In vitro* anti-inflammatory activity can be effectively assessed using UV-visible spectrophotometry. Standard assays such as protein denaturation inhibition and erythrocyte membrane stabilization rely on absorbance measurements at specific wavelengths (e.g., 560–660 nm) to quantify the protective effects of plant extracts against protein degradation and cell lysis. I*n vitro* anti-inflammatory activity of the herbal preparation HP-4 was evaluated by protein denaturation and erythrocyte membrane stabilization assays. HP-4 (100–500 µg/mL) significantly inhibited heat-induced denaturation of egg albumin, measured spectrophotometrically at 660 nm, and protected RBC membranes against hypotonic-solution-induced lysis in a concentration-dependent manner. These effects were comparable to those of acetylsalicylic acid (100–200 µg/mL), confirming HP-4's anti-inflammatory potential [29].

4.2.6.8 Confocal Microscopy

The concept of fluorescence confocal microscopy was introduced by Marvin Minsky in 1955. Unlike conventional fluorescence microscopy, which collects signals from the entire sample and produces blurred images, confocal microscopy achieves optical sectioning, sharper resolution, and three-dimensional reconstruction of thick specimens. It has various advantages that include improved resolution (axial and lateral), suppressed background fluorescence, higher signal-to-noise ratio, and multi-channel imaging abilities. These features make it an indispensable tool in areas of biomedical research, materials science, and dermatology research. In herbal drug research, it allows visualization of plant tissues, localization of various phytochemicals, evaluation of formulation integrity, and authentication of herbal drugs.

4.2.6.8.1 Principle

Fluorescence confocal microscopy is based on the principle of point illumination and spatial filtering to achieve optical sectioning. A focused laser excites fluorophores in the sample, causing them to emit light at a longer wavelength. The emitted fluorescence passes back through the objective and is directed toward a pinhole aperture placed in front of the detector. This pinhole eliminates out-of-focus light from above and below the focal plane, allowing only in-focus signals to be detected. As the laser scans across the specimen (x-y plane) and through its depth (z-axis), serial optical sections are captured, which can be reconstructed into high-resolution three-dimensional images [30, 31].

Fluorescence microscopy primarily follows the equation of fluorescence intensity, which is derived from the Beer-Lambert law of light absorption and the efficiency of fluorescence emission.

The equation is expressed as

$$I_f = 2.3 \times \varnothing \times X_{\epsilon} \times I_{ex} \times C \tag{4.4}$$

where

ϕ = quantum yield

ε = molar extinction coefficient

I_{ex} = excitation light intensity

C = fluorophore concentration

This equation describes how emitted fluorescence depends on fluorophore properties and excitation intensity. The fluorescence intensity (I_f) detected by a microscope is governed by excitation light, sample concentration, and intrinsic fluorophore properties.

4.2.6.8.2 Advantages

- **Improved optical sectioning:** By eliminating out-of-focus light, confocal microscopy reduces background interference from planes above and below the focal point, yielding images with improved sharpness and contrast compared to conventional widefield microscopy [32].
- **Three-dimensional visualization through optical sectioning:** It attains sequential optical slices that can be digitally combined to produce comprehensive three-dimensional reconstructions of thick specimens, eradicating the need for physical sectioning.
- **Improved axial resolution:** The use of point illumination and pinhole-based spatial filtering significantly enhances the resolution along the z-axis, allowing detailed imaging of thick or layered specimens.
- **Precise electronic zoom capability:** It offers an adjustable zoom and adjustable electronic zoom, allowing thorough visualization of smaller sections without switching objectives, thus refining on-screen resolution and magnification.
- **Quantitative and multi-channel imaging detection:** It permits reliable quantification of fluorescence intensity, simultaneous detection of multiple spectral channels, and time-gap acquisition, making them well suited for quantitative analysis and experiments concerning several fluorophores.
- **Enhanced signal-to-noise ratio:** Focused excitation, along with spatial filtering, minimizes background fluorescence, resulting in higher image contrast and better sensitivity in sensing weak signals.

4.2.6.8.3 Instrumentation

A fluorescence confocal microscope comprises the following parts, as shown in Figure 4.13 [33].

- **Light source**: The light source of a fluorescence microscope is an important component, which is responsible for providing the excitation light that is used to excite the fluorophores in the sample.

  The various types of light sources commonly used in fluorescence microscopy are as follows:

  a) **Mercury arc lamps**: They are the most commonly used light sources in fluorescence microscopy. They produce a broad spectrum of light that excites a wide range of fluorophores.

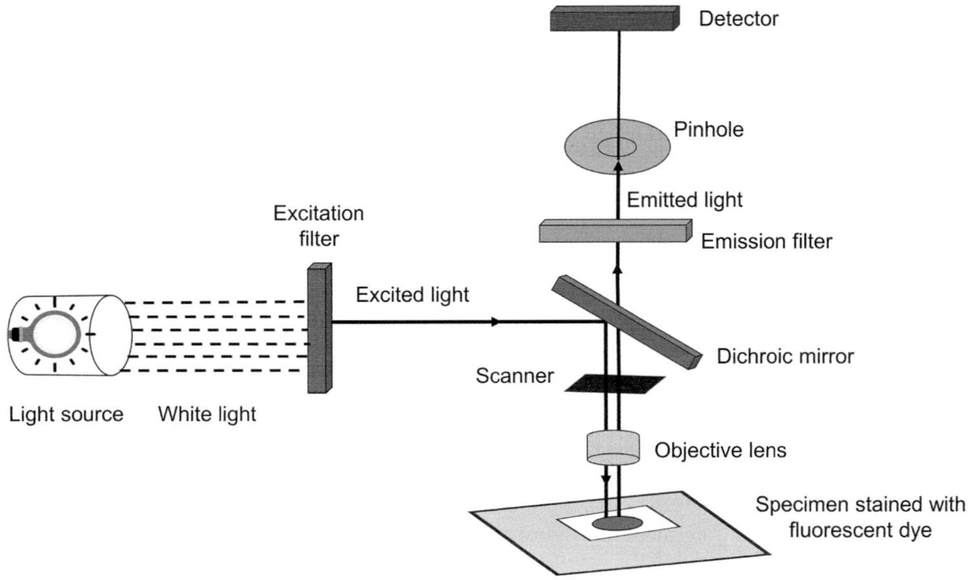

FIGURE 4.13 Schematic of a confocal fluorescence microscope

b) **Metal halide lamps**: They are a newer type of light source that produces a more focused spectrum of light than mercury arc lamps. They are often used to excite specific fluorophores and can provide higher intensity and more stable light output than mercury arc lamps.
c) **LED (light-emitting diode) source:** These light sources are becoming increasingly popular in fluorescence microscopy due to their long lifetime and low power consumption. They produce a narrow spectrum of light that can be used to excite specific fluorophores and can be easily controlled and modulated for advanced imaging applications.
d) **Laser light sources**: They produce a highly focused and coherent beam of light that can be used to excite specific fluorophores. They are often used in advanced imaging techniques, such as super-resolution microscopy, that require high-intensity, highly focused light.

- **Excitation filter**: It selects the precise wavelength of light needed to produce fluorescence to stimulate the fluorescent tags.
- **Dichroic mirror or beamsplitter**: It directs the excitation beam toward the sample and transmits emitted fluorescence.
- **Scanner/scanning mirror**: Scanning mirrors (galvanometer-driven) raster-scan the focused laser beam point by point across the specimen (x-y plane).
- **Objective lens**: High numerical aperture (NA) objectives focus the excitation light into the sample and collect emitted fluorescence. They ensure high spatial resolution.
- **Specimen stage:** It holds the sample and positions it precisely.
- **Pinhole aperture (confocal principle):** It is placed in front of the detector. It blocks out-of-focus light, allowing only in-focus fluorescence to reach the detector. It is also responsible for optical sectioning and improving contrast.
- **Emission filter**: It selects only the desired emission wavelengths to pass through and removes scattered laser light and background autofluorescence.
- **Detector:** It converts light signals into electrical signals. Photomultiplier tubes (PMTs) are common, while avalanche photodiodes (APDs) or hybrid detectors are used for higher sensitivity.

- **Computer and image processing software:** It controls the laser scanning, pinhole size, and detector sensitivity and reconstructs sequential point-by-point fluorescence signals into 2D or 3D images (optical sections).

4.2.6.8.4 Procedure

- In case of crude fresh herbal material (leaf, stem, root, seed, bark, etc.), it is first washed with distilled water to remove dust and debris.
- Then thin sections (~50–100 µm) are cut using a microtome or free-hand sectioning.
- The transverse sections are mounted on glass slides with a suitable mounting medium (e.g., glycerol).
- However, in the case of herbal drugs, it is first crushed and then dispersed in a small amount of mounting medium and then spread on a glass slide.
- Then the sections or dispersed herbal drugs are covered with a coverslip, and edges are sealed with nail varnish or mounting sealant if necessary.
- Sometimes, the herbal crude drugs and extracts can be directly utilized for imaging without fluorescent stains due to the natural autofluorescence property of phytochemicals.
- However, when natural autofluorescence is insufficient or when specific cellular structures need to be highlighted, fluorescent stains may be employed. Some of the commonly used dyes include Calcofluor white (binds strongly to cellulose and lignin), Acridine orange (stains nucleic acids), Fluorescein Isothiocyanate (conjugation with proteins or metabolites.
- Afterward, the confocal microscope is switched on and the lasers are allowed to stabilize.
- The excitation wavelength is selected based on autofluorescence or stains used.
- The dichroic mirrors and emission filters are adjusted for better signal separation.
- Then the glass slide is placed on the stage and moved from low (40×) to high (100×) magnification.
- The pinhole is adjusted for section thickness. The laser power, gain, and offset are optimized to get a better signal.
- The z-axis sections are captured and combined into z-stacks for 3D reconstruction, with multi-channel imaging used when multiple fluorophores are present.
- Advanced software is used for 3D reconstruction, intensity analysis, and co-localization.

4.2.6.8.5 Challenges

- **Interference due to autofluorescence:** Strong natural autofluorescence from phytoconstituents present in herbal drugs, such as chlorophyll and lignin, can suppress weak fluorescent signals, leading to bleed-through and making it tough to accurately differentiate or assign specific signals [34].
- **Spectral overlapping:** Significant spectral overlap requires the use of spectral imaging and linear unmixing, or else, overlapping emissions from endogenous plant metabolites and fluorescent dyes may cause inaccurate interpretation [35].
- **Photobleaching:** The continuous exposure to laser light during confocal imaging can result in photobleaching and the generation of reactive oxygen species, which may further harm living plant tissues and modify fluorescence intensity or dynamics, mainly in the case of time-based studies [36].
- **Limited imaging depth in opaque tissues:** In thick or opaque plant matrices, light scattering and absorption limit penetration depth, constraining visualization of internal structures and often demanding tissue cleaning or other imaging methods for in-depth analysis [31].
- **Refractive index variations:** The variations in refractive index across plant tissues, such as cuticle, cell walls, vacuoles, and mounting media, cause spherical aberrations, reducing the brightness and progressive loss of axial resolution with increasing depth of imaging [37].
- **Sample cleanup:** Sample cleaning is often required (using ClearSee/ClearSeeAlpha) to reduce autofluorescence by chlorophylls and enhance the transparency. However, sample

cleaning can be time consuming (days to weeks) and eventually may modify the chemistry or morphology of the sample [38].

- **Variability in quantification of fluorescence:** Accurate quantitative comparisons can be challenging as plant fluorescence is extremely influenced by the chemical environment, such as pH, quenching, binding, and unpredictable dye uptake, thus necessitating rigorous controls and standardized reporting [34].
- **Data complexity and calibration issues:** Handling multi-channel registration, detector linearity, and spectral system reliability introduces significant complexity, necessitating technical expertise along with stringent calibration and standardized controls [39].
- **Lack of standardized protocol for herbal drug authentication**: Even though microscopy is generally suggested for herbal drug authentication, standardized procedures and reference materials specific to confocal laser scanning microscopy remain inadequate and are not synchronized across various pharmacopoeias [40].

4.2.6.8.6 Applications

4.2.6.8.6.1 Localization of Phytochemicals The confocal microscopic technique was utilized to determine the spatial distribution of highly diffusible plant metabolites, like ions and small phenolic compounds, present in fruit tissues. Vidot et al. used a specialized cryo-laser scanning confocal microscopic method to successfully localize flavonoids, calcium, and potassium present in fruits such as apple and grape tissues, indicating the technique's capability for mapping of secondary metabolites in plant materials. It also confirmed that freezing temperature and rate of cooling significantly affected the thickness of the cell wall and morphology of tissues. The method reduced the artefacts due to photobleaching and phytocompound redistribution, thus permitting consistent localization of diffusible metabolites [41].

4.2.6.8.6.2 Mapping of Phytochemical Distribution In a study conducted by He et al., confocal laser scanning microscopy (CLSM), along with a metabolomics study, was utilized to investigate temporal changes of flavonoids in *Amygdalus pedunculata* leaves. The flavonoids were found to be mostly localized in palisade and spongy tissues, with higher intensity present in guard cells and nucleoli, and lowest in vascular bundles, by means of naturstoffreagenz A staining and autofluorescence detection. The technique also allowed precise mapping of flavonoid accumulation, confirming that the levels of flavonoid increased in the fruit development stage and decreased during the ripening stage [42].

4.2.6.8.6.3 Visualization of Secretory Structures In a study conducted by Gostin et al., CLSM was used to visualize trichome autofluorescence in the plant *Phlomis herba-venti*. It showed autofluorescence of chloroplasts in stalk, neck, and secretory cells, and yellow-green autofluorescence of secretion products in capitate and dendroid trichomes, demonstrating the presence of lipids, flavonoids, and other plant metabolites. It was further confirmed by histochemical analyses, which showed that secretion products comprised of phytoconstituents such as terpenes, phenolics, lipids and polysaccharides, with differences between types of trichomes [43].

4.2.6.8.6.4 Herbal Drug Authentication and Quality Control Liu et al. used confocal microscopy to build visual 3D models of Aconitum roots, allowing for precise identification of toxic species in Chinese herbal medicine. In the case of Aconitum, comparing the largest root cross-sections and exceptional internal structures based on their unique fluorescent signatures assured the safety and authenticity of the plant material [44].

4.3 CONCLUSION

Accurate analysis and standardization of phytopharmaceuticals remain at the heart of quality assurance in herbal medicine. With the increasing global interest in plant-based therapeutics, there's an increasing need for well-defined, reproducible methods that can validate both the identity and the consistency of herbal formulations. This chapter has emphasized the central role of instrumental techniques – especially HPLC, HPTLC, and LC-MS – in meeting that need. Each method serves a unique purpose. HPLC is particularly well suited for quantifying active compounds in complex matrices, offering reliability across a broad range of phytochemicals. HPTLC, by contrast, provides a more accessible platform for rapid screening, fingerprinting, and preliminary quality checks. Its visual output and relatively low operational costs make it a practical choice for routine analysis. LC-MS, with its ability to both separate and identify compounds with high specificity, plays a vital role in structural elucidation and trace-level detection, particularly in research and development settings. These techniques, with their contribution to building a scientific foundation for herbal products, bridge the gap between traditional use and modern expectations of quality, safety, and efficacy. These tools not just help identify the beneficial components but also detect adulterants, degradation products, and contaminants that may compromise the integrity of a formulation.

Standardization isn't a simple process, but requires a deep understanding of the botanical source, the extraction method, and the intended therapeutic use. Instrumental analysis, when applied accordingly, allows for this greater approach, providing data that can inform everything from batch-to-batch consistency to regulatory submissions. In industry, challenges like variability and a lack of uniform standards occur; however, the thoughtful integration of HPLC, HPTLC, and LC-MS into phytopharmaceutical workflows offers a path forward – not just for compliance, but for credibility. As the herbal sector continues to expand, these methodologies will remain indispensable, supporting not only product development but also the broader goal of ensuring that natural medicines meet the same scientific standards as their synthetic counterparts.

REFERENCES

1. Mohapatra, D., et al., Quality by design – based development and optimization of fourth-generation ternary solid dispersion of standardized Piper longum extract for melanoma therapy. *Drug Delivery and Translational Research*, 2023. 13(12): p. 3094–3131.
2. White, T.J., et al., Amplification and direct sequencing of fungal ribosomal RNA genes for phylogenetics. *PCR Protocols: A Guide to Methods and Applications*, 1990. 18(1): p. 315–322.
3. Drouet, S., et al., A critical view of different botanical, molecular, and chemical techniques used in authentication of plant materials for cosmetic applications. *Cosmetics*, 2018. 5(2): p. 30.
4. Selvakumari, E., et al., Application of DNA fingerprinting for plant identification. *JAIR*, 2017. 5(10): p. 149–151.
5. Ganie, S.H., et al., Authentication of medicinal plants by DNA markers. *Plant Gene*, 2015. 4: p. 83–99.
6. Sahu, U., et al., Development of simultaneous HPTLC method and validation for the quality assessment of ayurvedic formulation – ayush kvatha churna by using marker compound rosmarinic acid, trans-cinnamaldehyde and piperine. *Journal of Chromatographic Science*, 2025. 63(3): p. bmae019.
7. Meena, A.K., et al., Identification and estimation of bioactive constituents Negundoside, Berberine chloride, and Marmelosin by HPLC and HPTLC for development of quality control protocols for Ayurvedic medicated oil formulation. *Future Journal of Pharmaceutical Sciences*, 2021. 7(1): p. 171.
8. Sharma, B., A. Islam, and A. Sharma, HPTLC-MS: an advance approach in herbal drugs using fingerprint spectra and mass spectroscopy. *Traditional Medicine Research*, 2023. 8(2): p. 10.
9. Orsini, F., et al., HPTLC, HPTLC-MS/MS and HPTLC-DPPH methods for analyses of flavonoids and their antioxidant activity in Cyclanthera pedata leaves, fruits and dietary supplement. *Journal of Liquid Chromatography & Related Technologies*, 2019. 42(9–10): p. 290–301.
10. Orsini, F., et al., HPTLC, HPTLC-MS/MS and HPTLC-DPPH methods for analyses of flavonoids and their antioxidant activity in Cyclanthera pedata leaves, fruits and dietary supplement. *Journal of Liquid Chromatography & Related Technologies*, 2019. 42(9–10): p. 290–301.
11. Shivatare, R.S., D.H. Nagore, and S.U. Nipanikar, HPTLC'an important tool in standardization of herbal medical product: A review. *Journal of Scientific and Innovative Research*, 2013. 2(6): p. 1086–1096.
12. Purohit, D.C., et al., Rapid screening of undeclared hypoglycemics in counterfeit herbal antidiabetic products using HPTLC-MS. *Journal of Liquid Chromatography & Related Technologies*, 2022. 45(5–8): p. 100–106.
13. Sonia, K. and K. Lakshmi, HPTLC method development and validation: An overview. *Journal of Pharmaceutical Sciences and Research*, 2017. 9(5): p. 652.

14. Jain, D., et al., Isolation, development and validation of HPTLC method for the estimation of β-carotene from Gymnosporia senegalensis (Lam.) Loes. *Plant Physiology and Biochemistry*, 2023. 201: p. 107843.
15. Kagathara, C., K. Odedra, and N. Vadia, Development of HPTLC method for the simultaneous estimation of quercetin, curcumin, and ascorbic acid in herbal formulations. *Journal of the Iranian Chemical Society*, 2022. 19(10): p. 4129–4138.
16. Oladimeji, A.V. and M. Valan, HPLC techniques for phytochemistry. *IJCS*, 2020. 8(6): p. 2590–2596.
17. Bārzdiņa, A., et al., The potential use of herbal fingerprints by means of HPLC and TLC for characterization and identification of herbal extracts and the distinction of latvian native medicinal plants. *Molecules*, 2022. 27(8): p. 2555.
18. Raina, S., et al., Anticancer activity of cordia dichotoma against a panel of human cancer cell lines and their phytochemical profiling via HPLC and GCMS. *Molecules*, 2022. 27(7): p. 2185.
19. Patel, N.D., et al., HPLC fingerprint analysis coupled with multivariate analysis for quality assessment of Picrorhiza kurroa rhizomes. *Current Chromatography*, 2022. 9(1): p. 95–102.
20. Orth, H., C. Rentel, and P. Schmidt, Isolation, purity analysis and stability of hyperforin as a standard material from Hypericum perforatum L. *Journal of Pharmacy and Pharmacology*, 1999. 51(2): p. 193–200.
21. Sundhani, E., et al., Pharmacokinetic herb-drug interactions of glipizide with Andrographis paniculata (Burm. f.) and andrographolide in normal and diabetic rats by validated HPLC method. *Molecules*, 2022. 27(20): p. 6901.
22. Caldeirao, L., et al., Herbs and herbal infusions: Determination of natural contaminants (mycotoxins and trace elements) and evaluation of their exposure. *Food Research International*, 2021. 144: p. 110322.
23. Mandru, A., J. Mane, and R. Mandapati, A review on UV-visible spectroscopy. *Journal of Pharma Insights and Research*, 2023. 1(2): p. 091–096.
24. Pérez-Ràfols, C., N. Serrano, and J.M. Díaz-Cruz, Authentication of soothing herbs by UV – vis spectroscopic and chromatographic data fusion strategy. *Chemometrics and Intelligent Laboratory Systems*, 2023. 235: p. 104783.
25. Ajanal, M., M.B. Gundkalle, and S.U. Nayak, Estimation of total alkaloid in Chitrakadivati by UV-Spectrophotometer. *Ancient Science of Life*, 2012. 31(4): p. 198–201.
26. Chimagave, S.S., et al., Development and validation of stability indicating UV-spectrophotometric method for the estimation of hesperidin in bulk drugs, plant extract, Ayurveda formulation and nanoformulation. *Indian Journal of Pharmaceutical Education and Research*, 2022. 56(3): p. 865–872.
27. Saini, R., et al., Evaluation of polyphenolic content and anti-oxidant potential of poly-herbal formulation. *International Journal of Emerging Technologies*, 2019. 10(2): p. 243–8.
28. Joshi, P., et al., Antioxidant and anti-inflammatory activities of selected medicinal herbs and their polyherbal formulation. *South African Journal of Botany*, 2020. 130: p. 440–447.
29. Padmanabhan, P. and S. Jangle, Evaluation of in-vitro anti-inflammatory activity of herbal preparation, a combination of four medicinal plants. *International Journal of Basic and Applied Medical Sciences*, 2012. 2(1): p. 109–116.
30. Jayamohanan, R., Confocal microscopy – Working principle and applications in dermatology. *Journal of Skin and Sexually Transmitted Diseases*, 2023. 5(2): p. 81–89.
31. Elliott, A.D., Confocal microscopy: Principles and modern practices. *Current Protocols in Cytometry*, 2020. 92(1): p. e68.
32. Amos, W.B., J. White, and M. Fordham, Use of confocal imaging in the study of biological structures. *Applied Optics*, 1987. 26(16): p. 3239–3243.
33. Lichtman, J.W. and J.-A. Conchello, Fluorescence microscopy. *Nature*. 200: p. 5.
34. Czymmek, K.J., et al., Best practices in plant fluorescence imaging and reporting-a primer. *The Plant Cell*, 2025. 37(7): p. koaf143.
35. Garini, Y., I.T. Young, and G. McNamara, Spectral imaging: Principles and applications. *Cytometry Part A: The Journal of the International Society for Analytical Cytology*, 2006. 69(8): p. 735–747.
36. Thorn, K., A quick guide to light microscopy in cell biology. *Molecular Biology of the Cell*, 2016. 27(2): p. 219–222.
37. Diaspro, A., F. Federici, and M. Robello, Influence of refractive-index mismatch in high-resolution three-dimensional confocal microscopy. *Applied Optics*, 2002. 41(4): p. 685–690.
38. Kurihara, D., et al., ClearSee: A rapid optical clearing reagent for whole-plant fluorescence imaging. *Development*, 2015. 142(23): p. 4168–4179.
39. Zucker, R.M., et al., Reliability of confocal microscopy spectral imaging systems: Use of multispectral beads. *Cytometry Part A: The Journal of the International Society for Analytical Cytology*, 2007. 71(3): p. 174–189.

40. Ichim, M.C., A. Häser, and P. Nick, Microscopic authentication of commercial herbal products in the globalized market: Potential and limitations. *Frontiers in Pharmacology*, 2020. 11: p. 876.
41. Vidot, K., et al., Cryo-laser scanning confocal microscopy of diffusible plant compounds. *Plant Methods*, 2018. 14(1): p. 89.
42. He, Y., et al., Metabolomic and confocal laser scanning microscopy (CLSM) analyses reveal the important function of flavonoids in amygdalus pedunculata pall leaves with temporal changes. *Frontiers in Plant Science*, 2021. 12: p. 648277.
43. Gostin, I.N., Glandular and non-glandular trichomes from Phlomis herba-venti subsp. pungens leaves: Light, confocal, and scanning electron microscopy and histochemistry of the secretory products. *Plants*, 2023. 12(13): p. 2423.
44. Liu, C.C., et al., Identification of four Aconitum species used as "Caowu" in herbal markets by 3d reconstruction and microstructural comparison. *Microscopy Research and Technique*, 2015. 78(5): p. 425–432.

5 Instrumental Methodology for Structural Elucidation and Phytochemical Integrity Analysis of Herbal Drugs and Formulations

ABBREVIATIONS

Abbreviation	Description
APCI	Atmospheric Pressure Chemical Ionization
APPI	Atmospheric Pressure Photoionization
ATR	Attenuated Total Reflectance
$CDCl_3$	Deuterated Chloroform
DMSO-d_6	Deuterated Dimethyl Sulfoxide
EI	Electron Ionization
ESI	Electrospray Ionization
FT	Fourier Transformation
FTIR	Fourier Transform Infrared Spectroscopy
GC-MS	Gas Chromatography-Mass Spectrometry
HPLC-SPE-NMR	High-Performance Liquid Chromatography-Solid Phase Extraction-NMR Spectroscopy
ICH	International Council for Harmonisation
LC-MS	Liquid Chromatography-Mass Spectrometry
MAS	Magic Angle Spinning
NIST	National Institute of Standards and Technology
NMR	Nuclear Magnetic Resonance
OPLS-DA	Orthogonal Partial Least-Squares Discrimination Analysis
PCA	Principal Component Analysis
RF	Radiofrequency
SIM	Selected Ion Monitoring
USP	United States Pharmacopoeia

5.1 INTRODUCTION

The complexity of herbal drugs and formulations necessitates the use of advanced instrumental methodologies for structural elucidation and phytochemical integrity analysis. These approaches provide reliable insights into the molecular architecture of bioactive constituents, ensuring accurate identification, authentication, and standardization of herbal medicines. High-resolution spectroscopic

DOI: 10.1201/9781003540823-5

and chromatographic tools, such as infrared spectroscopy, nuclear magnetic resonance (NMR), mass spectrometry (MS), and hyphenated techniques like LC-MS and GC-MS (Figure 5.1), enable precise structural determination and quantification of active metabolites. These methodologies are necessary for detecting adulteration, monitoring batch-to-batch consistency, and ensuring the therapeutic efficacy of herbal formulations.

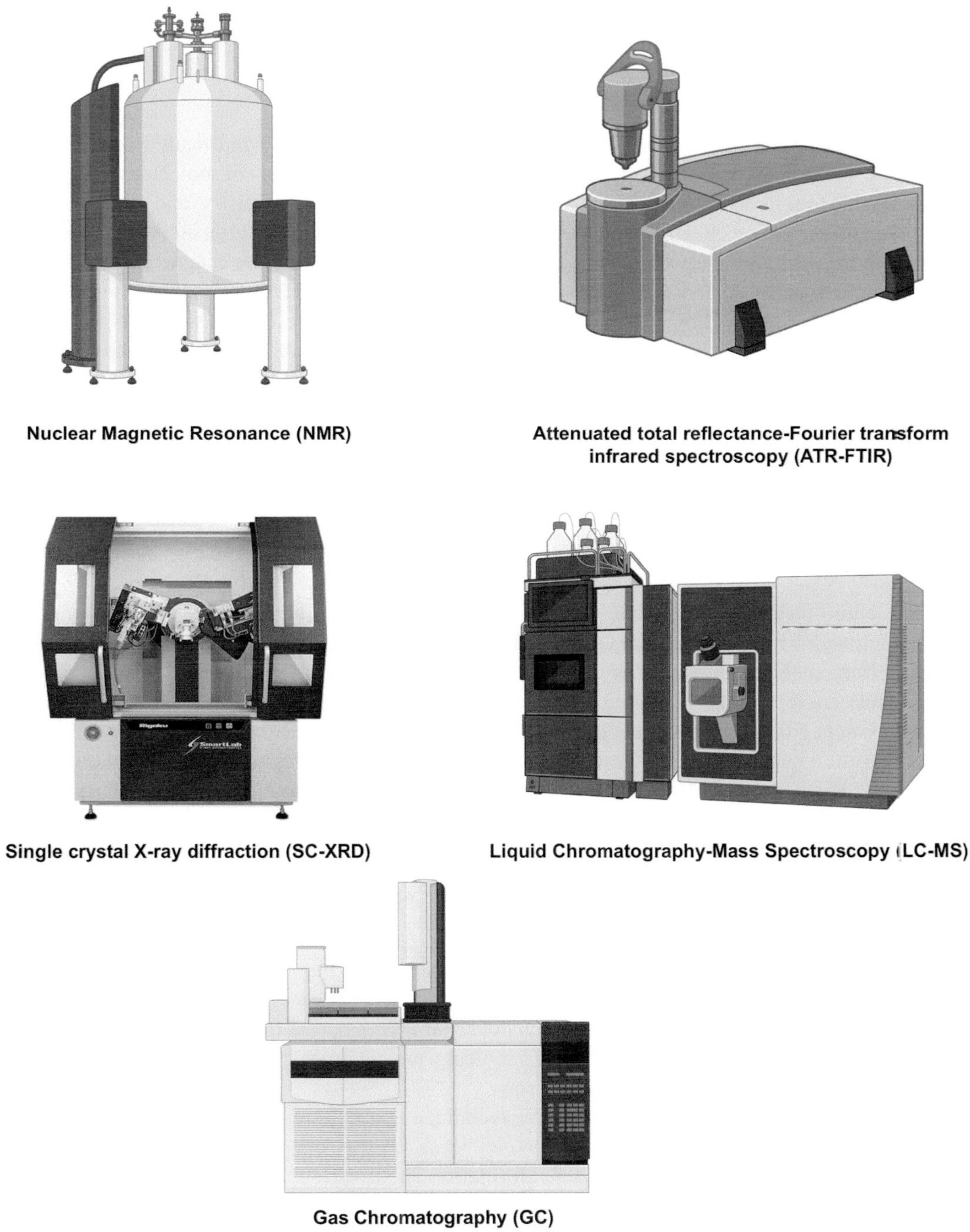

FIGURE 5.1 Instruments for structural elucidation and phytochemical integrity analysis of herbal drugs and formulations

Furthermore, the integration of these instrumental techniques into phytopharmaceutical research has strengthened quality control protocols and regulatory compliance. By establishing well-defined chemical markers and fingerprinting profiles, researchers and manufacturers can safeguard the phytochemical integrity of herbal products, thereby enhancing safety, reproducibility, and global acceptance. This chapter explores the instrumental strategies employed in the structural elucidation and quality assessment of herbal drugs and formulations. Emphasis is placed on the role of modern analytical tools in standardization, validation, and quality control, underscoring their significance in the development of scientifically credible and therapeutically effective phytomedicines.

5.2 NUCLEAR MAGNETIC RESONANCE

Nuclear magnetic resonance (NMR) spectroscopy is a non-destructive, sophisticated analytical tool utilized to study the structure, composition, and dynamics of compounds or molecules. NMR is an indispensable tool in diverse areas such as pharmaceuticals, organic chemistry, and natural product development, providing exhaustive information about the molecular structure, atomic connectivity, functional groups, and stereochemistry of a compound by analyzing the magnetic properties of atomic nuclei such as ^{1}H and ^{13}C. There are two main types of NMR, namely solution-state NMR and solid-state NMR (ssNMR). Solution-state NMR generates high-resolution spectra of the soluble compounds and is commonly used in cases of complex mixtures and small molecules. However, ssNMR evaluates insoluble, crystalline compounds using the magic angle spinning (MAS) technique to refine the NMR spectra resolution. Advanced techniques like 1D, 2D, and qualitative NMR (qNMR), when combined with chemometric tools (PCA), further enhance the applicability of NMR in precise structural elucidation, fingerprinting, and quality control of herbal drugs and other complex samples.

5.2.1 Principle

NMR is based on the principle that some atomic nuclei have an inherent intrinsic property called spin, along with a magnetic moment. When placed in a strong external magnetic field, these nuclei align their magnetic moments either in parallel or anti-parallel to the applied field, thereby producing discrete energy levels. When radiofrequency (RF) pulses are applied at a specific resonance frequency, depending on the magnetic field strength and type of nuclei, the nuclei absorb the energy, and a transition occurs between these spin states. As the nuclei return to their lower energy state, the emitted RF signals are detected by the instrument. The difference in resonance frequency happens due to the surrounding electronic environment of each nucleus, which provides detailed information about their molecular structure, chemical environments, connectivity, and dynamics. This phenomenon produces a spectrum that, upon interpretation, gives information about the nuclei or molecule [1–3].

5.2.2 Advantages

- **Simultaneous analysis:** NMR simultaneously analyzes herbal drugs or extracts containing complex phytochemicals without requiring pre-treatment; therefore, it can identify multiple phytoconstituents in a single run.
- **Non-destructive:** It is a non-destructive technique that requires minimal sample preparation, suitable for herbal drugs.
- **Structural elucidation:** It provides information about various functional groups present and the stereochemical configuration of the phytochemicals.
- **Solvent compatibility:** A wide range of deuterated solvents is compatible with NMR, which allows analysis of both polar and non-polar compounds.

- **Authentication and spectral fingerprinting:** The distinct NMR spectral fingerprints of the herbal drugs help in differentiating similar plant species, identifying possible adulterants, and evaluating batch-to-batch consistency for standardization and quality control purposes.

5.2.3 Instrumentation

A schematic representation of the NMR instrument is shown in Figure 5.2 [2, 4, 5]. The fundamental parts of an NMR spectrometer consist of the following:

- **Superconducting magnet:** A magnet is the main component of an NMR spectrometer. It contains liquid nitrogen or helium to maintain the low temperature to make the magnet superconducting in nature. It provides a homogeneous and highly stable magnetic field (generally 300–1200 MHz for proton NMR) required for alignment of the nuclear spin.
- **RF transmitter system:** It generates precise radiofrequency pulses, which are regulated by a pulse programmer to excite the nuclear spins and detect the emitted signal during relaxation.
- **Probe and sample tube:** The probe consists of RF coils, which transmit and receive the signals. The probe is designed to hold the NMR sample tube, and modern broadband probes enable the detection of multiple nuclei. The sample is placed in an NMR tube of about 5 mm thickness. The tube is spun at a controlled rate to minimize variations in the magnetic field, thereby improving the resolution of the spectra.
- **Shimming system:** Shimming is required for obtaining optimal peak and resolution. For shimming, small electromagnets known as shims are used to fine-tune the magnetic field.
- **Computer and software:** The NMR raw data acquisition, Fourier transformation, spectral processing, and analysis are carried out using specialized software platforms, for example, TopSpin (Bruker) and Mnova (Mestralab).

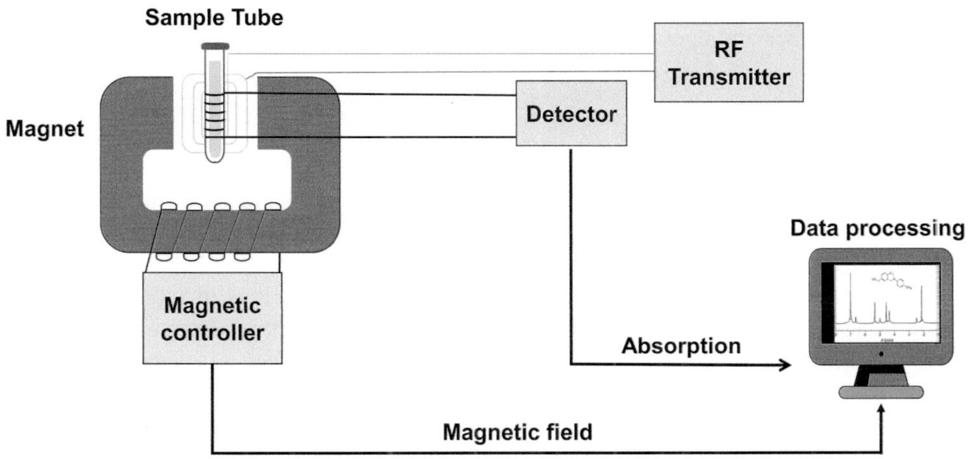

FIGURE 5.2 Schematic representation of an NMR spectrometer

5.2.4 Procedure

- Herbal drugs and extracts are homogenized by triturating or grinding to produce a fine powder.
- Then a small amount of powdered sample is dissolved in NMR-compatible deuterated solvents such as $CDCl_3$, DMSO-d_6, or D_2O, depending upon the compound of interest and also to avoid the proton signals from the solvent.
- The sample is filtered through a nylon membrane filter to remove particles that may interfere with the instrument and the spectra.
- A very small quantity of sample, usually 0.5–0.7 mL, is loaded into the NMR tube and capped properly, which is then placed in a strong magnetic field.
- The sample is irradiated with RF pulses that excite the nuclei of interest in the specimen.
- In the case of quantitative NMR, a known concentration of internal standard such as TMS (Tetramethyl silane) is added.
- Raw NMR data is processed to reduce noise and baseline drift, enhancing the signal-to-noise ratio. The free induction decay (FID) is then converted into a frequency-domain spectrum using the Fourier transformation (FT), accompanied by a series of processes, that is, apodization, phase and baseline correction, calibration, and peak analysis to improve signal-to-noise ratio. Several software tools like TopSpin (from Bruker) or Mnova (from Mestrelab) are used for data processing.

5.2.5 Challenges

- **Sensitivity:** NMR is inherently less sensitive than mass spectrometry, thus requiring relatively high concentrations of the sample and more time for complete analysis.
- **Sample preparation:** NMR requires a large amount of a soluble, pure sample for analysis. In the case of a solid sample, ssNMR is preferable, which is complex and time consuming. Additionally, nuclei such as ^{15}N and ^{13}C produce inherently weak signals, which require isotopic enrichment to improve the sensitivity of detection.
- **Data interpretation:** Herbal drugs and extracts contain a complex mixture of many phytochemicals, generating a complex NMR spectrum consisting of several overlapping peaks and signal multiplicity, which complicates the structure elucidation and requires expert interpretation.
- **Operating and maintenance cost:** The NMR spectrometer consists of superconducting magnets, which require liquid nitrogen or helium for cooling, necessitating proper continuous maintenance and operational costs.

5.2.6 Application to Herbal Drugs

In a study, the structure elucidation of organic compounds from natural sources was carried out using advanced 1D and 2D NMR techniques. The detailed structural analysis of compounds, mainly diterpenoids, triterpenoids, iridoids, and other secondary metabolites isolated from plants belonging to the Lamiaceae family, such as *Salvia*, *Ajuga*, *Teucrium*, *Sideritis*, *Nepeta*, and *Lavandula* growing in Turkey, was done. The characteristic signals and couplings in proton and carbon-13 NMR are used to identify different skeletons such as abietane diterpenes from *Salvia* species, ent-kaurane from *Sideritis*, iridoid monoterpenes-nepetalactones from *Nepeta* species, and neo-clerodane from both *Ajuga* and *Teucrium* species. The application of advanced 2D NMR techniques (COSY, NOESY, DEPT, HMQC, HMBC, HETCOR, and selective INEPT) further helped to establish the atom connectivity, stereochemistry, and elucidate the connection between diverse molecular fragments [6].

In another study, Radix *Polygoni multiflori*, a herb known to improve hypercholesterolemia, atherosclerosis, diabetes and other associated conditions, was studied. The herbal extracts were first fractionated into libraries of subfractions, and each fraction was tested with a target protein using ligand-observed NMR techniques for identification of the bioactive compounds. By combining NMR interaction data with high-performance liquid chromatography-solid phase extraction-NMR spectroscopy (HPLC-SPE-NMR), structural elucidation and lead optimization were established. This scheme identified a compound, 2,4-dihydroxy-6-[(1E)-2-(4-hydroxyphenyl) ethenyl] phenyl-β-D-glucopyranoside (TSG), from one subfraction, which binds to the human fatty acid binding protein 4 (FABP4). FABP4 is an important therapeutic target related to metabolic diseases; thus, this experimental approach revealed the bioactive compound, its molecular pathway, disease target, and provided atomic-level interaction details, which facilitated understanding the potent compound responsible for the therapeutic activities [7].

5.3 FOURIER TRANSFORM INFRARED SPECTROSCOPY

Fourier transform infrared spectroscopy, or FTIR, is the most widely used instrument in IR spectroscopy. It is used to identify functional groups in a sample, in solid, liquid or gaseous state, when an IR beam or radiation is passed through a sample. The sample absorbs some of the IR radiation while the rest is transmitted through it. The resultant spectrum depicts the %transmittance versus wavenumber (cm^{-1}), thus providing the molecular fingerprint of the sample. For the determination of functional groups in a molecule, it must be IR active. An IR active molecule is one which has a dipole moment. When the IR radiation interacts with the covalent bond of the materials having an electric dipole, the molecule absorbs energy, and the bond starts back and forth oscillation. Therefore, the oscillation which caused the change in the net dipole moment of the molecule should absorb IR radiation.

FTIR spectroscopy is a widely employed analytical technique that identifies the structural framework. Like fingerprints, no two compounds produce the same IR spectrum, resulting in precise qualitative identification of the sample. Additionally, peak intensities reflect the quantity of sample present, and with modern software, IR spectroscopy also supports accurate quantitative analysis.

5.3.1 PRINCIPLE

The basic principle of IR spectroscopy is the measurement of the amount of IR radiation, which is absorbed (or emitted) by a sample as a function of the wavelength [8]. When a sample is exposed to infrared radiation (IR), it absorbs the energy corresponding to specific vibrational transitions within its molecular bonds. These absorption events occur at characteristic frequencies that directly reflect the types of bonds and functional groups present in the molecule. The resulting spectrum is typically recorded in terms of wavenumbers (cm^{-1}), ranging from 4000 to 600 cm^{-1}, as wavenumbers are directly proportional to energy and provide a convenient scale for spectral interpretation. The IR region is generally divided into three regions: near IR (400–10 cm^{-1}), mid IR (4000–400 cm^{-1}), and far IR (14,000–4000 cm^{-1}). IR photons have sufficient energy to cause the vibration of groups of atoms with respect to the bonds that link them [9].

Before analyzing a sample, a background spectrum is collected to account for atmospheric interference, particularly from water vapor and carbon dioxide. This background is then subtracted from the sample spectrum to isolate the true absorbance characteristics of the analyte. The resulting absorption spectrum reveals distinct vibrational modes, such as stretching, bending, and twisting, corresponding to various bond types within the sample. By examining these vibrational patterns, one can reliably identify the functional groups present in the molecule, making FTIR a powerful tool for qualitative analysis [10].

5.3.2 Advantages

The advantages of FTIR include the following [1, 9]:

- **Simplicity:** The moving mirror in the interferometer is the only continuously moving part in the instrument. Thus, there is very little possibility of mechanical breakdown.
- **Internal calibration:** Determines frequencies by directly comparing the output of a visible laser, usually a HeNe laser, which serves as an internal reference for the interferometer's measurements. This approach, known as the Connes advantage, enhances frequency accuracy. The use of this laser reference is considered a major advantage of FTIR over conventional dispersive spectrometers.
- **Improved speed:** As all the frequencies are simultaneously measured in an FTIR spectrometer, the measurements are completed within seconds. This is known as the multiplex or Felgett advantage.
- **Improved sensitivity:** FTIR offers enhanced sensitivity due to highly sensitive detectors, increased optical throughput (Jacquinot advantage), and the ability to reduce noise through signal averaging from rapid scans.
- **Reduced interferences:** In FTIR, the sample is placed between the interferometer and detector to ensure only modulated IR source radiation reaches the detector, minimizing interference from unmodulated sample emissions.
- **Ease of measurement:** In the FTIR technique, all compounds show distinctive absorption or emission in the IR spectral range; hence, they can be examined by both qualitative and quantitative means.

5.3.3 Instrumentation

A schematic representation of the FTIR instrument is shown in Figure 5.3 [9, 11]. The major components of FTIR include the following:

- **IR radiation source:** The infrared source emits energy or radiation of all wavelengths in the infrared region from a glowing blackbody radiation source. The source is usually selected based on the desired spectral range (typically 4000–400 cm^{-1}); mercury lamps for the far infrared and glowers (Globar or Nernst glower) for the near infrared range are used. This beam then passes through an aperture, which controls the amount of energy to be incident on the sample.
- **Michelson interferometer:** The interferometer divides radiant beams, generates an optical path difference between the beams, and generates interference signals measured as a function of the optical path difference by the detector. As its name implies, the interferometer produces interference signals, which contain IR spectral information generated after passing through the sample. It consists of the following:
  a) **A mirror (fixed and moving):** The two mirrors are perpendicular to each other.
  b) **A beam splitter:** The beam splitter is a semi-reflecting device and is often made by depositing a thin film of germanium onto a flat potassium bromide (KBr) substrate. The beam splitter divides the amplitude of the beam: one part of the radiation beam goes to the moving mirror, and the other part to the fixed mirror. Radiation from the broadband IR source is collimated and directed into the interferometer and impinges on the beam splitter.
- **Sample compartment:** The beam enters the sample compartment, where it is transmitted through or reflected off the surface of the sample, depending on the type of analysis being carried out. This is where specific frequencies of energy, which are uniquely characteristic of the sample, are absorbed.
- **Detector:** The beam finally passes to the detector for final measurement. The detectors used are specially designed to measure the special interferogram signal. The detector used

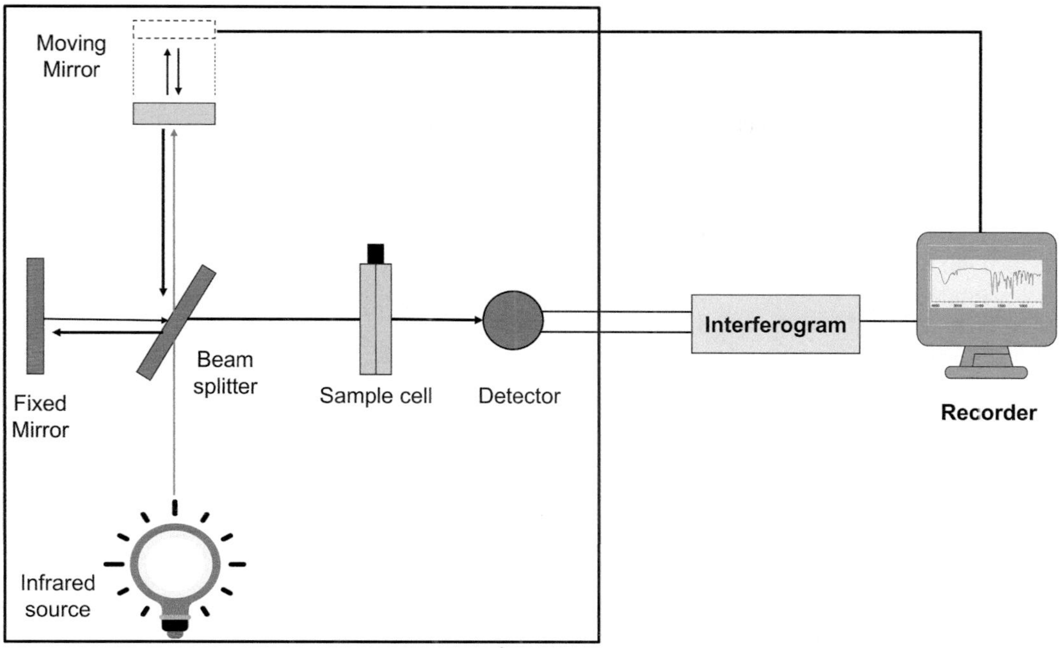

FIGURE 5.3 Schematic representation of FTIR

in an FT-IR spectrometer must respond quickly because intensity changes are rapid as the moving mirror moves quickly. Pyroelectric detectors or liquid-cooled photon detectors and photoelectric detectors such as mercury-cadmium-telluride (MCT) detectors are commonly used.

- **The computer**: The measured signal is digitized and sent to the computer where the FT takes place. The final IR spectrum of intensity versus frequency is then presented to the user for interpretation.

5.3.4 Procedure

- Herbal drugs or extracts are first properly dried to remove the water content, which may interfere during the analysis.
- The dried herbal drug or extract is then finely ground to obtain a uniform powder, which will provide better interaction with the IR beam.
- Further, a small amount of the powder (usually 1–2%) is mixed with dried KBr, then ground thoroughly, and then pressed to obtain a transparent pellet of optimum thickness for analysis. In advanced instruments, such as attenuated total reflectance-Fourier transform infrared spectroscopy (ATR-FTIR), samples in liquid, semi-solid, or fine powder are directly placed on ATR crystals (diamond, zinc selenide, germanium) for obtaining FTIR spectra.
- In certain cases, the herbal drug is first extracted using solvents such as ethanol or methanol and then the dried extract is used to enhance the detection of active phytocompounds.
- The obtained spectra show different absorption bands with functional groups such as C=O, C–O, C–H and O–H. The fingerprint region from 1500 to 500 cm^{-1} helps to distinguish compounds having structural similarity. The spectra are then matched or compared with reference spectral libraries for identification, whereas unmatched peaks may give an idea about new or unknown compounds or phytoconstituents.

5.3.5 Challenges

- **Interferences and baseline drift:** External factors such as atmospheric carbon dioxide and water vapor generates noise or undesirable absorption bands. Some herbal extracts or herbal drugs are hygroscopic in nature, which thus may affect the quality of the IR spectrum.
- **Sample preparation and sensitivity:** Solid samples can be analyzed by the KBr pellet preparation technique in FTIR, but in the case of a liquid sample, an ATR crystal is preferable for analysis. Also, the thickness of the KBr pellet, homogeneity, and method of preparation can significantly affect the spectra. For example, a KBr pellet too thick can lead to total absorption and distorted spectra.
- **Optics:** FTIR instruments have only a single beam, whereas dispersive instruments generally have a double beam.
- **Spectral overlapping:** Herbal drugs contain complex mixtures of herbs, which have many functional groups that may cause overlapping of the peaks and complicate the interpretation of spectra.
- **Quantitative analysis:** FTIR is an ideal technique for qualitative analysis, but without proper calibration using standards, it is not suitable for accurate quantitative analysis.
- **Spectral library:** FTIR depends on matching the known reference libraries for compound identification, so if the compound is unknown, then it is difficult to identify.

5.3.6 Application to Herbal Drugs

FTIR spectroscopy is a fast, non-destructive analytical tool commonly used for analyzing herbal drugs and extracts. It helps to identify the important functional groups present, allowing phytochemical profiling, quality control and authentication of herbal drugs. Each extract is known to have a unique fingerprint profile, useful for the detection of adulteration and thus confirming batch-to-batch consistency. FTIR also determines phytochemical integrity and drug-excipient compatibility, and when it is combined with chemometric tools, it aids in its broad applicability in semi-quantitative analysis, making it crucial in herbal drug research and development.

In a study, the drug-excipient compatibility of eugenol with the formulation excipients – chitosan, carbopol 940, polycaprolactone, sodium alginate, and tween 80 was analyzed using FTIR. The FTIR spectra of individual formulation components were compared with the spectra of physical mixtures. Eugenol showed characteristic peaks from 720 to 1250 cm^{-1} and at 1514, 1610, and 1637 cm^{-1}, which are related to C=C aromatic stretching. These characteristic peaks were retained intact in all the physical mixtures, therefore indicating the absence of any chemical interactions between eugenol and the excipients. Overall, the structural integrity of eugenol was confirmed by FTIR analysis, thus supporting its drug-excipient compatibility with the selected excipients for the formulation of nanoparticles and nanoemulsion gels [12].

In another study, ATR-FTIR was used to study extract-excipient compatibility in a *Piper longum* extract-loaded solid dispersion (SD). Figure 5.4 shows the FTIR of the solid dispersion, formulation components, and their physical mixture. In the case of the physical mixture of drug and polymers, the characteristic peak of the extract was found to be retained, representing the compatibility of the extract with the used excipient. However, in the case of SD, due to the complete molecular interaction, bonding among PLFEE and excipients, and coating by excipients, the characteristic peaks of PLFEE were found to disappear [13].

Recently, FTIR spectroscopy was utilized for phytopharmaceutical analysis to identify important phytochemicals present in the leaves of five medicinal plants, namely, *Azadirachta indica*, *Carica papaya, Ocimum sanctum*, *Murraya koenigii*, and *Ficus racemosa*. The characteristic absorption peaks and functional groups confirmed that important bioactive compounds such as azadiractolide, carpaine, eugenol, nimbolide, papain and ursolic acid were present in the selected plants,

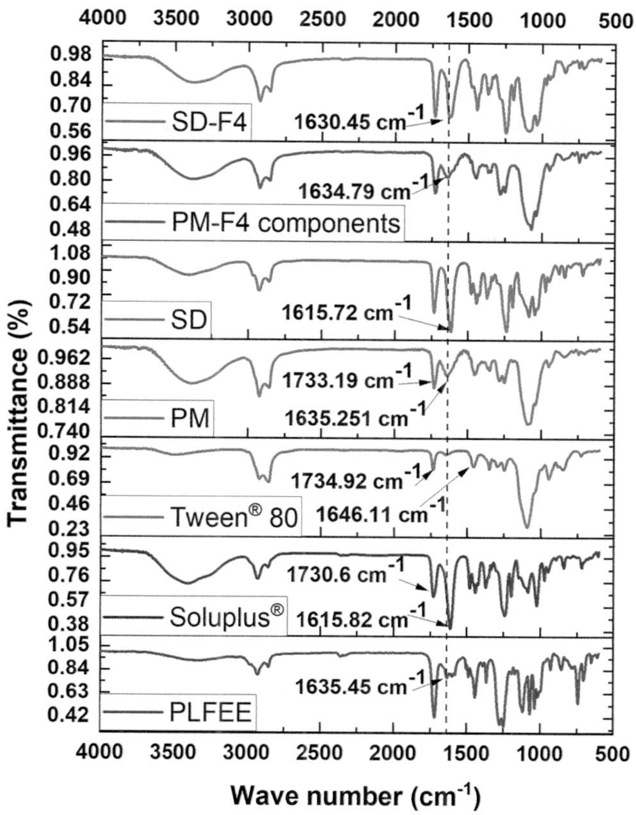

FIGURE 5.4 FTIR of extract, excipients, and solid dispersion for drug-excipient compatibility study

highlighting their diverse therapeutic potential, such as anti-diabetic, antioxidant, antimicrobial, and anticancer activities. FTIR, further coupled with PCA analysis, proved to be a non-destructive and rapid technique for distinguishing phytochemical fingerprint and evaluating its phytopharmaceutical importance [14].

In another study, structural differences between hydroxyvardenafil and hydroxythiovardenafil as adulterants were reported in a herbal dietary supplement. A strong absorption band of C=O stretch at 1703 cm^{-1} indicated the presence of the carbonyl group in the imidazotriazinone ring of hydroxyvardenafil, while hydroxythiovardenafil lacked this band and instead exhibited C=S stretches at 1482 and 1455 cm^{-1}, confirming the thio-substitution. Both compounds showed typical sulfonamide (SO_2) bands in the range of 1350 to 1280 cm^{-1} and 1170 cm^{-1}. The O–H stretch due to the hydroxyl group showed a broad band at around 3300 cm^{-1}, and bands in the 3000–2850 cm^{-1} and 950 cm^{-1} range are due to C–H stretching in the alkyl group and aromatic ring [15].

5.4 SINGLE CRYSTAL X-RAY DIFFRACTION

Single crystal X-ray diffraction (SC-XRD) offers a powerful and precise technique for investigating the phase composition and crystallinity of compounds by interpreting the X-ray diffraction patterns of a single crystal. It plays an important role in the pharmaceutical field in characterizing the solid-state properties of active pharmaceutical ingredients, plant bioactives, herbal drugs, and pharmaceutical excipients. It is widely used for investigating various polymorphic forms, distinguishing between amorphous and crystalline materials, identifying drug-excipient compatibility, ensuring batch-to-batch consistency, and stability of formulation.

5.4.1 Principle

In SC-XRD, X-rays interact with crystal substances to disclose structural insights. It involves interference in the crystal lattice, which can be destructive or constructive [16]. It is based on the diffraction of X-rays by the crystal lattice of the substance. When the X-ray interacts with a crystal, the scattering of light occurs as per Bragg's law [Equation (5.1)]. The law describes the relationship between the X-ray's wavelength and the interatomic distances within a crystal. The pattern of diffraction is unique for a crystalline substance, which is used for the identification and quantification of phases and to determine the degree of crystallinity.

$$n\lambda = 2d\sin\theta \tag{5.1}$$

where n is an integer/diffraction order, λ is the wavelength of the X-rays, d is the distance between lattice planes in the crystal, and θ is the diffraction angle.

5.4.2 Advantages

- **Non-destructive method**: It is a non-destructive method, making it suitable for quality control of herbal drugs.
- **Sensitivity:** It has high sensitivity for recognizing the crystalline phases.
- **Identification of polymorphism:** It is used to identify polymorphism, which can significantly alter drug solubility and therapeutic efficacy.
- **Quantitative mapping of electron density:** It allows mapping of electron density for the identification of atoms even in complex or heavy atom-containing structures.
- **Structural configuration:** It offers detailed information on crystal structure, lattice parameters, absolute stereochemistry, and unit dimensions.
- **Data processing:** It offers automated processing and interpretation of data.

5.4.3 Instrumentation

A schematic representation of the SC-XRD instrument is shown in Figure 5.5. The major components of the instrument include the following [16, 17]:

- **X-ray source:** Generates X-rays required for diffraction. Cu, Mo, Fe, and Co are mainly used as X-ray sources. Electrons are accelerated and struck on an anode/metal, generating X-rays.
- **Stage or sample holder:** It holds the sample.
- **Goniometer:** It rotates the sample holder and detector to follow Bragg's law. It controls the alignment of the X-ray source, sample holder, and detector. The θ represents the angle between the incident beam and the sample surface, whereas the 2θ represents the angle between the detector and the incident beam.
- **Detector:** It detects the X-ray and measures the intensity of the diffracted ray after exposure to the sample. Various detectors, such as proportional counters, scintillation counters, and solid-state detectors, are mainly used in XRD. They offer fast scanning and real-time acquisition of data.
- **Monochromator:** It produces monochromatic light by filtering the unwanted wavelengths (Kβ) from the X-ray. Graphite is mainly used as a monochromator.
- **System for data acquisition:** It acquires, analyzes (peak identification, crystallite size, phase analysis), and records the data using software.

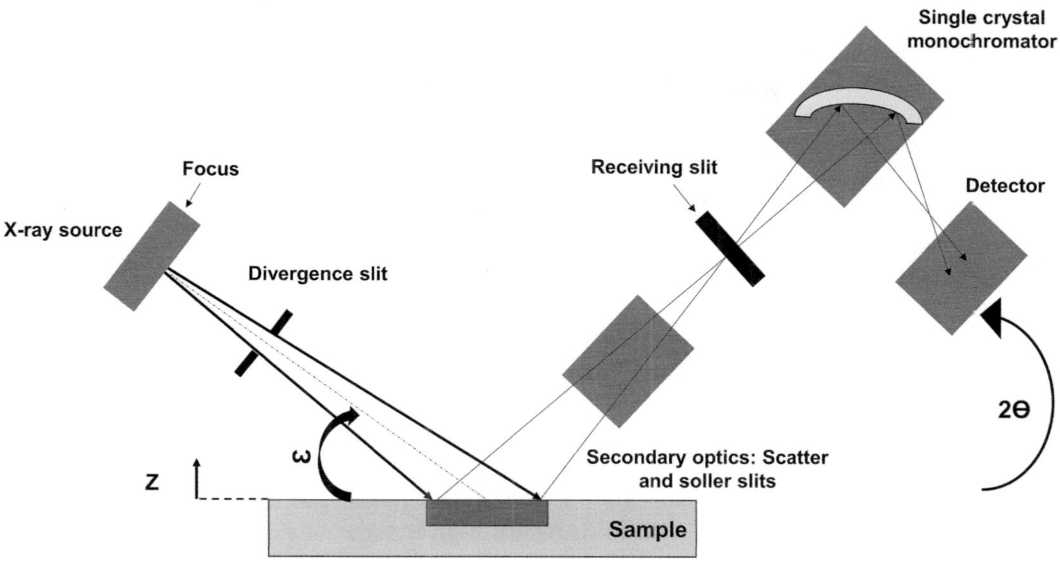

FIGURE 5.5 Schematic representation of SC-XRD

5.4.4 Procedure

- First, the herbal drug is finely ground into a powder. Then the compound of interest is isolated using chromatographic techniques to remove impurities which affect crystallization.
- Then the crystal is allowed to develop using different crystallization techniques such as slow evaporation and solvent layering.
- A defect-free, well-formed crystal, in the size range 0.1–0.5 mm, is selected using polarized light microscopy.
- The selected crystal is mounted on a glass fiber or loop using a suitable oil or adhesive oil; generally, paratone or epoxy is preferred.
- To get optimal diffraction, the crystal should be firmly fixed and aligned before analysis.
- The crystal is then mounted on an X-ray diffractometer. It is allowed to cool at a low temperature (typically ~100 K) by a cryostream to prevent thermal vibration and improve data quality.
- Afterward, the crystal is irradiated using monochromatic X-rays such as Mo Kα or Cu Kα radiation.
- The crystal is rotated and diffraction images are collected using a charge-coupled device (CCD) or complementary metal-oxide semiconductor (CMOS) detector.
- Further, data processing and interpretation of results are done using specialized software such as SHELX and Olex2.

5.4.5 Challenges

The analysis of phytochemicals in SC-XRD suffers from the following challenges [18–20].

- **Complexity and impurities:** The herbal drug contains a mixture of phytocompounds and excipients, which makes isolation of the desired compound and crystal growth difficult. Trace amounts of impurities can hamper nucleation formation, resulting in poor crystal quality.

- **Polymorphism:** Herbal drugs or extracts crystallize into several polymorphs or solvate forms, thus complicating reproducibility and structural assignment. Polymorph stability is known to shift over a period of time or due to slight changes in temperature or solvent.
- **Suitable single crystal:** Several phytochemicals present in herbal drugs and extracts are difficult to crystallize because they are found in low purity, exhibit flexibility, or have oily or amorphous nature, specifically when present in small quantities.
- **Sample quantity constraints:** Herbal phytochemicals are generally present at low quantities (mg or µg) in extracts, which is not sufficient for conventional crystallization screening.
- **Damage due to radiation:** High-flux modern synchrotrons can cause lattice damage due to radiation in small-molecule crystals, leading to loss of diffraction power at high angles, bond rearrangements, bond breakage, or changes in polymorphism. Therefore, it deteriorates the quality of data and can lead to incorrect structural interpretation.

5.4.6 Application to Herbal Drugs

In herbal drug research, SC-XRD appears as a powerful analytical technique used for well-defined structural interpretation of isolated phytochemicals. It gives accurate atomic-level information on molecular geometry, including angles, bond lengths, and absolute stereochemistry, which gives vital insight into the pharmacological role of chiral phytoconstituents. SC-XRD plays a key role in identifying polymorphic forms, confirming compound uniqueness after isolation, and studying co-crystal or compound-impurity connections. Its application confirms the structural accuracy and also contributes to standardization of herbal drugs during drug development.

In a recent study, the structural characterization of phytochemicals isolated from *Uncaria rhynchophylla* was done utilizing the SC-XRD technique to determine the absolute conformation of a newly identified indole alkaloid triglucoside and numerous other known alkaloids using very little sample (i.e., 50 µg). The novelty in the research was the utilization of the micro-drop crystallization technique, which allowed the development of single crystals of high quality obtained from minor phytoconstituents present in complex matrices of herbal extracts. This method allowed accurate 3D structural interpretation and validation of formerly projected stereochemistry, and thus established the usefulness of SC-XRD in standardization and quality control of herbal crude drugs [21].

5.5 LIQUID CHROMATOGRAPHY-MASS SPECTROMETRY

Liquid chromatography-mass spectrometry (LC-MS) is a widely used analytical tool that has become the most preferred technique for both qualitative and quantitative analysis. It offers fast screening and characterization of known and unknown phytoconstituents present in the extract of herbal drugs. It allows simultaneous separation, detection, and structural elucidation of the various compounds present in the sample. Due to its versatility and sensitivity, it has become an essential tool in natural product research, pharmaceutical analysis and metabolomics.

5.5.1 Principle

LC-MS is a hyphenated analytical technique that combines the chromatographic separation capabilities of liquid chromatography (LC) or HPLC with the mass identification or detection capabilities of a mass spectrometer (MS). LC first partitions a sample, which contains a complex mixture of several compounds based on their chemical nature (polarity, hydrophobicity, or molecular interactions) with the stationary and mobile phases. Each compound then elutes at a specific retention time from the chromatographic column. The eluted compounds from the LC are then transferred into a mass spectrometer via an ionization source, generally electrospray ionization (ESI) or atmospheric pressure chemical ionization (APCI), where they are converted into charged ions in a gaseous state.

These charged ions, based on their mass-to-charge (m/z) ratios, are then sorted and detected by the mass analyzer. The resultant mass spectrum based on the fragmentation pattern gives information about the molecular weight and structural framework of the compound.

5.5.2 Advantages

There are numerous advantages of LC-MS over other chromatographic techniques, which are as follows [22].

- **Improved sensitivity and selectivity:** It enables the detection of trace-level compounds present in nanogram to picogram levels, thus enhancing the sensitivity. Compounds eluting together, having poor chromatographic resolution, can be separated based on the m/z ratios.
- **Qualitative and quantitative analysis:** It allows simultaneous detection, confirmation of molecular weight, and quantification of several known and unknown compounds present in the herbal drug or extract in a single run, thereby enhancing the efficiency of analysis.
- **Sample preparation and method development:** Compared to other chromatographic techniques, LC-MS requires fewer sample pre-treatment and derivatization steps in the case of samples containing thermolabile and polar compounds. Also, the method development is comparatively rapid without the validation of retention time.
- **Versatility:** It can analyze a wide range of phytocompounds ranging from small molecules to large biomolecules having complex structural frameworks, different molecular weights and chemical nature, which are present in herbal drugs, extracts, and herbal formulations.
- **Accuracy:** LC-MS yields improved assay accuracy.
- **Structural interpretation:** LC-MS provides detailed structural information about the compound by applying controlled fragmentation, allowing identification of both known and unknown compounds by their molecular weight and structure analysis by matching them with a reference database or libraries such as NIST and METLIN.

5.5.3 Instrumentation

LC-MS instrumentation [22] typically consists of two main parts: the liquid chromatography (LC) system and the mass spectrometry (MS) system, connected via an interface for ionization (Figure 5.6).

- **Liquid chromatography system:** It separates different components present in a mixture based on the interaction with the stationary and mobile phases. It consists of the following:
  - **a. Solvent reservoir and pump:** The solvent reservoir holds the mobile phase solvents, generally methanol, water, acetonitrile, and so forth. The pump delivers the mobile phase solvents at a constant and precise flow rate, usually 0.1–1 mL per minute.
  - **b. Manual injector or autosampler:** It injects the sample either manually through a syringe or via an autosampler into the LC.
  - **c. Column:** The column consists of a stationary phase, which is packed with fine silica particles. The column or the stationary phase separates a mixture of compounds based on their size, polarity, or size. C-18 and C-8 reverse-phase columns are the most widely used columns.
  - **d. Column oven:** The column oven precisely controls the column temperature during the chromatographic separation or analysis, therefore improving the separation efficiency and reproducibility of chromatographic results.

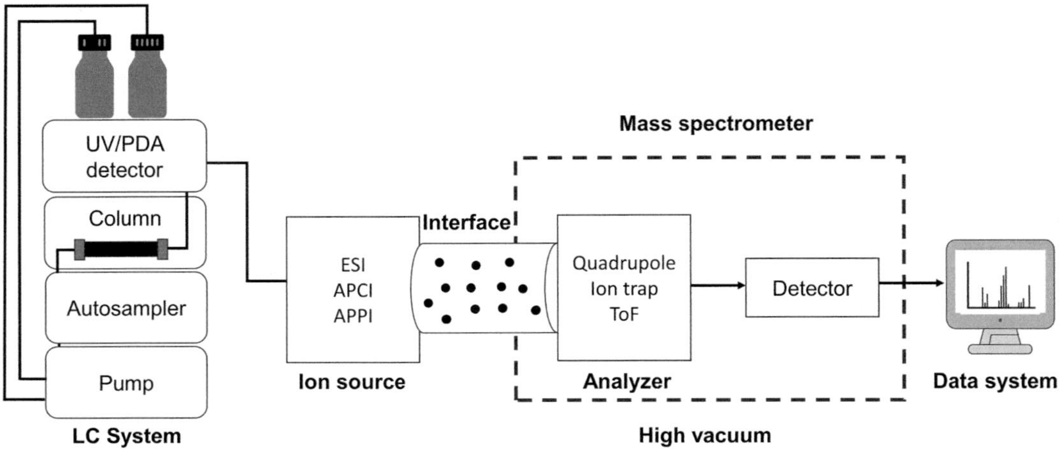

FIGURE 5.6 Schematic representation of LC-MS

- **Interphase:** The LC-MS interphase transfers the eluted sample from LC to the high vacuum MS by converting it into charged ions for further detection by the MS. The most commonly used ionization sources or interfaces are as follows:
  a. **Electrospray ionization (ESI):** It is a soft ionization method which generates multiply charged ions and is ideal for compounds which are large, polar, and thermally unstable.
  b. **Atmospheric pressure chemical ionization (APCI):** It is suitable for compounds that are polar to moderately polar and thermally stable, which have molecular weights generally less than 1500 Da.
  c. **APPI (atmospheric pressure photoionization)**: It is mostly used for non-polar compounds.
- **Mass spectrometer (MS) system:** The mass spectrometer or mass analyzer separates the ions depending on their mass-to-charge (m/z) ratios. Various types of mass analyzers are used based on their resolution level and accuracy. They are quadrupole, ion trap, time-of-flight (ToF), orbitrap, and magnetic sector analyzers.
- **Detector:** The detector detects charged ions and then converts them into electronic signals proportionate to the abundance of ions. The commonly used detectors include a Faraday cup, an electron multiplier, and so forth.
- **Computer software and data system:** The data system consists of a specific software program that controls various instrument parameters, obtains and stores data, and interprets the chromatograms and resultant mass spectra that help in the identification and quantification of compounds based on their m/z values and retention time.

5.5.4 Procedure

- A small amount of dried or fresh, finely powdered herbal drug or extract is accurately weighed and then extracted using a solvent suitable for extracting different phytoconstituents (phenols, flavonoids, alkaloids, etc.) at room temperature.
- Then the extract is filtered through a syringe filter to remove any residues from the drug or extract.
- An aliquot of the resultant filtered extract is pipetted and diluted using a suitable organic solvent.
- Finally, a required amount (usually 10–20 μL) of the sample is injected manually or via an autosampler into the LC column for analysis.

- Under optimized chromatographic conditions, the components present in the sample get separated based on their affinity toward the stationary phase and mobile phase.
- Then, separated compounds are first ionized using a suitable ionization source, and then, based on the m/z ratio, are detected by MS.
- The spectra or data generated by the MS are then analyzed to detect and quantify the phytocompounds present in the sample based on different m/z values and retention times.

5.5.5 Challenges

The LC-MS techniques have the following limitations [23–25].

- **Cost, maintenance, and expert personnel:** LC-MS equipped with high resolution systems such as Orbitrap, QToF are expensive and require maintenance, that is, daily calibration and proper cleaning. It also requires well-trained expert personnel for method development, troubleshooting, and complex data interpretation.
- **Contamination risk:** This sophisticated instrument is highly prone to contamination and carryover of the sample from the previous analysis.
- **Complex interpretation of data:** LC-MS generates vast, complex data, especially in the case of untargeted analysis, necessitating robust data processing. Since herbal drugs are very complex, the interpretation of LC-MS is very challenging for qualitative and quantitative estimation of numerous phytoconstituents.
- **Compound identification and interference issues:** Most herbal drugs contain structurally similar compounds, that is, isomers and compounds having different structures but similar masses; that is, isobars are difficult to differentiate using the LC-MS technique alone, creating errors in detection; therefore, further MS/MS fragmentation is required.
- **Standardization issues:** Standardization using LC-MS becomes challenging in the case of some herbal drugs containing individual bioactive phytocompounds, which lack reliable reference standards. This ultimately affects the accurate quantification and reliability of quality control results during the analysis of herbal drugs.

5.5.6 Application to Herbal Drugs

LC-MS has developed as an important, robust analytical technique for qualitative profiling and quantitative estimation of the phytochemicals owing to its variability and chemical complexity in the area of herbal formulation and herbal drug research. It ensures high sensitivity, selectivity, and capability for structural elucidation of complex mixtures commonly found in herbal matrices. It is useful for authentication, standardization and quality control of herbal drugs to ensure their purity as well as batch-to-batch consistency. It also differentiates overlapping between phytoconstituents and interprets their synergistic interactions required for rational development of herbal formulation and evaluation of its efficacy. Furthermore, after administration, herbal drug metabolites can be traced in biological fluids, thus proving their versatility in bioavailability studies.

In a recent study, Sun et al. investigated the effect of fresh and traditional post-harvesting methods on variations in the phytochemicals present in *Glycyrrhiza uralensis* (liquorice) using UPLC-MS and multivariate statistical analyses. A total of 38 compounds (flavonoids and saponins) showed significant variation in concentrations between the two methods, which can be regarded as important biomarkers to distinguish between different processed products. When compared with the traditional processing, fresh processing prevented the loss of hydrophilic compounds due to hydrolytic cleavage and also reduced the number of processing steps. Under the traditional processing method, flavonoids, mainly liquiritin and liquiritigenin, showed degradation and transformation due to the cleavage of the glycosidic bond. However, saponins such as glycyrrhizic acid were better preserved in the case of fresh processing, demonstrating their higher detoxifying potential. Thus, UPLC-MS combined with statistical tools such as principal component analysis (PCA) and orthogonal partial least-squares discrimination analysis (OPLS-DA) confirmed the phytochemical differences among the two processing methods [26].

In another study, anti-salmonella activity of hydroethanolic extract of *Detarium microcarpum* root bark was established based on the phytochemical analysis using LC-MS. The analysis revealed that the extract contained eight important phytocompounds such as flavonoids, saponins, triterpenes, coumarins, sterols and glucosides. Among these compounds, rhinocerotinoic acid and lupeol were isolated and rhinocerotinoic acid showed potent anti-salmonella activity. However, another identified compound, lupeol, did not exhibit significant antibacterial activity. Although rhinocerotinoic acid alone was found to be effective, the whole extract showed even better antibacterial activity, suggesting that multiple phytocompounds may work together in a synergistic manner to improve the efficacy of the extract [27].

5.6 GAS CHROMATOGRAPHY

Gas chromatography (GC) is a widely used analytical technique for the separation and quantification of volatile and semi-volatile compounds present in complex herbal mixtures. In the context of herbal drug analysis, GC plays an important role in the profiling of essential oils, detecting residual solvents, and assessing the quality and consistency for quality control of herbal formulations. The separation in GC is based on the affinity of the compounds toward the stationary phase of the column while being carried by an inert gas.

To increase the selectivity and reliability of this method, GC is often combined with either a headspace sampler (GC-HS) or a mass spectrometer (GC-MS). GC-HS is particularly useful for herbal matrices, as it allows for the extraction and analysis of volatile compounds directly from the vapor phase above the sample, reducing matrix interferences and the need for extensive sample preparation. This makes it ideal for analyzing delicate plant materials and formulations without degrading sensitive components.

The coupling of GC with mass spectrometry enhances the system's ability to not only quantify but also identify compounds through their mass spectra. This is especially beneficial when working with complex herbal mixtures where multiple volatiles may co-elute or exist at trace levels. Together, these techniques form a robust analytical platform for ensuring the safety, consistency, and regulatory compliance of herbal medicinal products, aligning with global quality standards.

5.6.1 Principle

GC-MS is a hyphenated analytical tool that utilizes GC for the separation of compounds and MS for compound identification. The sample is first vaporized in the GC and carried by an inert gas (commonly helium) through a capillary column. Inside the column, the volatile components are separated depending on their volatile nature and their affinity toward the stationary phase [28]. Each compound elutes from the column at a specific retention time (R_t), allowing detection of individual compounds.

The compounds separated from the column then enter the detector (usually MS), where ionization takes place using a suitable ionization technique. Most commonly, electron ionization (EI) at 70 eV is used for the fragmentation of the compounds. These fragmented ions are then sorted according to their mass-to-charge ratio (m/z) by a mass analyzer, such as a quadrupole or time-of-flight (ToF) system. The obtained GC chromatogram and mass spectra give structural information about the compounds, allowing both profiling and quantification [29].

5.6.2 Advantages

- **Analysis of volatile and semi-volatile compounds:** GC-MS is ideal for the analysis of compounds that are thermostable and vaporize readily. This makes it the method of choice for essential oils, aroma profiling, and volatile biomarkers [29, 30].

- **High sensitivity for trace-level detection:** Residual solvents, which are often present in very low concentrations in the extracts or herbal drugs, can be easily detected owing to the high sensitivity using GC-MS. It can identify impurities at parts-per-billion levels, thus complying with the ICH regulatory guidelines [28].
- **Accurate identification using mass spectra:** Volatile compounds often have similar retention times. The mass spectrometric detection in GC-MS allows precise differentiation based on unique fragmentation patterns [31].
- **Compliance with regulatory standards:** GC-MS is widely accepted by various pharmacopoeias (such as USP, EP) for residual solvent analysis in pharmaceuticals and herbal preparations, ensuring global regulatory compliance (USP, 2023).
- **Robust quantification:** GC-MS enables reliable quantification of residual solvents using internal standard calibration, improving accuracy and repeatability [32].
- **Library search capability enhances identification:** Mass spectral libraries such as NIST and Wiley facilitate rapid identification of unknown volatiles or solvent residues by comparison with reference spectra [33].
- **Minimal sample preparation:** Headspace sampling combined with GC-MS allows direct analysis of volatile constituents without complex extraction steps, preserving compound integrity [34].
- **Reduces matrix interference:** GC-MS, especially in selected ion monitoring (SIM) mode, minimizes interference from complex matrices like plant extracts, making it highly effective for herbal and botanical testing [35].

5.6.3 Instrumentation

A schematic representation of the GC-MS instrument is shown in Figure 5.7. The GC-MS system is composed of several integrated units, each performing a critical role in the analysis of volatile and semi-volatile compounds.

- **Gas flow control system:** An inert gas such as helium or nitrogen is used to carry vaporized analytes through the system. Flow rate and pressure are carefully regulated to ensure consistent transport and efficient separation. Capillary columns typically operate at lower flow rates than packed columns [28].
- **Sample introduction unit:** Samples are introduced using injection techniques such as split, splitless, or on-column modes. Splitless injection is commonly used for trace-level

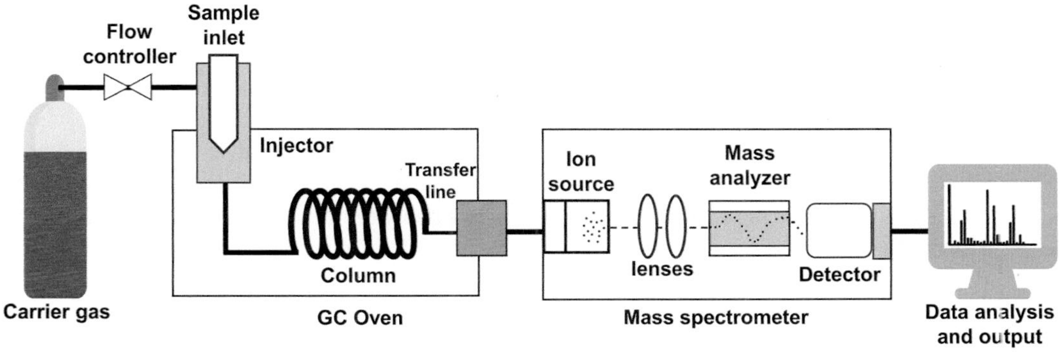

FIGURE 5.7 Schematic representation of GC-MS instrumentation

detection, ensuring that the entire sample reaches the column. Injection temperature and timing are critical for peak reproducibility.

- **GC column and oven:** The analytical column is placed in a temperature-programmed oven, where separation occurs based on the volatility and interaction of compounds with the stationary phase. Gradual heating helps elute analytes in order, improving resolution. Excess heat may cause column degradation and baseline drift.
- **Interface and transfer line:** A heated interface ensures smooth transfer of analytes from the GC to the MS without condensation or loss. This connection maintains the sample in its vapor phase until ionization.
- **Ionization and mass analysis:** In the mass spectrometer, analytes are typically ionized by electron ionization (EI), generating charged fragments. These ions are separated based on their mass-to-charge ratio using analyzers like quadrupoles or time-of-flight (ToF) systems. Each analyte produces a characteristic fragmentation pattern [31].
- **Detection and data processing:** The detector (usually MS) records ion intensities, producing a mass spectrum for each compound or analyte. The specialized software then matches these spectra with those of available reference libraries (e.g., NIST) for compound identification and also allows quantification based on peak areas and retention times [33].

5.6.4 Procedure

GC analysis involves the following procedure [35, 36]:

- The dried plant material or herbal extract is accurately weighed and extracted using a volatile, non-polar solvent such as hexane or dichloromethane, chosen based on the solubility and volatility of the target compounds.
- The extract is filtered to remove particulate impurities and concentrated under reduced pressure at low temperature to preserve heat-sensitive volatiles.
- If needed, the dried residue is reconstituted in the same solvent to achieve an appropriate concentration.
- An aliquot (typically 1–2 µL) is then introduced into the GC-MS system via split or splitless injection, depending on analyte levels.
- In residual solvent analysis, the autosampler collects the vaporized volatile sample from the headspace of vials and injected to MS system for qualitative and quantitative analysis.

5.6.5 Challenges

Despite its robustness and widespread use, GC-MS has several challenges, mainly when used for the analysis of complex plant matrices and natural products.

- **Sample volatility requirement:** GC-MS is inherently suitable for compounds that are volatile and also thermally stable. However, many plant-derived compounds, such as flavonoids, glycosides, and alkaloids, are non-volatile and thermolabile in their native forms, therefore limiting their direct analysis using GC-MS [35, 37, 38]. Water-soluble polar compounds create a problem during analysis as they bind irreversibly to the GC column material.
- **Derivatization requirement:** Polar and thermolabile compounds often require prior derivatization before GC-MS analysis. This step, though essential, is time consuming and moisture sensitive and may lead to inconsistent results if incomplete [37, 39].
- **Thermal degradation of sensitive compounds:** When exposed to high temperatures, some phytochemicals undergo degradation during analysis, and also after derivatization, thus leading to unwanted degradation products or artefact peaks that make sample interpretation difficult [40, 41].

- **Matrix interference:** Crude plant extracts are often chemically complex, which results in the simultaneous elution of compounds and interferes with analyte peak signals. Such matrix effects can interfere with both identification and quantification during analysis [41].
- **System contamination and maintenance:** Repeated analysis of concentrated or poorly purified plant samples can cause a build-up of non-volatile residues in the injector, column, or ion source. This contamination can decrease the instrument sensitivity, cause carryover of the residues, and increase the frequency of system maintenance requirements [29].
- **Quantification issues:** Accurate quantification needs known reference standards for each compound of interest, but these are often difficult to obtain or expensive for rare or complex plant secondary metabolites. This limits the quantitative reliability for quantitative analysis in untargeted metabolomics studies [39].

5.6.6 Application to Herbal Drugs

A study conducted by Khazan M *et al.* focused on the identification of synthetic adulterants in eight herbal weight loss supplements in Iran using GC-MS. The analysis revealed the presence of four synthetic drugs, which were present without disclosure on the product labels, namely sibutramine, phenolphthalein, bumetanide, and phenytoin. These substances are often linked with several health risks, such as cardiovascular and psychiatric disorders and renal and hepatic complications, along with carcinogenicity, thus raising further safety concerns. Similarly, other compounds such as caffeine, pseudoephedrine, theobromine, and amfepramone were detected, which were not listed on the product labels [42].

In another study, fingerprint profiling of an Unani polyherbo-mineral formulation, Safoof-e-Pathar Phori, was done for its authentication and standardization purposes. The formulation is used for its antilithiatic potential and contains four plants and two mineral constituents, namely, *Didymocarpus pedicellata*, *Dolichos biflorus*, *Rheum emodi*, *Raphanus sativus*, potassium carbonate, and potassium nitrate. GC-MS analysis of the hexane extract of the formulation and its plant components revealed the presence of various phytochemical compounds. A total of 51 compounds were detected, out of which 3-furancarboxylic acid, cyclooctene, stearic acid, and linoleic acid were among the major compounds identified. Comparative phytochemical profiling confirmed that most of these identified compounds were similar to those present in respective crude drugs, therefore confirming the phytochemical integrity and identity of the polyherbo-mineral formulation. The phytochemical identification was confirmed by matching the mass spectra with NIST and Wiley libraries [43].

In a recent study, the quality of walnut oil extracted by the cold press and solvent methods was thoroughly examined. The results showed that the yield of walnut oil extracted by the solvent method was higher (66%), compared to that of the cold press method (48%). However, the solvent method showed a decrease in several thermolabile compounds, whereas the cold press method preserved more phytoconstituents, polyphenols, and tocopherols. GC-MS analysis revealed a total of 60 compounds, mainly consisting of tocopherols, phytosterols, and fatty acids (PUFA, unsaturated fatty acids), which were detected in cold-pressed walnut oil. However, only 43 compounds, mostly fatty acids, were identified in the solvent-extracted oil. These results suggest that cold-pressed walnut oil preserves more nutritional components and bioactive phytocompounds [44].

5.6.7 Conclusion

The integration of these advanced analytical techniques, namely NMR, FTIR, Single-Crystal XRD, LC-MS, and GC-MS, offers a robust framework for accurate structural elucidation and phytochemical integrity assessment. These methodologies, through their unique capabilities, allow complete characterization of phytocompounds based on the molecular framework, stereochemistry, functional groups, and purity analysis. Such multi-instrumental approaches for phytochemical or herbal drug analysis improve the accuracy as well as reproducibility of results. Altogether, it supports reliability for authentication, standardization, and quality control of herbal drugs and extracts for phytopharmaceutical research and development.

REFERENCES

1. Vij, D., *Handbook of Applied Solid State Spectroscopy*. 2007, Springer Science & Business Media.
2. Lambert, J.B., E.P. Mazzola, and C.D. Ridge, *Nuclear Magnetic Resonance Spectroscopy: An Introduction to Principles, Applications, and Experimental Methods*. 2019, John Wiley & Sons.
3. Pérez Varela, I., G. Shear, and C. Cobas, Molecular melodies: Unraveling the hidden harmonies of NMR spectroscopy. *Molecules*, 2024. 29(4): p. 762.
4. Günther, H., *NMR Spectroscopy: Basic Principles, Concepts and Applications in Chemistry*. 2013, John Wiley & Sons.
5. Keeler, J., *Understanding NMR Spectroscopy*. 2010, John Wiley & Sons.
6. Topcu, G. and A. Ulubelen, Structure elucidation of organic compounds from natural sources using 1D and 2D NMR techniques. *Journal of Molecular Structure*, 2007. 834: p. 57–73.
7. Wan, H., et al., A NMR-based drug screening strategy for discovering active substances from herbal medicines: Using Radix Polygoni Multiflori as example. *Journal of Ethnopharmacology*, 2020. 254: p. 112712.
8. Bunaciu, A.A., H.Y. Aboul-Enein, and S. Fleschin, Application of Fourier transform infrared spectrophotometry in pharmaceutical drugs analysis. *Applied Spectroscopy Reviews*, 2010. 45(3): p. 206–219.
9. Dutta, A., Fourier transform infrared spectroscopy, in *Spectroscopic Methods for Nanomaterials Characterization*. 2017, Elsevier. p. 73.
10. Khan, S.A., et al., Fourier transform infrared spectroscopy: Fundamentals and application in functional groups and nanomaterials characterization, in *Handbook of Materials Characterization*. 2018, Springer. p. 317–344.
11. Jaggi, N. and D. Vij, Fourier transform infrared spectroscopy, in *Solid State Spectroscopy*. 2006, Springer. p. 411.
12. Pramod, K., et al., Unveiling the compatibility of eugenol with formulation excipients by systematic drug-excipient compatibility studies. *Journal of Analytical Science and Technology*, 2015. 6(1): p. 34.
13. Mohapatra, D., et al., Quality by design – based development and optimization of fourth-generation ternary solid dispersion of standardized Piper longum extract for melanoma therapy. *Drug Delivery and Translational Research*, 2023. 13(12): p. 3094–3131.
14. Dhillon, N., R. Gupta, and S. Agarwal, Analysis of therapeutic and phytopharmaceutical potential of selected plant species using FTIR spectroscopy and in silico methods. *Vegetos*, 2024: p. 1–12.
15. Jankovics, P., et al., Detection and structure elucidation of hydroxythiovardenafil as an adulterant in a herbal dietary supplement. *Journal of Pharmaceutical and Biomedical Analysis*, 2013. 74: p. 83–91.
16. Sghaier, H., Ionizing Radiation – Insights into Chemistry, Biology, and Applications. *BoD – Books on Demand*, 2025: p. 57.
17. Bunaciu, A.A., E.G. UdriŞTioiu, and H.Y. Aboul-Enein, X-ray diffraction: Instrumentation and applications. *Critical Reviews in Analytical Chemistry*, 2015. 45(4): p. 289–299.
18. McMonagle, C.J., et al., Lattice response to the radiation damage of molecular crystals: Radiation-induced versus thermal expansivity. *Structural Science*, 2024. 80(1): p. 13–18.
19. Song, J.-G., W.-C. Ye, and Y. Wang, Advanced crystallography for structure determination of natural products. *Natural Product Reports*, 2025. 42(3): p. 429–442.
20. Xiao, Y., et al., Current trends and advancements in crystallization and single-crystal structural analysis of small molecules. *Coordination Chemistry Reviews*, 2024. 517: p. 216035.
21. Koseki, Y., et al., Isolation of new indole alkaloid triglucoside from the aqueous extract of Uncaria rhynchophylla. *Journal of Natural Medicines*, 2025. 79(1): p. 28–35.
22. Kumar, P., S. Dinesh, and R. Rini, LCMS – a review and a recent update. *Journal of Pharmacy & Pharmaceutical Sciences*, 2016. 5(5): p. 377–391.
23. Gika, H.G., I.D. Wilson, and G.A. Theodoridis, LC – MS-based holistic metabolic profiling. Problems, limitations, advantages, and future perspectives. *Journal of Chromatography B*, 2014. 966: p. 1–6.
24. Kharyuk, P., et al., Employing fingerprinting of medicinal plants by means of LC-MS and machine learning for species identification task. *Scientific Reports*, 2018. 8(1): p. 17053.
25. Zhou, J.-L., L.-W. Qi, and P. Li, Herbal medicine analysis by liquid chromatography/time-of-flight mass spectrometry. *Journal of Chromatography A*, 2009. 1216(44): p. 7582–7594.

26. Sun, X., et al., Effects of different post-harvest processing methods on changes in the active ingredients of licorice based on LC-MS and plant metabolomics. *Phytochemical Analysis*, 2025. 36(2): p. 419–429.
27. Mbock, M.A., et al., In vitro and in vivo anti-salmonella properties of hydroethanolic extract of Detarium microcarpum Guill. & Perr. (Leguminosae) root bark and LC-MS-based phytochemical analysis. *Journal of Ethnopharmacology*, 2020. 260: p. 113049.
28. Poole, C.F. and S.A. Schuette, *Contemporary Practice of Chromatography*, Vol. 5. 2012, Elsevier.
29. Sparkman, O.D., Z. Penton, and F.G. Kitson, *Gas Chromatography and Mass Spectrometry: A Practical Guide*. 2011, Academic Press.
30. Adams, R.P. and O.D. Sparkman, Review of identification of essential oil components by gas chromatography/mass spectrometry. *Journal of the American Society for Mass Spectrometry*, 2007. 18(4): p. 803–806.
31. Gross, J.H., *Mass Spectrometry: A Textbook*. 2006, Springer Science & Business Media.
32. McMaster, Marvin C., *GC/MS: A Practical User's Guide*. 2008, John Wiley & Sons.
33. Kind, T., et al., FiehnLib: Mass spectral and retention index libraries for metabolomics based on quadrupole and time-of-flight gas chromatography/mass spectrometry. *Analytical Chemistry*, 2009. 81(24): p. 10038–10048.
34. Kolb, B. and L.S. Ettre, *Static Headspace-Gas Chromatography: Theory and Practice*. 2006, John Wiley & Sons.
35. Dettmer, K., P.A. Aronov, and B.D. Hammock, Mass spectrometry-based metabolomics. *Mass Spectrometry Reviews*, 2007. 26(1): p. 51–78.
36. Gupta, M., et al., A rapid and green GC-MS method for the sampling of volatile organic compounds in spices and flowers by concurrent headspace single-drop microextraction and solid-phase microextraction. *Analytical Methods*, 2023. 15(12): p. 1536–1545.
37. Villas-Bôas, S.G., et al., Mass spectrometry in metabolome analysis. *Mass Spectrometry Reviews*, 2005. 24(5): p. 10038–10048.
38. Fancy, S.-A. and K. Rumpel, GC-MS-based metabolomics, in *Biomarker Methods in Drug Discovery and Development*. Springer. p. 317–340.
39. Fiehn, O., et al., Metabolite profiling for plant functional genomics. *Nature Biotechnology*, 2000. 18(11): p. 1157–1161.
40. Bladt, S., *Plant Drug Analysis: A Thin Layer Chromatography Atlas*. 2009, Springer Science & Business Media.
41. Wagner, H. and S. Bladt, *Plant Drug Analysis: A Thin Layer Chromatography Atlas*. 1996, Springer.
42. Khazan, M., et al., Identification and determination of synthetic pharmaceuticals as adulterants in eight common herbal weight loss supplements. *Iranian Red Crescent Medical Journal*, 2014. 16(3): p. e15344.
43. Ahmad, W., et al., Comparative fingerprint profiling of Unani polyherbomineral (Safoof-e-Pathar phori) formulation by HPTLC, HPLC, and GC-MS. *Journal of AOAC International*, 2020. 103(3): p. 659–668.
44. Masoodi, L., et al., Quality analysis of walnut oil extracted by cold press and solvent methods for in vitro phytochemicals, antioxidant activity and GCMS analysis. *Measurement: Food*, 2025: p. 100223.

6 Instrumental Methodology for Crystallinity Analysis of Herbal Drugs and Formulations

ABBREVIATIONS

Abbreviation	Description
ASD	Amorphous Solid Dispersion
BSED	Backscattered Electron Detector
CCD	Central Composite Design
CDs	Carbon Dots
CMC	Carboxymethyl Cellulose
DSC	Differential Scanning Calorimetry
DVS	Dynamic Vapor Sorption
FEG	Field Emission Gun
FESEM	Field Emission Scanning Electron Microscopy
HPMC	Hydroxypropyl Methylcellulose
HPMCAS-L	Hydroxypropyl Methylcellulose Acetate Succinate
ITC	Isothermal Titration Calorimetry
NLC	Nanostructured Lipid Carrier
PLM	Polarized Light Microscopy
PVP	Polyvinylpyrrolidone
SAED	Selected Area Electron Diffraction
SED	Secondary Electron Detector
SEM	Scanning Electron Microscopy
ssNMR	Solid-State Nuclear Magnetic Resonance
TEM	Transmission Electron Microscope
TGA	Thermogravimetric Analysis
XRD	X-ray Diffraction

6.1 INTRODUCTION

Crystallinity plays an important role in determining the physicochemical and biopharmaceutical properties of herbal drugs and formulations, including solubility, stability, dissolution, and bioavailability. Phytoconstituents may exist in crystalline, amorphous, or polymorphic states, and these variations can significantly influence therapeutic efficacy and quality. Instrumental techniques (Figure 6.1) such as powder X-ray diffraction (PXRD), differential scanning calorimetry (DSC), thermogravimetric analysis (TGA), Fourier-transform infrared spectroscopy (FTIR), and Raman spectroscopy are widely employed for assessing crystallinity, phase transitions, and structural integrity [1]. These methodologies are essential for ensuring standardization, quality control, and reproducibility in phytopharmaceutical research. This chapter focuses on instrumental methodologies

DOI: 10.1201/9781003540823-6

FIGURE 6.1 Instruments for crystallinity analysis of herbal drugs and formulations

for crystallinity analysis, highlighting their principles, applications, and significance in maintaining the integrity and therapeutic reliability of herbal drugs and formulations.

6.2 X-RAY DIFFRACTION

X-ray diffraction (XRD) is a widely used technique to investigate crystallinity and phase composition of materials by analyzing the diffraction patterns of X-rays interacting with a sample. It plays an important role in the pharmaceutical field in characterizing the solid-state properties of active pharmaceutical ingredients, plant bioactives, herbal drugs, and pharmaceutical excipients. It is widely used for investigating various polymorphic forms, distinguishing between amorphous and crystalline materials, identifying drug-excipient compatibility, ensuring batch-to-batch consistency, and stability of formulation.

6.2.1 Principle

In XRD, X-rays interact with crystalline substances to disclose structural insights. It involves interference in the crystal lattice, which can be destructive or constructive [2]. It is based on the diffraction of X-rays by the crystal lattice of the substance. When the X-ray interacts with a crystalline material, the scattering of light occurs as per Bragg's law [Equation (6.1)]. The law describes the relationship between the X-ray's wavelength and the interatomic distances within a crystal. The pattern of diffraction is unique for a crystalline substance, which is used for the identification and quantification of phases and to determine the degree of crystallinity.

$$n\lambda = 2d\sin\theta \tag{6.1}$$

Here n is an integer/diffraction order, λ is the wavelength of the X-rays, d is the distance between lattice planes in the crystal, and θ is the diffraction angle.

6.2.2 Advantages of XRD

- It is a non-destructive method, making it suitable for quality control of herbal drugs.
- It has high sensitivity for recognizing the crystalline phases.
- It is used to identify polymorphism, which can significantly alter drug solubility and therapeutic efficacy.
- It is also useful for quantitative analysis by estimating the degree of crystallinity using peak integration.
- It offers detailed information on crystal structure, lattice parameters, and unit dimensions.
- Sample preparation is very easy.
- It offers automated processing and interpretation of data.

6.2.3 Instrumentation

A schematic representation of the XRD instrument is shown in Figure 6.2. The major components of an XRD instrument include the following [2, 3]:

- **X-ray source:** Generates X-rays required for diffraction. Cu, Mo, Fe, and Co are mainly used as X-ray sources. Electrons are accelerated and strike an anode/metal, generating X-rays.
- **Stage or sample holder:** It holds the sample.
- **Goniometer:** It rotates the sample holder and detector to follow Bragg's law. It controls the alignment of the X-ray source, sample holder, and the detector. In Figure 6.2, θ represents the angle between the incident beam and the sample surface, whereas the 2θ represents the angle between the detector and the incident beams.
- **Detector:** It detects the X-ray and measures the intensity of the diffracted ray after exposure to the sample. Various detectors, such as proportional counters, scintillation counters, and solid-state detectors, are mainly used in XRD. They offer fast scanning and real-time acquisition of data.
- **Monochromator:** It produces monochromatic light by filtering the unwanted wavelengths (Kβ) from the X-ray. Graphite is mainly used as a monochromator.
- **System for data acquisition:** It acquires, analyzes (peak identification, crystallite size, phase analysis), and records the data using software.

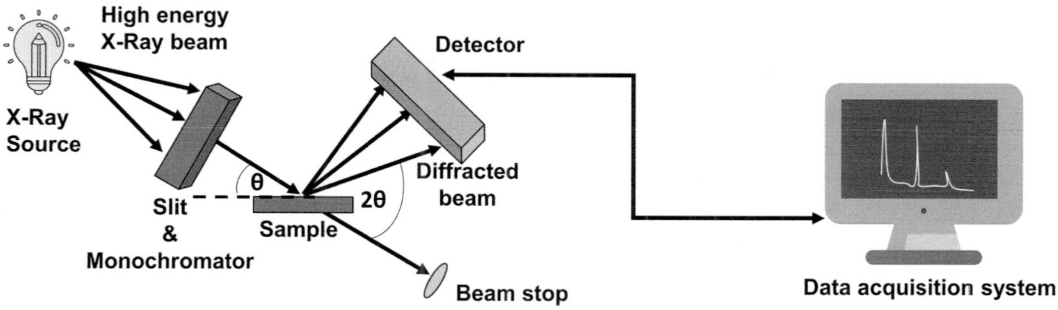

FIGURE 6.2 Schematic representation of XRD

6.2.4 Procedure

- Mostly, the herbal drug is finely ground into a powder. In the case of liquid extract, a film of extract is made on a clean glass section by the drop casting method and evaporating the solvents.
- Then the sample (powdered sample or thin film) is uniformly spread into a sample holder, slightly compacted to obtain a firm layer, and subjected to an XRD instrument.
- The sample is rotated over an angle (2θ), and the resulting diffraction peaks are recorded.
- Using various software, the diffraction pattern is analyzed and compared with standard databases to identify the crystalline phases and to calculate the degree of crystallinity.

6.2.5 Challenges

- **Complexity:** Herbal drugs contain a mixture of amorphous and crystalline materials, such as plant fibers, oil, or polysaccharides, due to which preparation of a homogenous herbal drug sample is difficult. The complex mixtures develop broad and overlapping diffractions, complicating the identification and quantification of the phase.
- **Incorrect interpretation:** Many of the herbal drugs contain a minimum or negligible number of bioactive components, which may appear as amorphous diffraction or very weak/not detectable crystalline diffractions, leading to false interpretation. The therapeutically inactive plant parts, such as cellulose and lignin, can dominate the diffraction of active components.
- **Issue with grinding:** Fine grinding is required to obtain a uniform bed of sample with a good diffraction pattern. However, during grinding, there may be a chance of amorphous modification, loss of volatile components, and degradation of bioactives due to the developed heat.
- **Moisture sensitivity:** Some of the plant extracts are highly hygroscopic, which causes broadening of the diffraction pattern, induces phase transition, and inconsistent reproducibility.
- **Quality and quantity of sample:** Obtaining sufficient, uniform samples with homogeneous particle size distribution is difficult in the case of low-yield extracts.

6.2.6 Application to Herbal Drugs

XRD is a major analytical technique for the investigation of the structure, composition, and properties of substances [2]. It provides comprehensive structural information, including atomic arrangements and lattice parameters, which are crucial for the characterization of materials [2]. The quantitative analysis of crystalline phases in a mixture allows the purity determination of pharmaceutical products [2]. It is also useful for the evaluation of material integrity and reliability by providing insights into stress–strain analysis, texture, and anisotropic properties [2].

XRD plays a vital role in the crystallinity analysis of herbal drugs, especially in identifying the polymorphism and quantifying the crystalline phases of plant bioactives. It can identify the amorphous and crystalline compounds of a plant bioactive, providing information on purity, stability, and solubility characteristics. Since most of the herbal drugs are multiconstituent complex mixtures of plant bioactives (extract/fraction), XRD provides information on the crystalline and amorphous properties, which may significantly affect the pharmacokinetic properties. It also analyzes the crystallinity of herbal formulations and excipients. It provides information on the drug/extract-excipient compatibility.

In a recent study, the crystallinity of crude peanut extract (CPE) was compared with CPE-loaded fast-disintegrating tablet formulation. The neat CPE demonstrated sharp peaks representing the crystallinity nature, whereas the formulation demonstrated a few characteristic peaks of CPE with reduced intensity, which is due to amorphous modification and complete incorporation of the CPE in the polymer matrix [4].

In another study, the crystallinity of ternary amorphous solid dispersions containing Indole2-carboxamide derivative (North 2) and curcumin (CUR) was studied by XRD and compared with the diffractions of neat compounds and drug-excipient physical mixture. Pure North 2 and CUR demonstrated sharp diffraction peaks representing their crystalline nature. The physical mixture of drug and excipients [polyvinylpyrrolidone (PVP), Eudragit EPO® (EPO), and hydroxypropyl methylcellulose (HPMC)] also demonstrated the retention of crystallinity. However, the ternary solid dispersion showed broad diffractions with complete absence of sharp crystalline peaks, demonstrating effective formation of amorphous solid dispersions [5].

Recently, the purity of curcumin in commercial turmeric powder was reported using a validated powder X-ray diffraction method. The developed PXRD provided elemental composition, crystallographic information, and 3D structure of curcumin, which was found to be suitable for the identification of curcumin in commercial turmeric powders and the possibilities of adulteration [6]. The crystalline properties of chitosan (CS)/agarose (AG) based on *Matricaria recutita* herbal extract (CH) bioscaffold for wound dressing application were reported using XRD. The XRD study reported evidence of strong hydrogen bonding among the polysaccharide polymers and the CH extract [7].

Fat from *Astrocaryum vulgare* (Tucumã fat) was explored as a pharmaceutical excipient for the formulation of a nanostructured lipid carrier (NLC). The physicochemical properties of fat in combination with carnauba wax were studied using XRD. The carnauba wax of varying proportions (10–50%) was added to prepare lipid mixtures to avoid the melting of fat at body temperature (37°C). Glyceride crystals are classified into α, β′, and β forms as per their interplanar distances. Crystals with a single strong line close to 4.15Å are classified as α. Crystals having two strong lines at lattice spacings of 4.2–4.3Å and 3.7–4.0Å are classified as β′, and crystals that do not fit into either classification are called β. The Tucumã fat showed two strong lines close to 4.2Å and 3.8Å, hence classified as β′. This crystalline form was reported to be ideal for the preparation of NLC as it is disorganized and stable one. The form α is more disorganized than β′, but is less stable and easily becomes the β form, which is more organized and does not accommodate the drug in the lipid matrix. Further, the lipid mixtures showed similar diffraction peaks to those of Tucumã fat and carnauba wax, suggesting the absence of polymorph formation. Based on the diffraction patterns, all mixtures are classified as the β′ form. Decreased diffraction intensity and broadening of the diffraction peaks are characteristic of less organized crystalline forms. All lipidic mixtures are found to be less crystalline than Tucumã fat and carnauba wax and are found to be suitable excipients for the development of NLC [8].

In a recent study, the crystallinity of amorphous solid dispersion (ASD) of *Withania somnifera* root powder extract was studied by PXRD. The diffractogram of the neat extract showed sharp diffraction peaks within 2θ of 10–40°, representing the crystalline nature of the contained phytoconstituents. The ASD was prepared with various hydrophilic polymers [hydroxypropyl methylcellulose acetate succinate (HPMCAS-L), polyvinylpyrrolidone (PVP K-30), and Soluplus®]. The Soluplus®-based ASD demonstrated short, sharp diffraction peaks similar to the extract, due to incomplete amorphous conversion of crystalline phytoconstituents. The HPMCAS-L-based ASD demonstrated short, sharp diffraction peaks, representing the presence of nanocrystallinity. In contrast, the ASD with PVP K-30 represented a broad diffraction pattern, representing the amorphous nature [9].

In another study, triptolide-loaded self-micelle solid dispersion was characterized for crystallinity using XRD and compared with physical mixture (Na_2GA+triptolide) and neat triptolide and excipient (Na_2GA). The triptolide showed characteristic crystalline peaks. The physical mixture also demonstrated crystalline peaks with a slight reduction of intensity due to the dilution effect of the excipient. In contrast, the diffraction pattern of SD represented a complete absence of sharp crystalline peaks, representing amorphous conversion [10].

NLCs of quercetin and piperine have been prepared with Compritol® 888 ATO as solid lipid, and squalene as liquid lipid to overcome the solubility and permeability issues. An XRD study was carried out to examine the crystallinity of drugs in the formulation. The study was carried out for individual phytoconstituents, blank NLCs, and drug-loaded NLCs. The XRD diffractogram of piperine and quercetin showed sharp diffraction peaks due to their crystallinity. However, the intensity was found to be reduced in the NLCs by 62.89% and 49.72% for piperine and quercetin, respectively, due to the amorphous modifications of drugs after incorporation into NLCs. The lyophilyzed blank NLCs and drug-loaded NLCs showed the crystalline diffractions due to the crystallization of sucrose, which was used as a cryoprotectant in NLCs [11].

In a recent study, rutin-loaded solid dispersion (SD) was prepared using Poloxamer 407 and Poloxamer 188 to improve aqueous solubility and oral bioavailability. The developed SD and its excipients were characterized for their crystallinity by XRD. Pure rutin showed sharp diffractions due to its crystalline property. The poloxamers 407 and 188 showed a few sharp diffraction peaks, representing their semi-crystalline structure. The physical mixture of rutin and excipients showed sharp crystalline peaks of rutin and weak peaks of excipients, due to the retention of the crystalline form of rutin. However, the crystalline peaks are found to be substantially reduced in the SD compared to the physical mixture. Almost all characteristic rutin diffraction peaks are found to be diminished due to amorphous modifications, which was reported for increasing dissolution of poorly soluble rutin [12].

In another study, the XRD of standardized ethanolic extract of *Piper longum* (PLFEE) excipients, their physical mixture (PM), and solid dispersion (SD) were studied to observe the crystallinity (Figure 6.3). The PLFEE showed sharp diffraction patterns, reflecting its crystallinity. The PM retained the sharp diffraction peaks of PLFEE and broad peaks of carrier matrix, signifying no alteration in the crystallinity of PLFEE. However, the decreased diffraction intensity of PLFEE in PM is due to the dilution effect of excipients. In contrast, XRD of the SD presented a typical halo pattern similar to excipients without any characteristic peaks of PLFEE, signifying the amorphous modification of PLFEE by SD [13].

6.3 DIFFERENTIAL SCANNING CALORIMETRY

Differential scanning calorimetry (DSC) has emerged as a powerful experimental technique for determining thermodynamic properties of biomacromolecules. It analyzes the heat flow associated with phase transitions in a material as a function of temperature or time [14, 15]. DSC is widely used to investigate the thermal properties of drugs, excipients, and formulations, including melting points, glass transition temperatures, polymorphic transitions, and crystallization. It offers valuable information on the stability, purity, and drug-excipient compatibility and optimal storage conditions of pharmaceutical products.

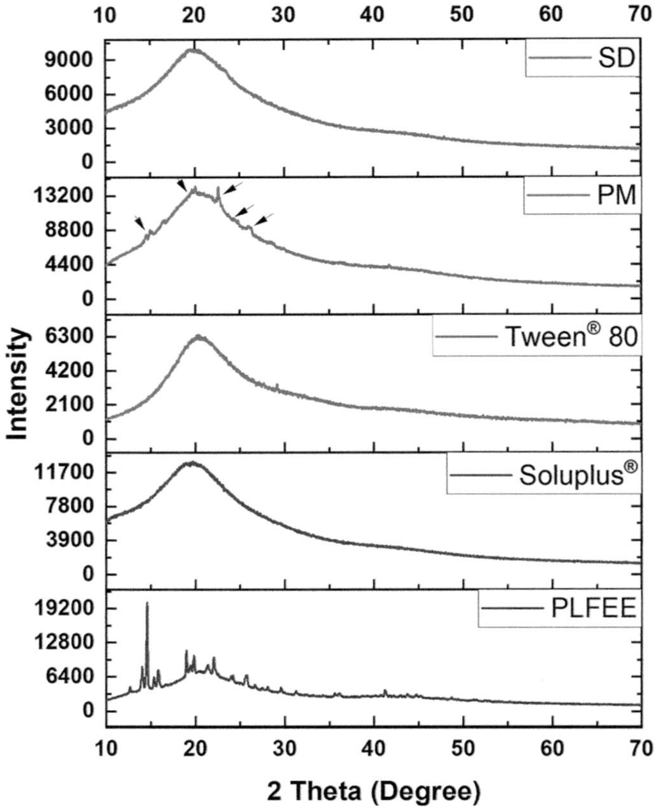

FIGURE 6.3 *X*-Ray diffractograms of extract, excipients, and solid dispersion for crystallinity analysis

PLFEE: *Piper longum* fruits ethanolic extract, PM: physical mixture of extract and polymer, SD: solid dispersion.

6.3.1 Principle

DSC is useful for studying heat flow into or out of a sample as it undergoes a controlled temperature condition [16]. When a sample is subjected to heating or cooling, the crystalline materials possess specific thermal transitions, such as crystallization or melting, which are not detected for amorphous substances. The temperature and the enthalpy changes provide valuable information on thermal properties, crystallinity, and thermal stability of a substance. The heat flow of the sample is compared with a reference (empty crucible), and the difference in the heat flow is plotted as a function of temperature to obtain a thermogram. The magnitude and sharpness of the peaks represent the crystalline properties of the material [16–21].

6.3.2 Advantages

- DSC directly detects the crystalline or amorphous state.
- It offers quantitative data (degree of crystallinity) on crystallinity and thermal properties.
- It identifies and detects phase transitions, such as melting, glass transition (Tg), and crystallization.
- It requires a small quantity of samples for analysis; hence, it is useful for expensive herbal drugs and phytomolecules.
- It offers an investigation of thermal properties.
- It is useful for the investigation of the stability and compatibility of drug formulations.

6.3.3 Instrumentation

A schematic representation of the DSC instrument is shown in Figure 6.4. The major components of DSC include the following [15, 21]:

- **Sample holder/pan:** It is made up of platinum or aluminum. It holds the sample, and a reference pan contains an inert substance (alumina).
- **Heating system/furnace:** Enables the simultaneous heating of both pans at a constant rate in a controlled environment.
- **Thermocouples or temperature sensor:** It monitors the temperature and measures the temperature difference between the sample and the reference.
- **Heat flow sensor:** It measures the energy input for the sample and reference.
- **Cooling system:** It uses liquid nitrogen or Peltier systems to study the thermal events at lower temperatures.
- **Data acquisition and integration system:** It sets the temperature, records the thermograms, and calculates the transition temperature and the enthalpy change.
- **Purge gas system:** It maintains an inert atmosphere for precise reading and prevents oxidation. Inert gases like helium and nitrogen are used as purge gases.

6.3.4 Procedure

- For the DSC study [19], a very small quantity of the sample (~5 mg) is placed in a sealed crucible.
- The crucible is subjected to the DSC instrument, and the sample is heated or cooled at a controlled rate to obtain a heat flow.
- The DSC thermogram is used to investigate the melting points, crystallization temperatures, and other thermal transitions.
- The thermal data is compared with reference standards to estimate the crystallinity and phase transitions of the sample.

6.3.5 Challenges

- **Complexity:** As the herbal extract is a mixture of phytoconstituents, each possesses its own thermal behavior. This makes it difficult to interpret the thermal behavior of active

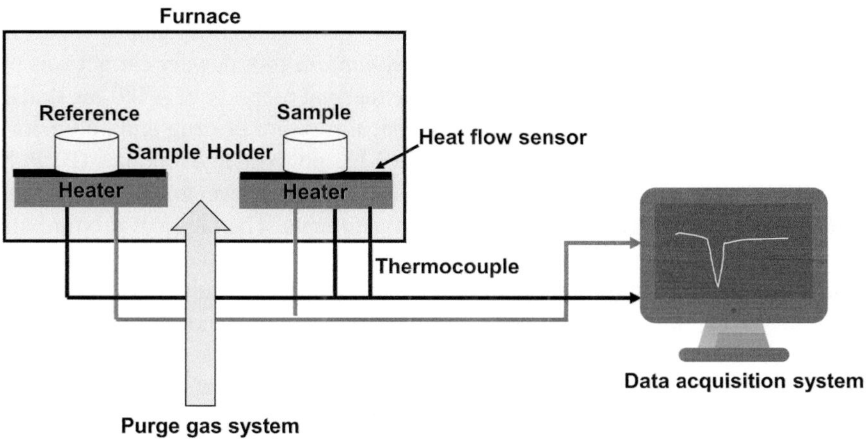

FIGURE 6.4 Schematic representation of DSC

phytoconstituent. Various phytoconstituents of a herbal extract or fraction may undergo thermal transitions in overlapping temperature ranges, making it difficult to assign peaks accurately.

- **Low concentrations of bioactive components:** Due to the very low concentration of active components, it is difficult to identify the thermal event of the compound of interest.
- **Moisture content:** Sometimes, herbal drugs are hygroscopic, which may produce an endothermic peak due to the moisture evaporation that may overlap with the endotherm of herbal drugs. Also, the moisture alters the stability of herbal drugs during heating.
- **Thermal degradation:** Most of the herbal drugs are thermolabile, which may degrade during the DSC study before the melting, which may cause difficulty in data interpretation.
- **Homogeneity:** A uniform sample is required to achieve consistent thermal behavior. However, most of the herbal drugs are fibrous, coarse, and oily in type, causing difficulty in obtaining a homogeneous sample.
- **Volatility:** Volatile components and essential oils may evaporate during the DSC, leading to the development of abnormal endotherms and pressure variations in the crucible.

6.3.6 Application to Herbal Drugs

In herbal drugs, DSC is used to investigate the thermal properties of bioactives, herbal extracts, fractions, and excipients [20]. The crystalline phytoconstituents show a sharp endothermic peak at their melting point, whereas the amorphous substances show a broad, holo/less-defined thermal transition. Such a study is especially helpful in the thermal analysis of herbal drug formulations for identifying drug-excipient compatibility, where the excipients may affect the crystallinity and stability of plant bioactives. It is useful to identify the melting point and heat of fusion, which helps to confirm the crystallinity of isolated plant bioactives. Impurities or mixed phases can alter thermal profiles, helping assessment of the purity and consistency of herbal extracts or powders. It is also useful to study polymorphic forms of plant bioactives, which may have different thermal transitions.

In one of the studies, crystallinity of ternary amorphous solid dispersions containing novel anti-tubercular agent Indole2-carboxamide derivative (North 2) and antioxidant compound curcumin (CUR) was studied by DSC and compared with the diffractograms of neat compounds. The solid dispersions are made using polyvinylpyrrolidone (PVP), Eudragit EPO® (EPO), and hydroxypropyl methylcellulose (HPMC). The pure compounds showed sharp crystalline peaks due to melting, whereas all the solid dispersions (North 2:CUR: PVP, North 2:CUR: EPO, and North 2:CUR: HPMC) showed no melting associated with CUR and North 2, but the appearance of a glass transition event (Tg), demonstrating effective formation of amorphous solid dispersions. DSC is also used to identify the miscibility between drugs and polymers. A significant reduction of the melting point of drugs in the presence of a polymer is an indication of good miscibility among them [5].

Amorphous solid dispersion (ASD) of *Withania somnifera* root powder extract was prepared to improve the saturation solubility and dissolution. The thermal property of ASD was studied by DSC and compared with that of the neat extract and physical mixture of drug and polymers [hydroxypropyl methylcellulose acetate succinate (HPMCAS-L), polyvinylpyrrolidone (PVP K-30), and Soluplus®]. DSC of pure extract showed two types of endothermic events (broad peak and multiple smaller peaks) due to the melting of various phytoconstituents. The physical mixture also demonstrated endothermic peaks, similar to those of pure extract. In contrast, except for Soluplus®-based ASD, all ASD formulations showed holo amorphous peaks (glass transition temperature/T_g) with complete absence of crystalline peaks, representing the amorphous modifications. The Soluplus®-based ASD demonstrated slight retention of crystallinity. The T_g for the ASD was found to be above the normal storage temperature (20–45°C), representing the stability of the developed ASD [9].

NLCs of plant bioactives have been prepared with Compritol® 888 ATO as solid lipid and squalene as liquid lipid to overcome the solubility and permeability issues of quercetin and piperine. A DSC study was carried out to examine the crystalline structure of drugs in the formulation. The

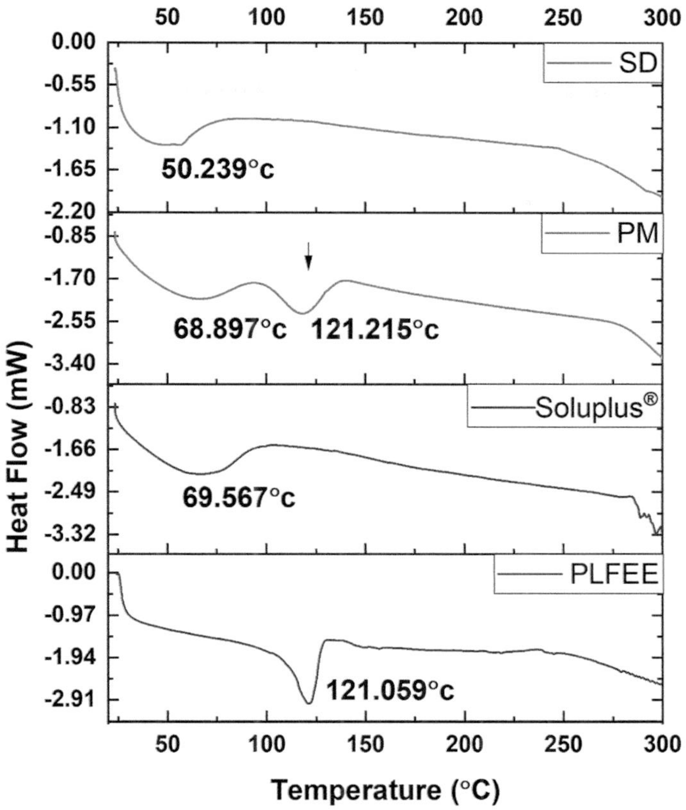

FIGURE 6.5 DSC thermograms of extract, excipients, and solid dispersion for crystallinity analysis

PLFEE: *Piper longum* fruits ethanolic extract, PM: physical mixture of extract and polymer, SD: solid dispersion.

study was carried out for individual phytoconstituents, blank NLCs, and drug-loaded NLCs. The DSC thermogram of piperine and quercetin showed sharp endothermic peaks due to the melting of the drugs, reflecting crystallinity. However, the thermogram of NLCs showed the absence of sharp endothermic peaks and successful conversion of both the phytocompounds into an amorphous form while encapsulated in a NLC matrix [11].

In another study, DSC of standardized ethanolic extract of *Piper longum* (PLFEE), excipients, their physical mixture (PM), and its solid dispersion (SD) was studied to observe the crystallinity (Figure 6.5). The PLFEE exhibited a sharp endothermic peak, signifying its crystallinity, attributed to the melting of numerous phytoconstituents. The excipient showed a broad endothermic peak due to its glass transition temperature, representing its amorphous nature. The PM retained the endothermic peak of PLFEE and the excipient. The presence of two distinct endotherms in the PM reveals the phase separation state in which the crystalline PLFEE and amorphous CMs coexist in their original state. In contrast, the thermograms of SD showed a complete absence of the PLFEE endothermic peak, signifying a complete transition of crystalline extract into its amorphous state [13].

6.4 POLARIZED LIGHT MICROSCOPY

Polarized light microscopy (PLM) is a non-destructive microscopy technique that uses polarized light to observe optically birefringent materials. PLM is widely used to identify and characterize crystalline or amorphous materials, detect polymorphs, evaluate particle size, and shape in the pharmaceutical field. It provides a rapid and visually intuitive strategy for investigating the microstructure of pharmaceutical substances.

6.4.1 Principle

PLM utilizes polarized light to analyze the optical properties of crystalline substances. When a crystalline material is observed under polarized light, it shows optical birefringence/double refraction, due to its unique internal structure. The crystals look like brightly colored materials under polarized light. The color and intensity of the birefringence depend on the crystal structure and orientation. In contrast, the amorphous material does not possess any optical birefringence; thus, it can be differentiated from crystalline materials. PLM is useful for studying the crystallinity and crystal morphology [22, 23].

6.4.2 Advantages

- It is a non-destructive and an easy technique.
- It offers detailed information on crystal size, shape, and optical properties.
- It can be used to directly observe the crystalline structures in herbal samples.

6.4.3 Instrumentation

A schematic representation of the PLM instrument is shown in Figure 6.6. The main components of PLM include the following:

- **Light source:** An LED or a halogen lamp is used to provide bright light for illumination.
- **Polarizer:** It is mainly made up of a polaroid sheet or a Nicol prism and is placed below the sample stage, which converts ordinary light into plane-polarized light.

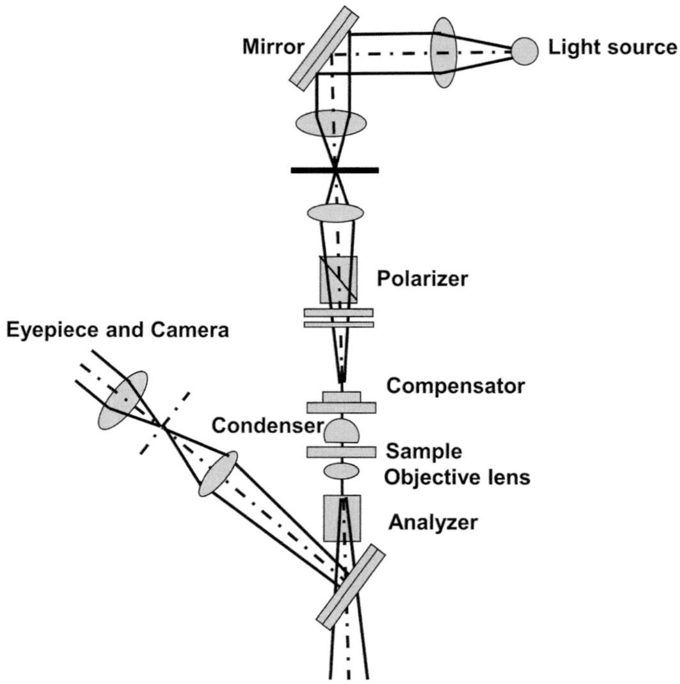

FIGURE 6.6 Schematic representation of PLM

- **Condenser:** It mainly focuses the polarized light onto the specimen. Strain-free condensers are used to prevent artifact birefringence.
- **Stage:** A rotatable stage that allows 360° rotation is used to observe changes in optical birefringence. Usually, a thin glass slide with a cover slip is used as a sample and placed on the stage.
- **Objective lenses:** Strain-free objective lenses are mostly used to avoid internal birefringence from the lenses.
- **Analyzer:** It is a second polarizer that is positioned above the objective. It is perpendicular to the first polarizer. The cross-polarization makes the crystalline birefringent materials to be observed as bright photomicrographs on a dark background.
- **Compensators or retardation plates:** This is an optional component often inserted between the analyzer and the polarizer to measure birefringence and enhance contrast.
- **Eyepieces and camera:** These are used for observing, capturing, and performing digital analyses.

6.4.4 Procedure

- The PLM sample is prepared as a thin section or placed on a microscope slide.
- The sample is subjected to PLM and viewed under a polarized light at various angles.
- The obtained image is analyzed for birefringence properties.
- The crystal sizes, shapes, and morphology are recorded.
- The birefringence property of the herbal extract/bioactive is compared with that of the herbal formulation to observe the changes in crystallinity.

6.4.5 Challenges

- **Complexity:** Since the herbal drugs contain a mixture of crystalline and amorphous materials, the overlapping data cause difficulty in interpreting the birefringence property. Multiple phytoconstituents may possess similar birefringence, making it difficult to distinguish the optical properties of individual components.
- **Low crystallinity:** Some of the herbal drugs contain a very low proportion of crystalline materials, leading to undetectable or very weak optical birefringence.
- **Difficulty in sample preparation:** Some of the herbal extracts are sticky, and herbal drugs are fibrous in nature. The hygroscopic nature of plant extracts also causes difficulty when mounted on a microscope.
- **Interference:** The cellulose, lignins, and starch may appear as bright birefringence, which may dominate the active phytoconstituents and interfere with their optical properties.
- **Qualitative measurement:** PLM is a qualitative analysis technique, which can't quantify the degree of crystallinity; hence, it needs supportive analytical tools like XRD and DSC.

6.4.6 Application to Herbal Drugs

PLM is mainly useful for the investigation of the morphology and birefringence of crystalline substances in herbal drugs and formulations. The crystal size, shape, and distribution can be investigated and quantified in powdered herbal drugs, offering a detailed understanding of their crystalline properties. It helps to identify the existence of amorphous and crystalline phases in powdered herbal samples. It analyzes the polymorphic forms based on variances in optical properties.

The PLM of standardized ethanolic extract of *Piper longum* (PLFEE) and its solid dispersion (SD) was studied to observe the crystallinity and morphology (Figure 6.7). The solid dispersion of PLFEE was prepared using Soluplus® and Tween® 80 and optimized using central composite design (CCD)-based quality-by-design approach. The PLFEE represented the intense birefringence, reflecting the presence of crystalline phytoconstituents. The physical mixture of drug and polymer

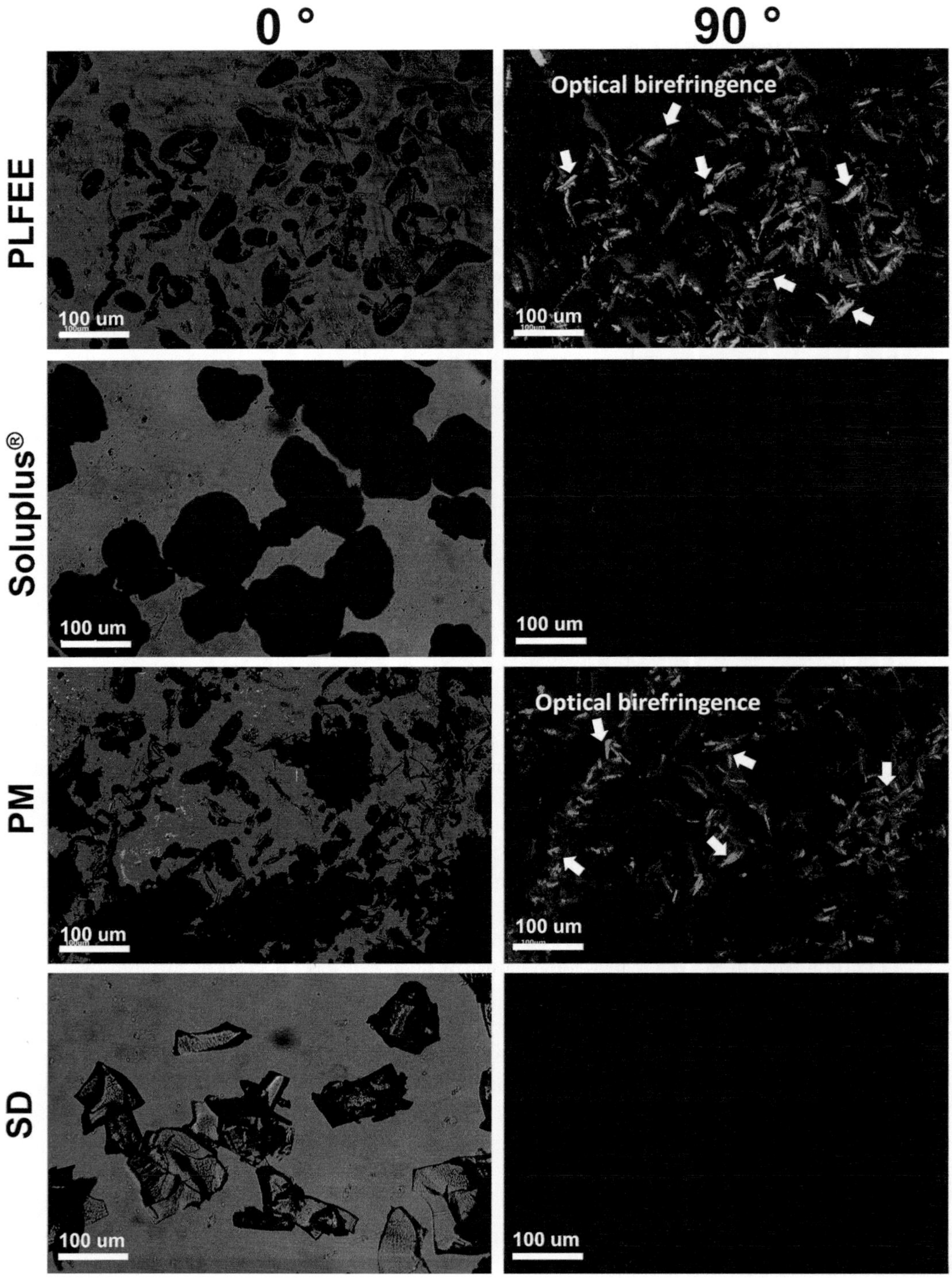

FIGURE 6.7 PLM photomicrographs of extract, excipients, and solid dispersion at 10X magnification for crystallinity analysis

PLFEE: *Piper longum* fruits ethanolic extract, PM: physical mixture of extract and polymer, SD: solid dispersion.

also represented optical birefringence, demonstrating the retention of crystallinity of PLFEE in PM. In contrast, the SD demonstrated no birefringence, which is due to complete amorphous modifications [13].

In a recent study, triptolide-loaded self-micelle solid dispersion was prepared to improve oral bioavailability and antitumor activity. The solid dispersion was prepared using the sodium salt of glycyrrhizic acid (Na_2GA) and was characterized for crystallinity using PLM. The triptolide and physical mixture (Na_2GA+triptolide) demonstrated birefringence properties. In contrast, the birefringence was found to have completely disappeared in solid dispersion, which is similar to the birefringence of Na_2GA, representing amorphous modification [10].

Rutin-loaded solid dispersion (SD) was prepared using Poloxamer 407 and Poloxamer 188 to improve dissolution and oral absorption. The developed SD and its excipients were investigated by PLM for studying changes in crystal morphology and optical properties. The miscibility and uniformity of the rutin in the excipient matrix were investigated, as phase separation leads to recrystallization of drugs in the SD. In all SD formulations, the rutin crystals were found to be fully miscible with the Poloxamer 407 and 188, indicating that RUT is homogenously dispersed in the excipient matrices [12].

6.5 THERMOGRAVIMETRIC ANALYSIS

Thermogravimetric analysis (TGA) is used to analyzes the temperature-based weight changes of substances. It provides information on thermal stability and the composition of materials. It is an important tool for studying the thermal properties of herbal drugs, especially crystallinity, thermal degradation, and moisture content. The thermal properties of active components of herbal drugs, excipients, and drug products can be studied to develop a rational herbal formulation with excellent drug-excipient compatibility and stability [15, 20, 24].

6.5.1 Principle

TGA is based on the principle of measuring the changes in weight of a material as a function of temperature or time under a controlled environment (air or nitrogen). During analysis, the sample is heated at a constant rate, and the loss of weight is recorded as a function of temperature. The loss of weight occurs due to loss of moisture, thermal decomposition of phytoconstituents, and volatilization of organic solvents in a herbal extract. The TGA data is represented as a thermogram, where the percentage weight loss is plotted against temperature or time. Different stages of weight loss represent different thermal events like water loss, thermal degradation, and so forth [25].

6.5.2 Advantages

TGA possesses the following advantages [20]:

- **Quantitative**: TGA offers quantitative data on percentage weight loss, which can be used to calculate percentage moisture content, percentage volatile compounds, and so forth.
- **Low amount of sample**: It usually requires a very low quantity of samples (~10 mg).
- **Versatility**: It can be used to investigate a wide variety of samples, such as extracts, bioactives, excipients, and herbal formulations.
- **Sensitivity**: It is very sensitive, so it can detect even a minute loss of weight.
- **Temperature control**: It allows precise control over a wide range of temperatures, making it ideal for analyzing the thermal properties of herbal drugs for understanding their thermal stability and crystallinity.
- **Does not require sample preparation**: TGA is a simple, fast, and cost-effective analytical method that does not need any sample preparation step or specialized equipment.

6.5.3 Instrumentation

A schematic representation of the TGA instrument is shown in Figure 6.8. The main components of TGA include the following [15, 24]:

- **Microbalance:** It uses an ultrasensitive balance that measures weight changes in the microgram range in real time.
- **Furnace:** It offers a wide range of temperatures (room temperature up to ~1400°C) to investigate the temperature-dependent weight loss of samples.
- **Sample holder or pan:** It is used to hold the sample in the furnace. These are made up of inert materials (alumina, platinum) that avoid the reaction of the material with the pan and withstand high temperatures.
- **Thermocouple or temperature controller:** It precisely controls and monitors the temperature to investigate the sample under isothermal and dynamic modes.
- **Purging gas:** Nitrogen and argon are mostly used as a purging gas to provide an inert atmosphere.
- **Software and data acquisition tools:** These acquire the data, process it, and analyze it to generate real-time thermal events.

6.5.4 Procedure

- The herbal sample (~10 mg) should be finely powdered to achieve uniformity and uniform heating.
- The sample is placed in a platinum or aluminum pan and subjected to the furnace of the TGA instrument.
- The study is carried out under isothermal and dynamic mode (5–20°C/min) within a wide range of temperature (room temperature to 1400°C) as per the material.
- The analysis is carried out in an inert atmosphere/nitrogen gas to prevent possible oxidation or other reactions with atmospheric air.
- The weight is monitored over the temperature range, and the percentage weight loss is calculated.

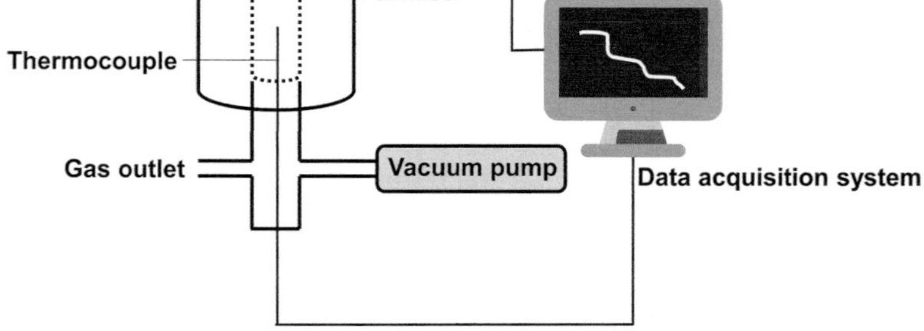

FIGURE 6.8 Schematic representation of TGA

6.5.5 Challenges

- **Complexity:** Herbal drugs are multicomponent mixtures, which leads to overlapping thermal events, making it difficult to assign the degradation of specific phytoconstituents. Different phytoconstituents may degrade within the same temperature range, causing a broad thermal weight loss event.
- **Moisture content:** Most of the herbal drugs are hygroscopic in nature, which show an initial weight loss due to the evaporation of moisture.
- **Presence of volatile components:** Some of the herbal drugs contain volatile oils and components, which may ambiguous the thermal decomposition data.
- **Heterogeneity:** Herbal samples are heterogeneous in nature, causing difficulty in obtaining reproducible data.
- **Interactions:** Since the herbal drugs are a multicomponent system, they may interact among themselves to produce an altered thermal pattern.
- **Residual materials:** Herbal drugs often contain residual substances, which may be obtained as inorganic salts, charred organic, and ash after the TGA study.

6.5.6 Application to Herbal Drugs

TGA offers important information on the thermal properties and crystallinity of herbal drugs. Crystalline substances possess a different thermogram compared to amorphous materials. Crystalline materials possess greater thermal stability, whereas the amorphous substances lose weight rapidly and possess broader thermal transitions. It is also useful for identifying the moisture content of the extract. Most herbal drugs contain water in their crystalline structure. The level of water can be quantified from the percentage weight loss due to dehydration to distinguish crystalline drugs from amorphous. The transition temperature of the crystalline substance can be identified. Different polymorphs possess unique thermal patterns. TGA enables the detection of different polymorphic forms of phytoconstituents. In addition to crystallinity and polymorphism, TGA helps to study the thermal stability of herbal drugs and formulations. It is also useful in identifying the thermal profile of excipients used in herbal formulation, stability, and drug-excipient compatibility [20, 26]. It is also used for the investigation of volatile components, the investigation of major components, moisture content of herbal drugs [20].

The TGA of *Piper longum* extract, physical mixture (drug+Soluplus®+Tween® 80) of drug and polymer, and its solid dispersion was made to observe the thermal properties, degradation patterns, and crystallinity (Figure 6.9). The extract showed two-step degradation patterns. The physical mixture demonstrated a degradation pattern almost similar to Soluplus® due to its higher proportion. In contrast, the solid dispersion showed a four-step degradation pattern, which is completely different from crystalline extract and physical mixture, demonstrating the existence of an entirely new amorphous phase in solid dispersion [13].

6.6 DYNAMIC VAPOR SORPTION

Dynamic vapor sorption (DVS) is used to measure the desorption and sorption of moisture in a sample as a function of temperature and relative humidity. It is a gravimetric tool that analyzes the amount and kinetics of solvent sorption by a sample [1]. It is widely used to study moisture uptake, hygroscopicity, and crystallinity of drugs, herbal drugs, and formulations. DVS is very sensitive and can be used for the quantification of very low amounts of amorphous content in predominantly crystalline materials [1]. Understanding the interaction of herbal drugs with moisture at various humidity levels, DVS provides information on crystallinity, stability, and hygroscopicity. Such information is useful for herbal drug packaging to ensure their stability and efficacy.

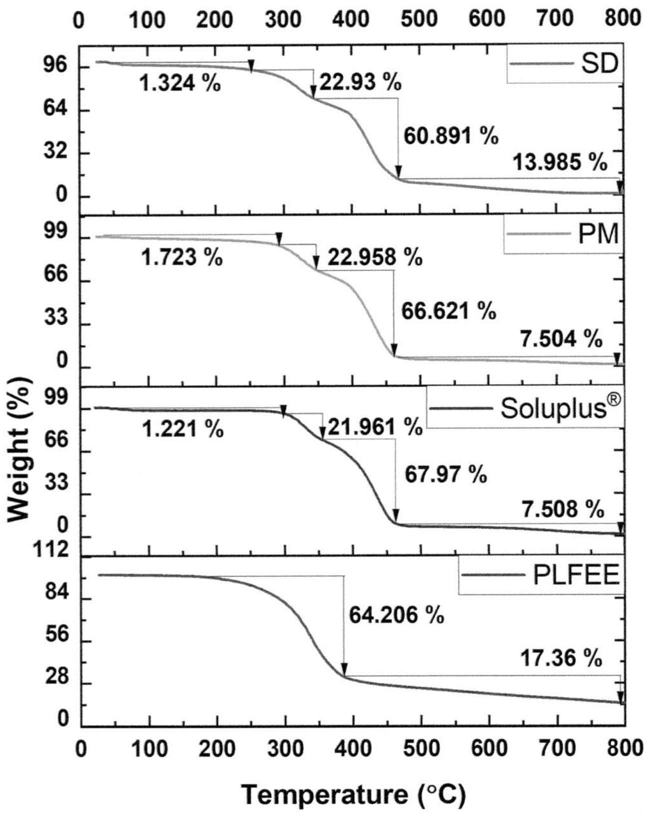

FIGURE 6.9 TGA thermograms of extract, excipient, and solid dispersion

PLFEE: *Piper longum* fruits ethanolic extract, PM: physical mixture of extract and polymer, SD: solid dispersion.

6.6.1 Principle

It is based on the principle of recording the alteration of the mass of a material during exposure to various levels of humidity or vapor pressure. The sample is exposed to a controlled environment, where humidity is provided incrementally and weight is measured simultaneously. The change in mass occurs due to desorption and sorption of moisture. During increased humidity, hygroscopic materials absorb moisture and gain weight, whereas during a decrease in the relative humidity, the absorbed moisture is lost, leading to a decrease in weight. The weight is monitored over several cycles of absorption and desorption. The generated data is used to study various aspects of solvent sorption isotherms. The isotherms are mainly divided into two components: sorption during increasing humidity events and desorption during decreasing humidity events. The difference in uptake of water vapor between the sorption and desorption isotherms is known as "hysteresis" [1]. By studying the mass change, DVS offers information on how the material interacts with vapor, providing information on amorphous and crystallinity properties [1].

6.6.2 Advantages

- **High sensitivity**: DVS can identify very small alterations in weight due to changes in moisture loss and uptake, making it very sensitive to study small changes in the structure of materials under very low humidity conditions.

- **Quantitative study**: It provides quantitative information on moisture uptake and release, which can be used to study crystallinity, moisture sorption capacity, and stability of drugs.
- **Real-time monitoring**: It provides real-time monitoring of the material's behavior with respect to humidity.
- **Sample versatility**: A wide variety of samples can be analyzed by DVS, including API, herbal drugs, powders, granules, formulations, and excipients.

6.6.3 Instrumentation

A schematic representation of the DVS instrument is shown in Figure 6.10 [1]. The major components include the following:

- **Sample chamber:** It is a temperature-controlled chamber that provides precise and uniform exposure to relative humidity and temperature.
- **Sample and reference pans:** These are mostly hemispherical quartz crucibles, used because of their chemical inertness and ease of cleaning. Metal-coated pans are also used for their antistatic abilities.
- **Ultra-microbalance/microbalance:** It is the heart of this instrument and can measure very small changes in sample mass (lower than 1 part per million) with respect to humidity.
- **Humidity generator:** Precisely provides relative humidity via humidity generator and mass flow controllers (MFLs). The MFL generates the required humidity by mixing dry and saturated vapor gas flows in the precise proportions.
- **Temperature controller:** Maintains constant temperature.
- **Gas flow controller:** It carries dry and humid air to the sample.
- **Temperature and humidity probes:** They accurately measure temperature and humidity.
- **Data acquisition system:** It collects the data, processes, and analyzes the change in mass with respect to humidity. The sorption, kinetics, and hysteresis can be recorded by the integrated system.

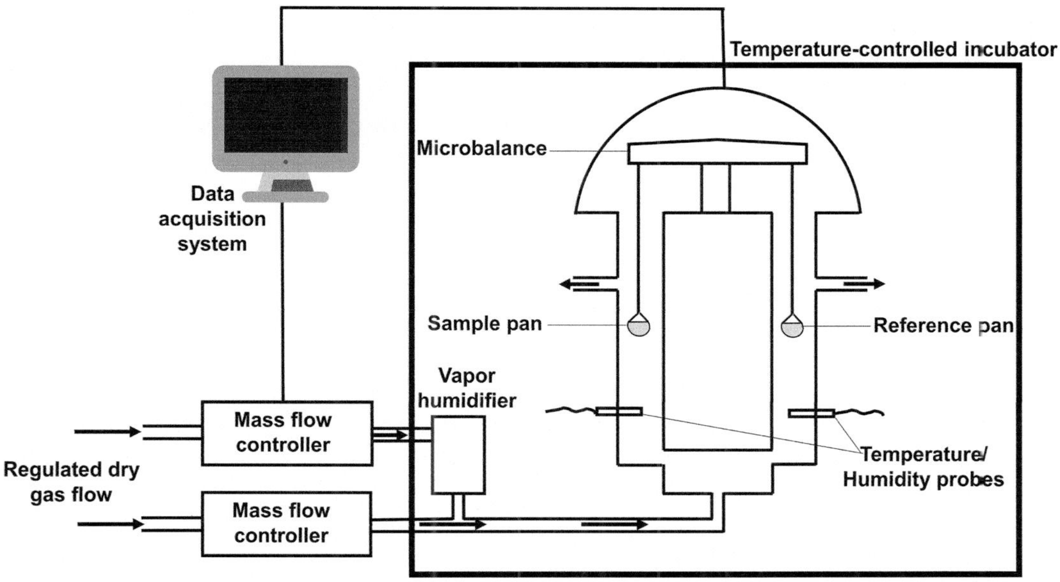

FIGURE 6.10 Schematic representation of DVS

6.6.4 Procedure

DVS analysis includes the following steps [1]:

- The sample is weighed accurately to obtain the initial mass.
- The sample is loaded into the pan inside the balance, which is placed in the sample chamber.
- Then, air or dry nitrogen is blown over the sample, and the relative humidity is gradually increased from 0% to 90% RH.
- The instrument waits until the mass equilibrium at each RH is achieved.
- Then, the relative humidity is decreased to produce a desorption curve.
- A typical DVS analysis involves the following: Desorption (exposure of sample to low humidity (e.g., 0–10% RH) to remove water in the sample, absorption (gradual increase in humidity in steps: for example, 10%, 20%, 30%, up to 90% RH) to record how the sample absorbs moisture at different levels, and desorption (back) (further reduction of humidity and monitoring the moisture loss).
- The process is repeated for multiple cycles to study the reversibility of the sorption and desorption phenomenon.
- The change in mass is recorded and monitored due to absorption or desorption of moisture.
- The change in weight as a function of humidity is plotted to generate sorption and desorption curves.
- The moisture uptake, rate of moisture absorption, equilibrium moisture content, phase change, and crystallization are estimated.

6.6.5 Challenges

- **Complexity:** Herbal drugs are multiconstituent-based complex materials, and each component has different hygroscopic properties, which creates difficulty in interpreting sorption profiles. The geographical, seasonal, and processing variability also leads to variation in composition that affects the properties of herbal drugs.
- **Variability:** Herbal drugs need pre-treatments, such as grinding, drying, and so forth, that may alter the properties and variability in the data among batches.
- **Structural heterogeneity:** Plant materials possess irregular structure, porosity, and surface area that affect moisture absorption and non-uniform DVS data.
- **Degradation:** Some phytoconstituents are unstable in the presence of water, which may lead to degradation and hydrolysis during the experiment.
- **Volatility:** Volatile components of the plant may evaporate at higher temperatures during the study.

6.6.6 Applications

The hysteresis and the rate of moisture absorption provide information on the crystallinity of the herbal drug. Crystalline and amorphous forms of the same chemical entity differ in their interaction with water [1]. Amorphous substances mainly show more significant moisture uptake and broader sorption profiles compared to crystalline substances due to their high surface area and disordered structure. Water adsorption onto crystalline substances occurs as a monolayer and multilayer, whereas in the case of amorphous solids, it occurs in two different thermodynamic states, namely, "bound state" and "solvent-like state" of water. The adsorption of moisture occurs in the case of crystalline materials, whereas adsorption and absorption occur in the case of amorphous materials. So amorphous materials sorb relatively more water compared to crystalline materials [1]. Thus, DVS can differentiate between amorphous and crystalline materials based on the amount and rate of moisture absorption [1]. Water mainly interacts with the surfaces of solid substances through the formation of hydrogen bonds. The adsorption at the surface of a solid mainly depends on the vapor pressure, binding energies, and temperature between the solid and vapor [1].

DVS also differentiates various polymorphic forms as they possess different moisture sorption profiles. DVS is helpful for studying the moisture sensitivity of herbal drugs. Most of the herbal drugs are sensitive to moisture. DVS evaluates the sensitivity of herbal drugs to different humid conditions, which is crucial for stability studies and predicting suitable storage conditions. Herbal drugs and formulations are sensitive to excess moisture. DVS helps to study the crystallinity and stability of herbal drugs due to moisture uptake, which is essential for shelf life and stability prediction.

DVS was used to study the physicochemical properties of *Adansonia digitata* L. Mucilage (ADM). The water sorption characteristic of ADM powder was studied using DVS, with humidity increased from 10% RH steps to 100% RH during the sorption phase and decreased similarly for the desorption phase at a temperature of 21°C. A marked hysteresis was observed between 10% and 80% RH in the corresponding water sorption isotherm, which represents the extent of water absorption by the ADM. Amorphous substances exhibit a considerable bulk absorption of water, which in some cases is accompanied by swelling effects. Such a phenomenon was reported in the moisture sorption isotherm of ADM, in which swellings were noticed at 0.05% at 344 min and 59.86% after 2488 min. The water sorption study confirmed the partial amorphous property of ADM, which was in accordance with the broad DSC thermogram exhibiting a T_g [27].

A DVS study was performed for co-crystal berberine-ibuprofen (BI) at 25°C using an RH cycle of 40-0-95-0-40% to investigate the relationship between humidity and BI stability. The DVS curve showed that when humidity increased from 40% to 95% and subsequently decreased to 10%, no significant alteration in the weight of BI was observed. When the humidity was below 10%, the BI was found to be partially dehydrated. However, when the humidity was further increased above 30%, the dehydrated substance reverted to the crystalline form of BI [28].

The hygroscopicity of the synephrine dry powder inhaler was evaluated using DVS at 25°C. At 90% RH, the synephrine powder showed a moisture absorption weight gain of 0.30%, representing slightly hygroscopic nature hygroscopic weight gain of < 2% but ≥ 0.2%. Under conventional storage conditions (60% RH), the percentage moisture absorption and weight gain are less than 0.2%, representing a non-hygroscopic nature, meeting the requirements of the production, transport, and storage of dry powder inhalers. It was also found to be suitable for pulmonary drug delivery via the respiratory system in the presence of moisture, which will not cause agglomeration and deposition of particles [29].

6.7 MICROCALORIMETRY

Microcalorimetry is an analytical technique used to analyze the heat flow associated with chemical and physical processes occurring in a material, such as dissolution, crystallization, and phase transitions. The phase behavior and crystallinity of a material can be studied using microcalorimetry. It provides quantitative, real-time data on the heat flow that is useful for analysis of the physical properties and to optimize formulation and storage conditions [30]. It is used to study the thermal properties of herbal drugs, including stability and crystallinity of bioactives, drugs, and excipients.

6.7.1 Principle

Microcalorimetry is based on the principle of analyzing the heat flow (release or absorption) during a physical or chemical alteration in a material. The amount of heat absorbed or released is related to the energy involved in the process during phase transitions (melting, crystallization). Crystalline substances release heat during transition from an amorphous or liquid state to a crystalline state. The heat of crystallization is measured by microcalorimetry. During the dissolution or melting of crystalline material, it absorbs heat, and the endothermic event is measured by microcalorimetry. They are very sensitive to evaluating very small changes in heat flow. A variety of calorimeters are used, such as adiabatic, isothermal, and Tian-Calvet calorimeters [30–32].

6.7.2 Advantages

- **High sensitivity**: It can detect very minute changes in heat flow, hence being useful for crystallization studies of small amounts of materials.
- **Quantitative**: It provides quantitative information on the heat flow during phase transition, which can be useful to analyze the amount of heat absorbed or released, crystallization enthalpy, and other thermodynamic parameters regarding crystallinity.
- **Real-time monitoring**: It provides real-time information, which is useful for studying phase transitions, such as melting and crystallization.
- **Wide applicability**: It applies to a wide range of samples, including pure API, phytoconstituents, and complex herbal drugs or formulations.

6.7.3 Instrumentation

A schematic representation of the microcalorimetry is shown in Figure 6.11. The important components of a microcalorimeter include the following [30]:

- **Sample and reference cells:** In the microcalorimeter, a sample cell is available to load the sample, whereas the reference cell serves as a blank or control. Both cells are kept in the same environmental conditions. The change in the heat is measured with respect to a reference.
- **Thermopile:** The thermopile is also known as a heat flux sensor that detects minute changes in heat and generates an electrical signal that is proportional to the heat flow of the sample.
- **Thermostatted enclosure:** It maintains a constant temperature for the analysis and ensures that any heat changes observed in a sample are not due to ambient fluctuation.
- **System for data acquisition:** It converts the signals from the sensor to digital data and generates a real-time heat flow chart.
- **Gas flow and humidity control:** For isothermal titration calorimetry (ITC), a controlled gas or vapor environments are provided.

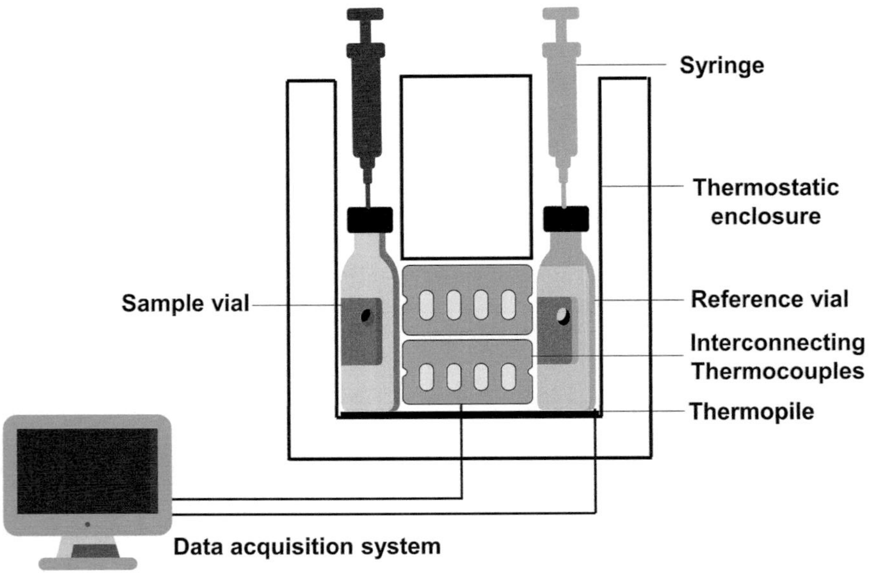

FIGURE 6.11 Schematic representation of microcalorimetry

6.7.4 Procedure

- An accurately weighed, uniformly ground sample is placed in the sample container. A finely powdered sample is used to offer uniform heat exchange to study the crystalline materials. Sometimes, the hygroscopic samples are preheated to remove the moisture.
- The sample is filled into the sample cell, and the reference cell is filled with inert substances.
- Then the sample is subjected to a controlled temperature environment with heating at a constant rate (1–10°C/min), and the heat flow is recorded. The temperature range is selected depending on the expected crystallization or phase transition.
- The microcalorimeter measures the heat absorbed or released by the sample continuously.
- The thermogram is represented as the heat flow with respect to temperature or time.
- Crystallization is an exothermic process that produces heat and generates a positive peak, whereas the melting is an endothermic event that generates a negative peak.
- The thermogram is analyzed to observe heat changes with respect to melting or crystallization.
- By comparing the thermal patterns of crystalline and amorphous materials, the phase transitions and crystallinity can be studied.
- The thermogram provides information on crystallization, transition temperature, and polymorphism.

6.7.5 Challenges

- **Complexity:** The thermal events generated from microcalorimetry may be the summation of multiple overlapping thermal events, making interpretation difficult.
- **Low amount of active constituents:** The active components in a herbal extract or product are very low, which may mask thermal events may be masked by the highly abundant other components.
- **Reproducibility:** The batch-to-batch variation in the extraction, harvesting time, and processing of herbal drugs leads to variation in the contents, which leads to inconsistent data reproducibility.
- **Volatile components and moisture:** The herbal drugs often contain moisture and volatile components, which may evaporate during analysis, leading to altered thermal events.

6.7.6 Applications

It is useful for crystallization and phase transition behavior. When the drugs or phytoconstituents crystallize, they release heat, the amount of which is correlated with the crystallization process. Different polymorphic forms of the same compound possess different crystallization behaviors, including heat release or absorption [33]. Microcalorimetry distinguishes between polymorphs based on their thermal signatures. Crystalline materials possess a well-defined structure, in which the crystallization event is exothermic, whereas amorphous materials possess a disordered structure, producing an altered thermal event without any crystallization peak compared to the crystalline material. Microcalorimetry is also used to study the stability of herbal drug products under different temperature conditions, by identifying any transformation or degradation that may affect the drug's crystallinity. Based on the changes in crystallinity and stability, the interaction among the herbal drugs and excipients can be studied in a herbal product by microcalorimetry. Isothermal microcalorimetry was used for quantifying the amorphous content of lactose as a model compound. It showed a prolonged exotherm followed by a sharp crystallization exotherm. The length of time prior to the onset of crystallization correlated with the mass of amorphous material [32].

6.8 SELECTED AREA ELECTRON DIFFRACTION

Selected area electron diffraction (SAED) is a crystallographic technique performed within a transmission electron microscope (TEM) to study the crystallinity and lattice parameters of a sample at the nanometer scale. It offers information on the crystal structure of a substance by analyzing the diffraction pattern generated by electrons transmitted through a selected microscopic area of the sample. In the pharmaceutical field, the SAED is used for investigating nanoscale crystalline structures, polymorphs, and poorly crystalline or nanocrystalline materials.

6.8.1 Principle

In the SAED study, an electron beam is focused on a small area of the sample under TEM investigation, and the resulting diffraction pattern is captured. This pattern comprises detailed information about the crystal structure, such as orientation, lattice spacing, and symmetry. Crystalline material shows a dotted SAED pattern, whereas the amorphous material shows a diffused diffraction pattern. The SAED is especially useful for samples with very fine structures, offering high-resolution information on crystallinity at the atomic level. The SAED of nanocrystals provides ring patterns that can be used to identify texture and discriminate nanocrystalline from amorphous phases [34].

6.8.2 Advantages

- Provides atomic-level resolution of crystal structures.
- Ideal for studying nano-sized or submicron crystalline particles.
- Useful for studying the crystallinity of highly complex or heterogeneous samples.

6.8.3 Instrumentation

A schematic representation of the SAED instrument is shown in Figure 6.12 [35]. The major components include the following:

- **Transmission electron microscope:** It acts as a core instrument for SAED. It enables TEM imaging and diffraction from the same region of the specimen. It has a stable electron beam and a well-aligned electron column.
- **Sample holder:** The sample holder accurately controls the position, orientation, and tilt of the sample.
- **Selected area aperture:** It is otherwise known as the SA aperture, which is a mechanical aperture (diameter: 0.2–5 μm) kept in the image plane of the objective lens. It is used to select the region of interest of the sample. The electrons diffracted from a selected area of the sample produce a diffraction pattern.
- **Objective lens:** It allows a focused beam of electrons on the specimen and plays a dual role in imaging and producing the initial diffraction pattern.
- **Diffraction camera and detector:** Photographic film is mostly used to capture the diffraction patterns. The latest instruments use CMOS and CCD cameras with high resolution and sensitivity, which are capable of real-time observation.
- **Vacuum system:** It is required to reduce the electron scattering by air and offers a clean beam path that protects sensitive samples.
- **Electron gun:** Mainly, a field emission gun (FEG) or a thermionic (LaB_6) device is used as an electron gun, which provides a coherent, intense, and stable electron beam.
- **Controller and software:** They allow beam alignment, aperture positioning, and photo acquisition. They calibrate the diffraction pattern and measure the interplanar spacings.

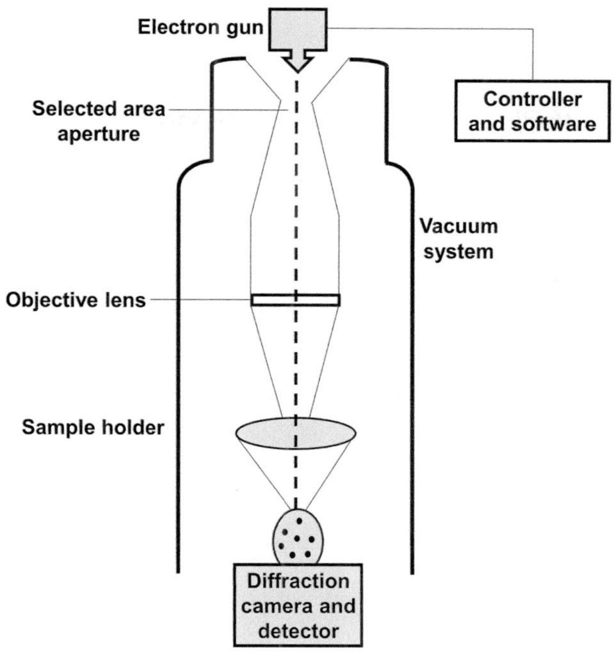

FIGURE 6.12 Schematic representation of SAED

6.8.4 Procedure

- The sample is prepared as an ultra-thin film on a TEM grid by the drop-casting or dipping method [35].
- During TEM analysis, the electron beam is focused on a small area of the nanoparticle.
- The SAED diffraction pattern is captured and analyzed to determine lattice spacing and crystal orientation.
- The crystallinity and phase structure of the sample are interpreted from the diffraction pattern.

6.8.5 Application to Herbal Drugs

During TEM analysis, the additional SAED tool is used to study the crystalline properties of nanostructures/nanoformulations. Herbal drugs and formulations may contain nano-sized polymorphic forms of crystals that can be resolved by SAED analysis from the diffraction patterns. It investigates the atomic arrangement of phytoconstituents at the molecular level. The dotted SAED pattern represents crystallinity, whereas the diffused pattern represents an amorphous nature.

SAED analysis was used to verify the crystallinity of plant-derived carbon dots (CDs). The CDs synthesized from leaves and stems of *Tinospora cordifolia* [36, 37] and leaves of *Piper longum* [38] represented a diffused ring SAED pattern, representing an amorphous nature, which was in accordance with XRD results.

In a recent study, solid dispersion of *Piper longum* was reported for melanoma therapy. The SAED analysis of nanomicellar structures during the solubilization of solid dispersion in the dissolution media was investigated to evaluate the crystallinity. The dissolution samples during HR-TEM analysis reflected a diffused ring pattern due to the retention of the amorphous structure, which is in accordance with XRD, DSC, and PLM outcomes (Figure 6.13) [13].

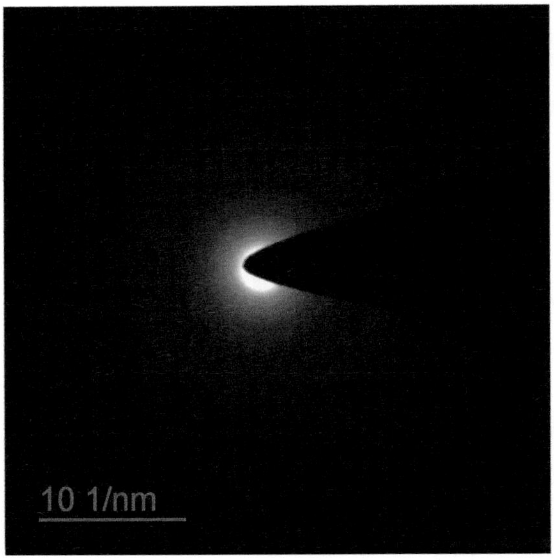

FIGURE 6.13 SAED pattern solid dispersion sample during dissolution

In other studies, SAED was used to characterize the crystallinity of green-synthesized metallic nanoparticles, such as silver nanoparticles from *Alysicarpus vaginalis* extract [39], *Withania coagulans* extract [40], *Mitragyna parvifolia* extract [41], and gold nanoparticles from *Mikania micrantha* extract [42]. All the metallic nanoparticles displayed crystalline properties due to the dotted SAED patterns.

In a recent study, the transferosomes of *Piper longum* for melanoma therapy were investigated by SAED analysis for their crystallinity. The transferosomes are developed using a central composite design (CCD)-based quality-by-design approach. The optimized formula showed an amorphous nature due to the diffused ring pattern of SAED, which was in accordance with the PLM, XRD, and DSC results [43].

6.9 SCANNING ELECTRON MICROSCOPY

Scanning electron microscopy (SEM) is a characterization technique that captures images and analyzes a specimen by scanning an accelerated electron beam, followed by selectively collecting and recording secondary electrons, back-scattered electrons, and other signals arising from the beam and specimen interactions. SEM permits the observation and characterization of materials at the micro- to nano-scale [44].

6.9.1 Principle

SEM is based on scanning a specimen with a focused high-energy beam of electrons. The electrons interact with the atoms of the sample and produce secondary electrons, backscattered electrons, and X-rays. The signals are collected from such interactions, which are used to generate high-resolution photomicrographs. The images are used to study the surface morphology, composition of the sample, and texture [35, 45–47].

6.9.2 Advantages

- **High resolution**: It provides high-resolution images from the nanometer to the micrometer range.

- **Surface morphology**: SEM offers detailed surface morphology of the sample and crystallinity. It enables the visualization of morphology via three-dimensional images of the sample and crystals.
- **Non-destructive**: It is a non-destructive technique that allows the repeated analysis of the same sample.
- **Elemental analysis**: In many cases, SEM is integrated with energy dispersive X-ray spectroscopy (EDX), which offers elemental analysis of crystals and the identification of trace elements.
- **Quick and efficient**: SEM study is relatively fast, offering immediate visual results of surface morphology and crystals in the specimen.
- **Versatility**: It can be applied to a wide variety of materials, including powder, film and compact, making it suitable for the identification of drug crystals at various stages of formulation and development.

6.9.3 Instrumentation

A schematic representation of the SEM instrument is shown in Figure 6.14. The major components of SEM include the following [35, 44–46]:

- **Electron gun/electron source**: It generates a high-energy electron beam. Field emission guns and thermionic guns (e.g., LaB_6, tungsten filament) are mainly used as electron sources in SEM.
- **Condenser lenses**: They control and focus the intensity of the electron beam.
- **Objective lens**: It further focuses the electron beam onto the specimen surface as a fine spot.
- **Scanning coils**: It scans the electron beam across the surface of the sample.
- **Sample chamber**: It holds the specimen on a stage that can be rotated and tilted. It is mainly kept in a vacuum to avoid scattering of electrons by air.

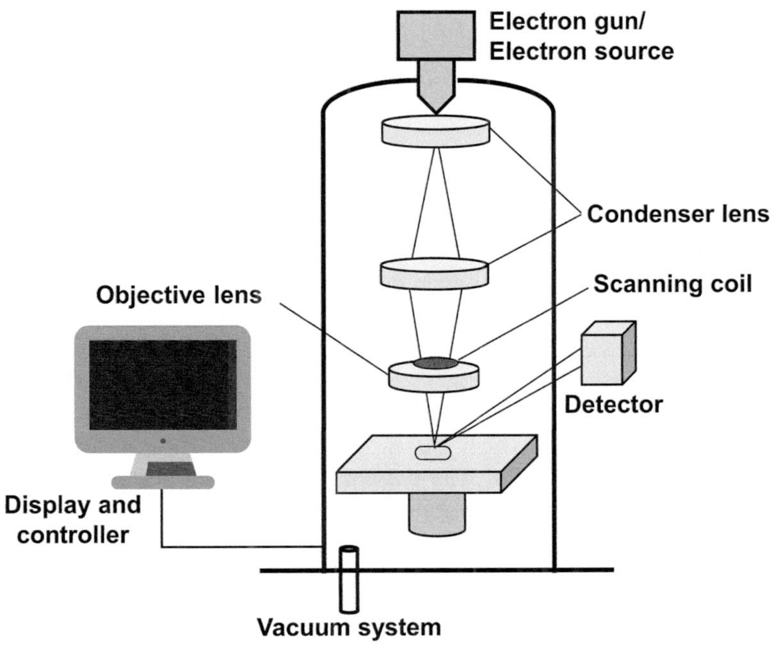

FIGURE 6.14 Schematic representation of SEM

- **Vacuum system**: It uses a combination of rotary and turbo molecular pumps that maintain a very high vacuum in the specimen chamber and SEM column.
- **Display and controller**: The controller allows the focus, magnification, and contrast, whereas the display allows the visualization of the generated images.
- **Detector**: It collects the secondary, backscattered emitted electrons and X-rays to generate an image. Secondary electron detector (SED), backscattered electron detector (BSED), and X-ray detectors (EDS/EDX) are mainly used in SEM. The SED is used for topographical imaging, and the BSED is used for contrast based on atomic number, whereas EDS/EDX is used for elemental analysis using energy-dispersive X-ray spectroscopy.
- **Additional agents**: Sometimes, cryo-SEM and environmental SEM components are attached to capture thermally unstable samples and imaging at low vacuum or weight conditions.

6.9.4 Procedure

SEM analysis involves the following steps [35, 45]:

- The samples are dried at room temperature or below 40°C to remove moisture and mounted on a sample stub using double-sided carbon adhesive tape.
- The samples are made conductive by coating with gold or palladium using a sputter coater to prevent the accumulation of charge during SEM analysis.
- Then the specimen is placed in the SEM chamber under vacuum to avoid the interference of air with the electron beam.
- Then the electron beam is focused on the specimen, and the secondary electrons are generated, which are collected by a detector to generate the surface morphology of the specimen.
- From the SEM photomicrographs, the morphology, shape, size, and uniformity of the sample or crystals can be studied. Multiple crystal structures of a sample represent the polymorphism.
- Sometimes, the EDX analysis is performed to analyze the elemental composition of a sample, which identifies the presence of possible contaminants or specific elements that may be a part of the crystalline structure.

6.9.5 Challenges

- **Hygroscopic:** Herbal drugs are mostly hygroscopic in nature, which require dehydration before SEM analysis, since incomplete drying can merge the delicate structures with altered morphology.
- **Nonconductive:** Herbal drugs are mostly nonconductive, which accumulate charge under the electron beam and produce photomicrographs with bright spots, poor resolution, and distortion. The coating with conductive materials (gold, platinum, palladium) can mask the surface.
- **Complexity:** Herbal drugs and extracts are heterogeneous in nature, containing starch, fiber, and crystals with variable morphology, which creates difficulty during image analysis.
- **Moisture content and volatility:** The presence of high moisture and volatile components in the herbal drugs causes outgassing under vacuum, leading to a change in chamber pressure and consistent imaging. Pre-treatment of herbal drugs may alter the original morphology.
- **Degradation:** Herbal drugs are sensitive to electron beams, leading to degradation and burning.

6.9.6 Application to Herbal Drugs

- The SEM images are visualized to investigate the surface morphology and crystallinity of the herbal drug. The crystallinity of the drugs can be interpreted based on the presence of well-defined, distinct crystals. However, the presence of disordered structure/lack of well-defined crystalline structures represents amorphous forms.
- The shape and size distribution of the crystals provide information on their behavior toward solubility, dissolution, and absorption.
- The plant bioactives may exist as different polymorphic forms, and each possesses a distinct morphology. SEM analysis provides insights into various crystal morphologies as a representative of various polymorphic forms having different pharmacokinetic properties.
- SEM analysis is useful for evaluating the quality/uniformity of herbal drugs and formulations by analyzing the crystalline forms of plant bioactives and detecting the presence of agglomerates/potential contaminants.
- SEM is also useful for investigating the drug-excipient interaction study. The interactions between the drug and excipient cause the alteration in the crystalline forms. Such changes can be identified by SEM.

Grapefruit flavonoid naringenin was reported to improve the dissolution and oral bioavailability via the development of a 4th-generation solid dispersion. Crystallinity analysis was studied using DSC, XRD, and SEM studies. SEM analysis of neat naringenin revealed its prismatic-shaped crystalline structure. However, the crystalline structure was not observed in the optimized solid dispersion. The solid dispersion demonstrated irregular spherical-shaped particles with a rough surface, representing amorphous modifications, which is in accordance with DSC and XRD results [48].

Amorphous solid dispersion of berberine was reported for improved bioavailability and hypoglycemic effect compared to neat berberine. Crystallinity analysis was carried out using SEM, DSC, and XRD. Pure berberine showed agglomerated bundles of crystalline fine needles with smooth surfaces. The physical mixture of drug and excipients (sodium caprate and PEG6000) also showed the existence of crystalline aggregates. In contrast, the solid dispersion showed the absence of crystalline agglomerates, which are in accordance with the XRD and DSC outcomes [49].

In another study, the solid dispersion of kaempferol was reported for improved oral bioavailability. The crystallinity was analyzed using DSC, XRD, and SEM analyses. DSC and XRD of the solid dispersion revealed its amorphous nature. The SEM photomicrographs of kaempferol represented irregular crystalline structures. Poloxamer 407, as an excipient, represented spherical, smooth-surfaced particles with its amorphous nature. The physical mixture of the drug and Poloxamer 407 showed the presence of crystalline structures in dispersed form around the Poloxamer 407. The solid dispersion demonstrated a wrinkled and irregular surface with a complete absence of drug crystals, representing amorphous modifications [50].

Solid dispersion of poorly water-soluble piperine was prepared with polyvinylpyrrolidone (PVP) and Eudragit to enhance dissolution rate and antioxidant activity. The crystallinity was evaluated using field emission scanning electron microscopy (FESEM). Pure piperine showed a random crystal shape with varying lengths and widths. The prepared solid dispersions showed disordered amorphous structures [51].

In a recent study, rutin-based solid dispersion (SD) was prepared using Poloxamer 407 and Poloxamer 188 to overcome poor aqueous solubility and permeability during oral administration. The developed SD and its excipients were characterized for their crystallinity and morphology using SEM analysis. Pure rutin showed a fragmentary-shaped crystal structure with irregular morphology. The poloxamer 407 and 188 particles showed spherical and round-shaped structures. The physical mixture of rutin and excipients showed the most scattering shapes of the rutin crystals, indicating that all the rutin still existed in a crystalline form. The rutin-loaded SD showed a small portion of irregular-shaped rutin crystals on the smooth surfaces of SD matrices due to amorphous modifications. On increasing the concentration of rutin in the SD from 20% to 40%, this morphology was found to be more pronounced [12].

In a recent study, HR-SEM analysis was carried out to study the crystallinity and morphology of the solid dispersion of *Piper longum* standardized extract (Figure 6.15). The extract and excipient showed irregular surface morphology. The physical mixture showed agglomerated particles of extract and excipient. The morphology of individual formulation components was not observed in the solid dispersion due to the complete molecular interaction between the extract and excipient, owing to the creation of new surface morphology during solvent evaporation. When the solution of extract and solid dispersion was dried separately on a glass slide and observed after 14 days, PLFEE showed crystalline blocks representing natural crystal growth. In contrast, the solid dispersion dried on the glass slide appeared as a smooth surface without any crystallinity, representing the retention of amorphous property and inhibition of nucleation and crystal growth during storage [13].

6.10 RAMAN SPECTROSCOPY

Raman spectroscopy is a non-destructive analytical tool used to investigate molecular vibrations and offer detailed information about the bonding, chemical structure, and crystallinity of substances. It involves the scattering of monochromatic light by molecules, resulting in the shifting of the light frequency. The shifting provides the information on the vibrational modes of the molecules. It is an important tool for crystallinity analysis of herbal drugs [52, 53].

6.10.1 Principle

Raman spectroscopy is based on the principle of inelastic scattering of light (Raman effect). When a specimen is exposed to a monochromatic light (from a laser), most of the photons are scattered elastically (known as Rayleigh scattering), but a small fraction of the scattered photons undergoes an energy shift. The shifting is ascribed to the interaction among photons and the vibrational modes of the molecules in the substance. The Raman shift (Δv) offers information about the vibrational modes of the sample. The quantity of energy lost or gained by the photons depends on the vibrational state of the involved molecules [52, 53].

6.10.2 Advantages

Raman spectroscopy has the following advantages for the analysis of herbal drugs [52, 53].

- **Non-destructive:** It is a non-destructive technique.
- **Easy sample preparation:** The preparation of sample doses does not require complex steps.
- **Wide applicability:** It applies to solid, liquid, and semisolid samples; hence, it is useful for powdered, semisolid, and liquid plant extracts. It can be used for the analysis of inorganic compounds.
- **High sensitivity:** It is highly sensitive to molecular structures and provides detailed information on chemical compositions.
- **Low water sensitivity:** It allows the analysis of aqueous or damp materials. As water molecules do not show strong Raman scattering, they can be used for the analysis of wet samples.
- **Ability to distinguish among polymorphs:** It has the capacity to differentiate among various polymorphs.
- **Highly efficient and quick:** It provides results within a few minutes, which is useful for high-throughput screening. It provides sharper and better-resolved peaks.

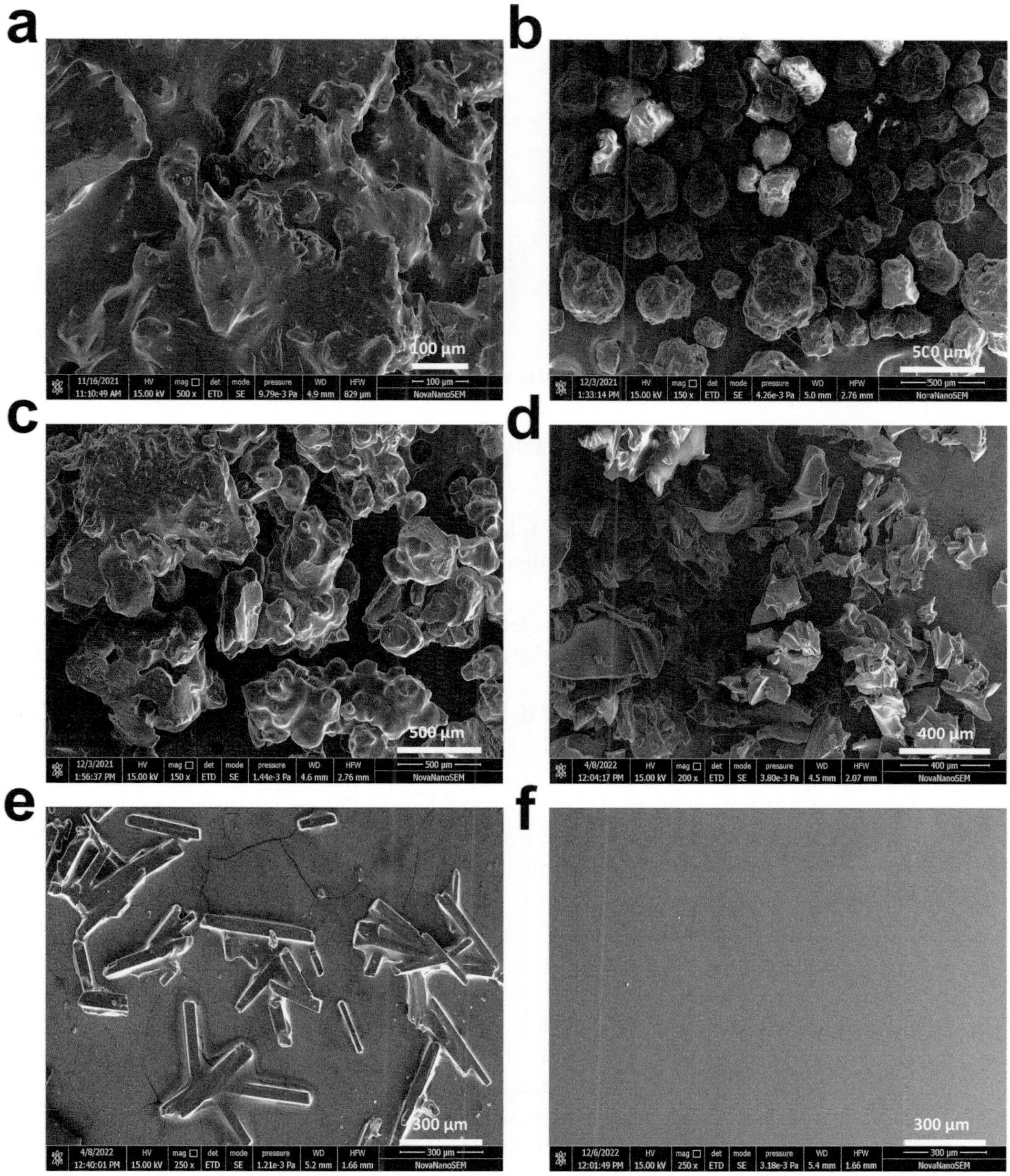

FIGURE 6.15 HRSEM results of extract, excipient, and solid dispersion. (a) PLFEE, (b) Soluplus®, (c) PM, (d) SD, (e) PLFEE dried on a slide, and (f) SD dried on a slide

PLFEE: *Piper longum* fruits ethanolic extract, PM: physical mixture of extract and polymer, SD: solid dispersion.

6.10.3 Instrumentation

A schematic representation of the Raman spectroscopy instrument is shown in Figure 6.16 [53]. The major components of Raman spectroscopy include the following:

- **Laser source:** It provides the monochromatic light required for excitation of the sample.
- **Sample holder:** This is used to hold the samples under investigation.

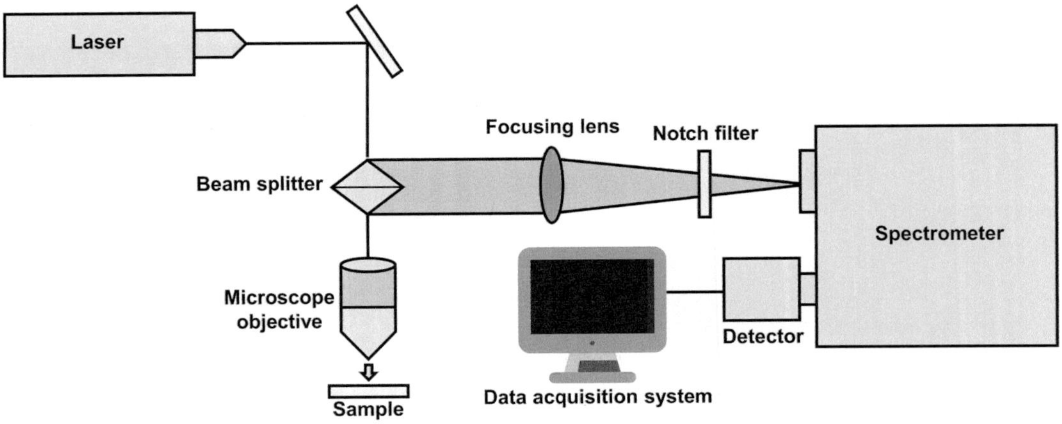

FIGURE 6.16 Schematic representation of Raman spectrometer

- **Optical system/light collection optics:** It helps focus the laser light onto the sample and collect the scattered light. The scattered light from the sample is then focused on the wavelength selector.
- **Wavelength selector**: The wavelength selector filters the reflected and scattered light so that only the Raman light is allowed to the detector. They limit the radiation absorbed by a sample to a certain wavelength range.
- **Spectrometer:** It disperses the scattered light to generate a Raman spectrum.
- **Detector:** After going through the wavelength selection, the light enters the detector. Then the collimating lens and the transmission grating, respectively, collimate and disperse optical information into a spectrum. After a focusing lens has linearly focused the spectrum onto a linear image sensor, each pixel of light is transformed into an electrical signal. Following the conversion of the electrical signal by an analogue-to-digital converter, the data is stored.

6.10.4 Procedure

- A very small quantity of powder sample is placed on a microscope glass slide or sample holder. The powdered sample is compressed into a pellet to acquire an improved signal. In the case of a liquid sample, the sample is placed in a cuvette and subjected to the instrument.
- The sample is exposed to a monochromatic laser that causes molecular vibration of the sample and leads to scattering of light.
- Then the scattered light is collected by a lens and passed through a filter to remove Rayleigh/elastic-scattered light. The remaining Raman scattering or inelastically scattered light is directed into a spectrometer.
- Then the spectrometer disperses the light into its constituent wavelengths, producing a Raman spectrum.
- The spectrum shows the Raman shift on the x-axis (cm^{-1}) and the intensity of the scattered light on the y-axis.
- The peaks of the Raman spectrum are selective to a particular vibration specific to the chemical structure.
- The spectra are compared with the reported data to identify the molecular structure of the compound, and the degree of crystallinity is measured by analyzing the intensity and position of representative peaks related to the amorphous and crystalline phases.
- The intensity and shift in the peaks are also specific to certain polymorphic forms, thus distinguishing various polymorphic forms.

- Comparison among the Raman spectra of various batches helps to understand batch-to-batch consistency.

6.10.5 Challenges

The analysis of herbal drugs and formulations by Raman spectroscopy faces the following challenges [53]:

- **Fluorescence interference:** Many herbal drugs contain pigments and other fluorescent compounds that can fluoresce under excitation by a laser. The produced fluorescence frequently suppresses the weak Raman signal, making it difficult for spectral interpretation.
- **Complex composition:** Herbal drugs are a complex mixture of phytoconstituents that produce overlapping spectra. Such spectra create difficulty in assigning the specific peaks to individual constituents.
- **Weak Raman spectra:** Some phytoconstituents of herbal drugs possess weak Raman scattering cross-sections, leading to low intensity. For this, higher laser power is required, which may degrade the sample.
- **Thermal sensitivity and degradation:** Sometimes, the focused laser beams can cause local heating, leading to charring or degradation of plant samples.
- **Surface roughness:** The irregular surface textures scatter light and reduce the signal-to-noise ratio, hence making it difficult to obtain a consistent spectrum.
- **Requirement of a highly intense laser source:** A laser source of high intensity is required to produce Raman scattering of optimum intensity with a high signal-to-noise ratio.

6.10.6 Application to Herbal Drugs

Based on the vibrations due to the crystalline and amorphous forms, Raman spectra can identify and quantify the crystallinity. Bulk-scale Raman measurements are regularly used for real-time examination of the polymorphic form of API and excipients, assessing the content uniformity and chemical alterations or particulate interactions in pharmaceutical dosage forms [54].

It can distinguish among the amorphous, crystalline, and polymorphic forms of herbal drugs and ensure the batch-to-batch consistency. Raman spectroscopy offers advantages for polymorph screening due to high sensitivity to crystalline geometry and requires little or no sample preparation, thus preserving crystals in their original form [52]. The spectral fingerprint is specific to a particular compound, which can be used for the identification of phytoconstituents [52]. It is also used to authenticate herbal drugs and to detect adulterations [52]. It is useful for the geographical origin identification of raw herbal drugs [52]. The spectral data of the drug-excipient mixture can be used to understand the drug-excipient compatibility.

Raman spectroscopy has been used for the chemical profiling of ginsenoside Rg3, a major component of ginseng. The results revealed that two isomers (20-(R)-Rg3 and 20-(S)-Rg3) of ginsenoside Rg3 exhibit clear differences in peak intensity for Raman bands. Based on such differential chemical profiles, the authors were able to distinguish between the isomers. These outcomes pave the way to establish clear relations among physicochemical properties and crystalline forms [55].

6.11 SOLID-STATE NUCLEAR MAGNETIC RESONANCE

Solid-state nuclear magnetic resonance (ssNMR) is an important analytical tool that provides detailed information on the atomic-level structure of molecules, including identification of amorphous and crystalline phases. It is a highly useful technique for the crystallinity study of herbal drugs, providing detailed insights into the molecular structure, amorphousness, and polymorphism of phytoconstituents. It ensures the consistency and stability of herbal formulations [56–62].

6.11.1 Principle

ssNMR is based on the principle of interaction among atomic nuclei (mainly carbon and hydrogen) and the provided external magnetic field. The sample is subjected to the instrument in a solid form, where the nuclei are exposed to the high magnetic field. By the influence of radiofrequency pulses, these nuclei resonate at characteristic frequencies, and the resulting signals offer information on the local environment of the nuclei, which can reveal structural details, such as molecular conformations, crystallinity, and polymorphs [57, 62, 63].

6.11.2 Advantages

- **Non-destructive:** ssNMR is a non-destructive technique, and the sample can be reused.
- **Detailed information on structure:** It provides detailed insights into the local atomic environment in a solid sample, enabling detailed structural analysis and crystallinity study of herbal drugs.
- **No need for sample preparation:** Compared to other analytical techniques, ssNMR does not need sample preparation, making it suitable for easy and high-throughput analysis.
- **Ability to differentiate crystalline and amorphous forms:** ssNMR can discriminate between amorphous and crystalline phases of herbal drugs. It is also useful to identify various polymorphs in herbal drugs.
- **Quantitative analysis:** It quantifies the proportions of amorphous and crystalline phases in a sample, offering an accurate measurement of crystallinity.
- **Comprehensive data:** ssNMR offers comprehensive spectral data, including insights into local molecular environment, molecular dynamics, and interactions within the sample, which are important to understand the behavior of herbal drugs in both solid and solution states.

6.11.3 Instrumentation

A schematic representation of the ssNMR instrument is shown in Figure 6.17. The major parts of ssNMR include the following [64]

- **Equipment for sample preparation:** The rotor packing tools are used for uniform and efficient loading of crystalline and powdered specimens. Sometimes a centrifuge is used for making pellets of biological samples.
- **Magnet:** It provides a stable and strong magnetic field (300–1200 MHz) to align nuclear spins.
- **Probe:** It includes a magic angle spinning (MAS) probe and a static probe. MAS spins the sample at the magic angle to average out anisotropic interactions, whereas the static probe is used when MAS is not feasible.
- **MAS spinners and rotors:** The rotors are a hollow cylinder that hold solid specimens, and the spinner spins the rotor with the help of nitrogen or pressurized air.
- **Radiofrequency (RF) transmitter and receiver:** The RF coils provide pulses and detect NMR signals. High-power RF amplifiers are required to excite a broad range of frequencies.
- **Tuning and matching circuits:** It optimizes signal quality for specific nuclei (e.g., ^{15}N, ^{13}C, ^{31}P).
- **Decoupling system:** It applies continuous and pulsed RF fields to remove dipolar couplings between abundant and rare spins (e.g., ^{1}H decoupling during ^{13}C acquisition).
- **Data acquisition and processing system:** Software controls pulse sequences, data acquisition, processing, and analysis, whereas the digitizer converts the analogue NMR signals to digital form.

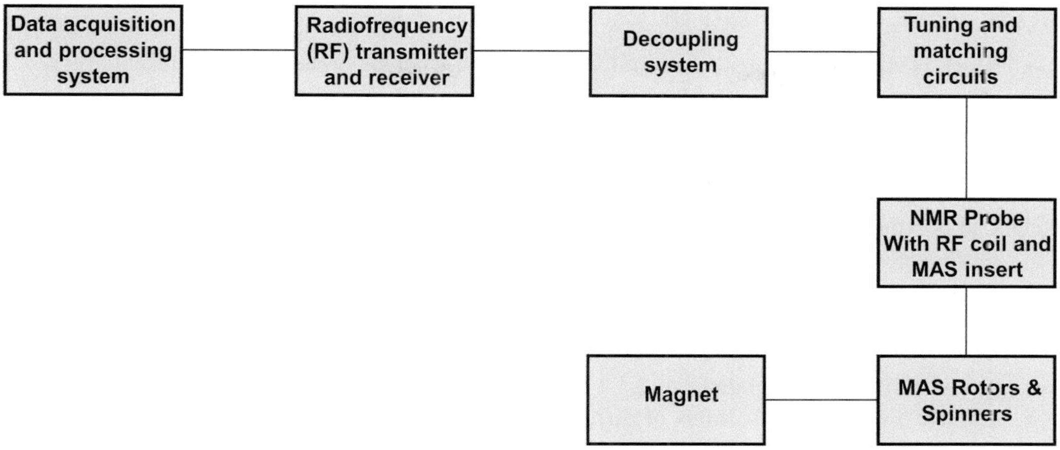

FIGURE 6.17 Schematic representation of ssNMR

6.11.4 Procedure

- A herbal drug is mainly prepared by triturating or grinding to produce a fine powder that ensures homogeneity of the sample.
- Then the sample is packed into the NMR rotor, and the NMR probe is placed in a strong magnetic field (~7 to 15 Tesla for ssNMR).
- Then the sample is irradiated with RF pulses that excite the nuclei of interest in the specimen.
- Cross-polarization is used to enhance signal intensity for certain nuclei, whereas MAS is used to average out anisotropic interactions to improve spectral resolution and reduce line broadening.
- NMR acquires data based on the resonance frequencies of the nuclei in the sample.
- Then, the raw NMR data is processed to remove noise and baseline drift and improve the signal-to-noise ratio.
- Peak fitting and analysis are performed to identify key resonances and determine the crystallinity of the sample by analyzing the peak sharpness and line width.

6.11.5 Challenges

- **Complexity of sample:** Due to the multicomponent system, complex spectra occur for herbal drugs, making data interpretation difficult.
- **Heterogeneity:** Herbal drugs often contain both crystalline and amorphous compounds, complicating the data interpretation.
- **Presence of moisture content:** The presence of moisture in the herbal drugs affects MAS efficiency and the quality of the signal.
- **Non-uniform rotor packing:** The presence of irregular particles in the herbal drugs leads to non-uniform rotor packing, which results in poor spinning and artifacts.
- **Lack of reference data:** Due to the limited availability of reference ssNMR spectral libraries for phytochemicals and the presence of novel phytomolecules, it is difficult to interpret the data.

6.11.6 Application to Herbal Drugs

- Sharp NMR peaks represent crystalline materials, whereas broad peaks represent amorphous materials. The loss of long-range molecular order in amorphous materials induces detectable differences in the chemical environment of some or all the nuclei with respect to the corresponding crystalline forms [56, 59, 62].
- The chemical shifts and coupling patterns are used to identify polymorphs of a compound. The polymorphism of a drug can arise from differences in the conformational properties of the molecules or in their packing modes in the crystal lattice. In both cases, the local environment nuclei will be different in different polymorphs, which are recognized by ssNMR [56, 65].
- Using the relative intensity of peaks for amorphous and crystalline compounds, the degree of crystallinity is calculated.
- ssNMR distinguishes solvates of different solvents and different solvates with the same chemical compositions, solvated, and desolvated forms. Compared to the nonsolvated forms, the presence of the solvent clearly causes an alteration to the local environment of the nuclei, by direct or indirect interaction, leading to modifications of the properties of the crystal lattice [56].
- By investigating the position and line width of the ssNMR peaks, the purity and crystallinity of the phytoconstituents can be evaluated.
- The NMR spectra are compared with the reference compound to identify the specimen.
- The ssNMR is also useful for investigating the interaction among herbal drugs and excipients of a formulation for drug-excipient compatibility studies.

Cellulose is an important pharmaceutical excipient. The crystallinity index is an important parameter used to recognize cellulose as a pharmaceutical and biomedical material. Since cellulose is an insoluble polymer, ssNMR is used for structural analysis. Solid-state 13C CPMAS NMR spectroscopy and XRD are mostly used to investigate the crystallinity of cellulose [65].

^{1}H, ^{13}C, and ^{17}O ssNMR was used to study two uncommon polymorphs of curcumin. Curcumin possesses three polymorphs (forms I, II, and III) in the solid state. A complete set of solid-state ^{1}H, ^{13}C, and ^{17}O NMR data was reported for form III of curcumin. The stability of form III was also investigated by the ssNMR. All the polymorphs differ in their spectral pattern, representing the suitability of the technique for the identification of polymorphs [66].

Piperine is one of the bioactive phytoconstituents that is widely used in the pharmaceutical industry as a bioavailability enhancer of drugs. In a recent study, ssNMR was used to study the polymorphic form of piperine. The method was found to be suitable for testing the quality of dietary supplements containing piperine as an absorption enhancer [67].

6.12 HOT STAGE MICROSCOPY

Hot stage microscopy combines optical microscopy with the ability to regulate the temperature of the specimen. It permits real-time investigation of the morphology, physical properties, phase transitions, melting, and crystallinity of herbal drugs and specimens under controlled temperature conditions. It provides valuable information on the polymorphism and stability of phytoconstituents [68–70].

6.12.1 Principle

It involves the integration of an optical microscope with a heating/cooling stage to investigate the specimen under controlled temperature. The specimen is kept on a microscopic slide, and the temperature is varied. The alterations in the physical appearance of the specimen, such as melting, morphological changes, or crystallization, are observed in real time using the hot stage microscope.

6.12.2 Advantages

- **Real-time investigation:** Hot stage microscopes allow real-time investigation of physical alterations, phase transitions, melting, crystallization, and other thermal properties of the sample.
- **Detailed morphological study:** It provides detailed morphology of the crystalline structures and alteration during heating or cooling, which are important for investigating the crystallinity and polymorphism of herbal drugs.
- **Cost-effective:** Hot stage microscopy is of relatively low cost compared to other high-end techniques for crystallinity analysis, such as XRD and DSC.
- **Easy-to-use and sample preparation:** It is relatively easy to investigate and does not require tedious sample preparation.

6.12.3 Instrumentation

A schematic representation of a hot stage microscope is shown in Figure 6.18 [71]. The major components include the following:

- **Heating and cooling stage:** The heating and cooling stages provide a controlled temperature (–190°C to 400°C) to the sample under investigation. The Peltier system and liquid nitrogen are used for cooling.
- **Sample holder:** Usually, a microscopic slide is used for sample holding, and a cover slip is used to prevent the sample from evaporating.
- **Temperature controller:** These are used to precisely control the temperature of the stage with the use of electronic devices and programmed software.
- **Recording and analysis system:** Image analysis tools and capture software are used for recording and analyzing the phase changes.
- **Additional accessories:** Sometimes, humidity controllers and gas purging systems (N_2, CO_2) are used to study hygroscopic samples and to avoid oxidative degradation at very high temperatures.

6.12.4 Procedure

- The sample is powdered and placed on a microscope slide with a cover slip.
- Then the slide is positioned on the hot stage of the microscope, and the temperature is controlled precisely at a controlled rate.

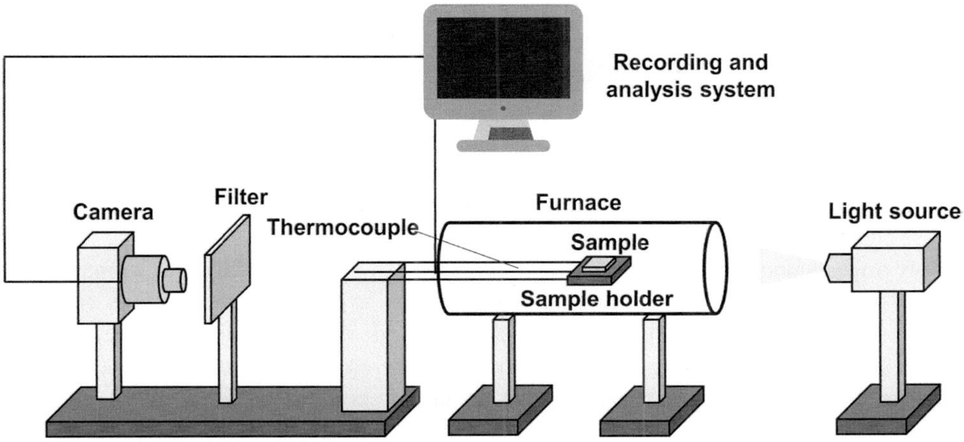

FIGURE 6.18 Schematic representation of a hot stage microscope

- Then the specimen is visualized under the optical microscope for changes in morphology (e.g., crystallization, melting).
- The images and videos are captured at various stages of heating or cooling to examine any changes in the physical state of the sample.
- The temperature at which changes occur is correlated with crystallinity and stability information.

6.12.5 Challenges

- **Complexity:** The overlapping thermal events in complex herbal drugs make it difficult to identify transitions related only to crystalline phases. Some of the phytoconstituents may degrade over a range of temperatures. The amorphous and semi-crystalline phytocomponents may mask the crystalline transitions.
- **Low crystallinity:** As many bioactive phytoconstituents are present in very low concentrations, it is difficult to observe thermal events due to the dilution effects.
- **Visual challenges:** The investigation of phase transitions can be subjective and vary from person to person.
- **Decomposition:** Sometimes, thermolabile herbal drugs may thermally decompose before melting, which creates difficulty in investigating.

6.12.6 Application to Herbal Drugs

Crystalline and amorphous forms can be easily identified using a hot stage microscope equipped with a polarizing filter [68]. Hot stage microscopy allows the visual observation of the sample during thermal studies. During an experiment, polymorphs may often be observed converting to other forms before melting; hence, it may be used to distinguish between polymorphs that have a small difference in melting points. It also checks the stability of a polymorph at a particular temperature [68]. It has been frequently used for cocrystal screening and characterization to interpret ambiguous data obtained from TGA and DSC [68]. Hot stage microscopy is used for particle size distribution and characterization of bioactives in the final formulation [68]. It can be used in addition to DSC to detect possible drug-excipient compatibility. The visual capabilities offer an advantage as the interactions can be visually observed, which was limited in DSC studies [68]. Hot stage microscopy is useful for investigating the crystallinity of herbal drugs. It helps to identify the crystal habits (size, shape) and monitor polymorphism in herbal drugs. Some herbal drugs may undergo phase transitions from a crystalline to an amorphous phase. The temperature at which these transitions occur can provide the stability and crystallinity information of phytoconstituents in herbal drugs. By using hot stage microscopy, different polymorphic forms of a phytoconstituent can be observed as they transform under varying temperature conditions. It also allows real-time observation of alterations in crystallinity, which can be correlated with the stability of herbal drug formulations. If a herbal drug crystallizes or melts at a different temperature than expected, this could indicate instability or degradation.

The thermal properties of *Kaempferia parviflora* extract and its solid dispersions were studied by hot stage microscopy by observing the changes in morphology. The solid dispersion was prepared to improve the solubility and dissolution rate of *Kaempferia parviflora* extract using polyethylene glycol-polyvinyl alcohol graft copolymer and hydroxypropyl methylcellulose. Compared to neat extract, the solid dispersions showed altered thermal events at different temperatures, representing modifications in crystallinity [72]. In another study, the solvate formation of bis(demethoxy)curcumin was studied by hot stage microscopy. Hot stage microscopy allowed the visual observation of crystal structures under microscopy and their thermal properties [73].

6.13 CONCLUSION

The crystallinity of herbal drugs significantly influences the solubility, dissolution, bioavailability, therapeutic efficacy, and stability. Analytical techniques, such as XRD, PLM, DSC, and SAED, provide complementary insights into the crystalline attributes of herbal drugs. XRD and DSC are especially useful for bulk analysis of crystallinity, whereas PLM and SAED are suitable for studying the morphology and atomic structure of crystalline substances. By employing a combination of the aforementioned analytical techniques, the crystallinity, quality, and consistency of herbal drugs and herbal formulations can be investigated. Such tools help overcome the challenges of polymorphism, crystallization, and stability, which are essential for developing high-quality herbal formulations.

REFERENCES

1. Sheokand, S., S.R. Modi, and A.K. Bansal, Dynamic vapor sorption as a tool for characterization and quantification of amorphous content in predominantly crystalline materials. *Journal of Pharmaceutical Sciences*, 2014. 103(11): p. 3364–3376.
2. Sghaier, H., Ionizing Radiation – Insights into Chemistry, Biology, and Applications. *BoD – Books on Demand*, 2025. p. 57.
3. Bunaciu, A.A., E.G. UdriŞTioiu, and H.Y. Aboul-Enein, X-ray diffraction: Instrumentation and applications. *Critical Reviews in Analytical Chemistry*, 2015. 45(4): p. 289–299.
4. Aswathy, K., et al., Formulation and in-vitro characterization of fast-disintegrating herbal extract sublingual immunotherapy tablet for peanut-induced allergic asthma. Saudi *Journal of Biological Sciences*, 2022. 29(3): p. 1283–1297.
5. Borde, S., et al., Formulation and characterization of ternary amorphous solid dispersions of a highly potent anti-tubercular agent and curcumin. *Journal of Drug Delivery Science and Technology*, 2021. 64: p. 102564.
6. Paranthaman, R., J. Moses, and C. Anandharamakrishnan, Powder X-ray diffraction conditions for screening curcumin in turmeric powder. *Journal of Food Measurement and Characterization*, 2022: p. 1–9.
7. Manoharan, S., P. Balakrishnan, and L.K. Sellappan, Fabrication of highly flexible biopolymeric chitosan/agarose based bioscaffold with Matricaria recutita herbal extract for antimicrobial wound dressing applications. *International Journal of Biological Macromolecules*, 2024. 281: p. 136195.
8. Gomes, A.T.A., et al., Tucumã (Astrocaryum vulgare) fat: An Amazonian material as a pharmaceutical input for lipid nanoparticle production. *Journal of Thermal Analysis and Calorimetry*, 2022. 147(1): p. 355–365.
9. Dudhat, K., et al., Application of amorphous solid dispersion technology for improving the physicochemical properties, saturation solubility, and in vitro dissolution of withania somnifera methanolic root powder extract. *Journal of Pharmaceutical Innovation*, 2023. 18(3): p. 1338–1349.
10. Zhu, D., et al., Mechanochemical preparation of triptolide-loaded self-micelle solid dispersion with enhanced oral bioavailability and improved anti-tumor activity. *Drug Delivery*, 2022. 29(1): p. 1398–1408.
11. Chaudhari, V.S., U.S. Murty, and S. Banerjee, Nanostructured lipid carriers as a strategy for encapsulation of active plant constituents: Formulation and in vitro physicochemical characterizations. *Chemistry and Physics of Lipids*, 2021. 235: p. 105037.
12. Yusuf, H., et al., A lyophilized surfactant-based rutin formulation with improved physical characteristics and dissolution for oral delivery. *Saudi Pharmaceutical Journal*, 2023. 31(6): p. 1077–1083.
13. Mohapatra, D., et al., Quality by design – based development and optimization of fourth-generation ternary solid dispersion of standardized Piper longum extract for melanoma therapy. *Drug Delivery and Translational Research*, 2023. 13(12): p. 3094–3131.
14. Haines, P., M. Reading, and F. Wilburn, Differential thermal analysis and differential scanning calorimetry, in *Handbook of Thermal Analysis and Calorimetry*. 1998, Elsevier. p. 279–361.
15. Bair, H.E., et al., Thermomechanical analysis (TMA) and thermodilatometry (TD), in *Thermal Analysis of Polymers: Fundamentals and Applications*. 2009, John Wiley & Sons, Inc. p. 319–385.
16. Gill, P., T.T. Moghadam, and B. Ranjbar, Differential scanning calorimetry techniques: Applications in biology and nanoscience. *Journal of Biomolecular Techniques: JBT*, 2010. 21(4): p. 167.

17. Singadi, R., et al., Application of differential scanning calorimetry to study the interpretation on herbal medicinal drugs: A review. *TMR Pharmacology Research*, 2022. 2(2): p. 6.
18. Leyva-Porras, C., et al., Application of differential scanning calorimetry (DSC) and modulated differential scanning calorimetry (MDSC) in food and drug industries. *Polymers*, 2019. 12(1): p. 5.
19. Chiu, M.H. and E.J. Prenner, Differential scanning calorimetry: An invaluable tool for a detailed thermodynamic characterization of macromolecules and their interactions. *Journal of Pharmacy and Bioallied Sciences*, 2011. 3(1): p. 39–59.
20. Pirsaheb, M., et al., Thermal analysis techniques in herbal medicine: A comprehensive review on unveiling integrity and quality for future perspectives. *Journal of Analytical and Applied Pyrolysis*, 2023. 175: p. 106192.
21. Spink, C.H., Differential scanning calorimetry. *Methods in Cell Biology*, 2008. 84: p. 115–141.
22. Carlton, R.A., *Pharmaceutical Microscopy*. 2011, Springer Science & Business Media.
23. Di Maggio, R.M. and F.A. da Silva Salvador, Optical microscopy applied to forensics, in *Mineralogical Analysis Applied to Forensics: A Guidance on Mineralogical Techniques and Their Application to the Forensic Field*. 2022, Springer. p. 1–36.
24. Alexander, K.S., A.T. Riga, and P.J. Haines, Thermoanalytical instrumentation and applications, in *Ewing's Analytical Instrumentation Handbook*, Fourth Edition. 2019, CRC Press. p. 433–490.
25. Zainal, N.F.A., et al., Thermal analysis: Basic concept of differential scanning calorimetry and thermogravimetry for beginners. *Chemistry Teacher International*, 2021. 3(2): p. 59–75.
26. Sampaio, T.P.D., et al., Biological evaluation and compatibility study of oral mucoadhesive hydrogel formulations using Schinopsis brasiliensis Engler. extract and excipients. *Journal of Thermal Analysis and Calorimetry*, 2023. 148(1): p. 141–158.
27. Mahmud, H., et al., Extraction and physicochemical characterization of Adansonia digitata L. Mucilage. *Nigerian Journal of Pharmaceutical Research*, 2023. 19(1): p. 23–35.
28. Wang, M., et al., A co-crystal berberine-ibuprofen improves obesity by inhibiting the protein kinases TBK1 and IKKε. *Communications Biology*, 2022. 5(1): p. 807.
29. Ke, J., et al., Preparation, quality evaluation and preliminary pharmacokinetic-pharmacodynamic studies of synephrine dry powder inhaler. *Drug Delivery*, 2025. 32(1): p. 2486346.
30. Calvet, E. and H. Prat, *Recent Progress in Microcalorimetry*. 2016, Elsevier.
31. Russel, M., et al., Different technique of microcalorimetry and their applications to environmental sciences: A review. *Journal of American Science*, 2009. 5(194208): p. 12.
32. Gaisford, S., Isothermal microcalorimetry for quantifying amorphous content in processed pharmaceuticals. *Advanced Drug Delivery Reviews*, 2012. 64(5): p. 431–439.
33. Briggner, L.-E., et al., The use of isothermal microcalorimetry in the study of changes in crystallinity induced during the processing of powders. *International Journal of Pharmaceutics*, 1994. 105(2): p. 125–135.
34. Nayfeh, M.J.F., A.o.N.S.i. Plasmonics, and fullerines, Chapter 7 – Characterization and simulation technologies of nanomaterial, in *Micro and Nano Technologies*. 2008, Elsevier. p. 153–167.
35. Inkson, B.J., Scanning electron microscopy (SEM) and transmission electron microscopy (TEM) for materials characterization, in *Materials Characterization Using Nondestructive Evaluation (NDE) Methods*. 2016, Elsevier. p. 17–43.
36. Mohapatra, D., et al., Carbon dots from an immunomodulatory plant for cancer cell imaging, free radical scavenging and metal sensing applications. *Nanomedicine*, 2021. 16(23): p. 2039–2059.
37. Mohapatra, D., et al., Tinospora cordifolia leaves derived carbon dots for cancer cell bioimaging, free radical scavenging, and Fe 3+ sensing applications. *Journal of Fluorescence*, 2022: p. 1–18.
38. Mohapatra, D., et al., In vitro cancer cell imaging, free radical scavenging, and Fe3+ sensing activity of green synthesized carbon dots from leaves of Piper longum. *Journal of Cluster Science*, 2023. 34(3): p. 1269–1290.
39. Mishra, K.N., et al., Bio-fabrication of silver nanoparticles using alysicarpus vaginalis extract: Preparation, characterization and comparative in vitro antibacterial evaluations. *ChemistrySelect*, 2023. 8(24): p. e202301113.
40. Chethan, H., et al., Formulation development and evaluation of hydrogel containing silver nanoparticles with withania coagulans aqueous extract. *Indian Journal of Pharmaceutical Sciences*, 2023. 85(4).
41. Dwivedi, A., et al., Microwave-assisted green synthesis of silver nanoparticles using Mitragyna parvifolia bark extract and their biological activities: An economical and environment-friendly approach. *Green Chemistry Letters and Reviews*, 2024. 17(1): p. 2395919.
42. Das, S., et al., Microwave-based rapid green synthesis of gold nanoparticles utilizing a notorious weed (Mikania micrantha) extract and evaluation of their biomedical properties. *Waste and Biomass Valorization*, 2024: p. 1–18.

43. Mohapatra, D., et al., Quality-by-design-based development of ultradeformable nanovesicular transgelosome of standardized Piper longum extract for melanoma. *Nanomedicine*, 2023. 18(14): p. 963–985.
44. Ni, C., Scanning electron microscopy (SEM), in *Encyclopedia of Tribology*. 2013, Springer. p. 2977–2982.
45. Zhou, W., et al., Fundamentals of scanning electron microscopy (SEM), in *Scanning Microscopy for Nanotechnology: Techniques and Applications*. 2006, Springer. p. 1–40.
46. Datye, A. and A. DeLaRiva, Scanning electron microscopy (SEM), in *Springer Handbook of Advanced Catalyst Characterization*. 2023, Springer. p. 359–380.
47. Abd Mutalib, M., et al., Scanning electron microscopy (SEM) and energy-dispersive X-ray (EDX) spectroscopy, in *Membrane Characterization*. 2017, Elsevier. p. 161–179.
48. Khan, A.W., Kotta, S., Ansari, S.H., Sharma, R.K., & Ali, J, Enhanced dissolution and bioavailability of grapefruit flavonoid Naringenin by solid dispersion utilizing fourth generation carrier. *Drug Development and Industrial Pharmacy*, 2014. 41(5): p. 772–779.
49. Zhaojie M, M.Z., Shengnan W, Xiaojia B, Hatch GM, Jingkai G, Li C, Amorphous solid dispersion of berberine with absorption enhancer demonstrates a remarkable hypoglycemic effect via improving its bioavailability. *International Journal of Pharmaceutics*, 2014. 467(1–2): p. 50–9.
50. Colombo, M., et al., Solid dispersion of kaempferol: Formulation development, characterization, and oral bioavailability assessment. *Aaps Pharmscitech*, 2019. 20: p. 1–9.
51. Dewi, R.T., et al., Enhancement of dissolution rate of piperine using solid dispersion approach and evaluation of its antioxidant activity. *Trends in Sciences*, 2025. 22(1): p. 8757–8757.
52. Chen, D.-D., et al., Raman spectroscopy in quality control of Chinese herbal medicine. *Journal of the Chinese Medical Association*, 2017. 80(5): p. 288–296.
53. Bandyoypadhyay, R., et al., Application of NIR and Raman spectroscopy for quality evaluation and gradation of medicinal plants: Possibilities and challenges. *Evidence-Based Validation of Herbal Medicine*, 2022: p. 703–721.
54. Khanal, D., W.-R. Ke, and H.-K. Chan, Raman spectroscopic evaluation of crystallinity, chemical composition and stability of pharmaceutical powder aerosols. *International Journal of Pharmaceutics*, 2022. 611: p. 121341.
55. Qu, X.-B., et al., Raman spectroscopy study on the structure of ginsenoside Rg3. *Europe PMC*, 2008. 28(3): p. 569–571.
56. Geppi, M., et al., Solid-state NMR studies of pharmaceutical systems. *Applied Spectroscopy Reviews*, 2008. 43(3): p. 202–302.
57. Bugay, D.E., Solid-state nuclear magnetic resonance spectroscopy: Theory and pharmaceutical applications. *Pharmaceutical Research*, 1993. 10(3): p. 317–327.
58. Reif, B., et al., Solid-state NMR spectroscopy. *Nature Reviews Methods Primers*, 2021. 1(1): p. 2.
59. Xu, Y., et al., The role of solid-state nuclear magnetic resonance in crystal engineering. *CrystEngComm*, 2016. 18(28): p. 5236–5252.
60. Brown, S.P., Applications of high-resolution 1H solid-state NMR. *Solid State Nuclear Magnetic Resonance*, 2012. 41: p. 1–27.
61. Mananga, E.S., et al., Advances in theory of solid-state nuclear magnetic resonance. *Journal of Nature and Science*, 2015. 1(6): p. e109.
62. Müller, K. and M. Geppi, *Solid State NMR: Principles, Methods, and Applications*. 2021, John Wiley & Sons.
63. Bryce, D.L., NMR crystallography: Structure and properties of materials from solid-state nuclear magnetic resonance observables. *International Union of Crystallography Journal*, 2017. 4(4): p. 350–359.
64. Rosay, M., M. Blank, and F. Engelke, Instrumentation for solid-state dynamic nuclear polarization with magic angle spinning NMR. *Journal of Magnetic Resonance*, 2016. 264: p. 88–98.
65. Paradowska, K. and I. Wawer, Solid-state NMR in the analysis of drugs and naturally occurring materials. *Journal of Pharmaceutical and Biomedical Analysis*, 2014. 93: p. 27–42.
66. Dai, Y., et al., Solid-State 1H, 13C, and 17O NMR characterization of the two uncommon polymorphs of curcumin. *Crystal Growth & Design*, 2020. 20(11): p. 7484–7491.
67. Siudem, P., Ł. Szeleszczuk, and K. Paradowska, Solid state NMR study of the bioenhancer piperine, supported by GIPAW DFT calculations. *Journal of Pharmaceutical and Biomedical Analysis*, 2024. 243: p. 116095.
68. Kumar, A., P. Singh, and A. Nanda, Hot stage microscopy and its applications in pharmaceutical characterization. *Applied Microscopy*, 2020. 50(1): p. 12.
69. Panna, W., P. Wyszomirski, and P. Kohut, Application of hot-stage microscopy to evaluating sample morphology changes on heating. *Journal of Thermal Analysis and Calorimetry*, 2016. 125(3): p. 1053–1059.

70. Vitez, I.M., et al., The evolution of hot-stage microscopy to aid solid-state characterizations of pharmaceutical solids. *Thermochimica Acta*, 1998. 324(1–2): p. 187–196.
71. Reynaert, C., E. Śnieżek, and J. Szczerba, Corrosion tests for refractory materials intended for the steel industry – A review. *Ceramics-Silikáty*, 2020. 64(3): p. 278–288.
72. Weerapol, Y., et al., Improved dissolution of Kaempferia parviflora extract for oral administration by preparing solid dispersion via solvent evaporation. *Asian Journal of Pharmaceutical Sciences*, 2017. 12(2): p. 124–133.
73. Yuan, L. and H. Lorenz, Solvate formation of bis (demethoxy) curcumin: Screening and characterization. *Crystals*, 2018. 8(11): p. 407.

7 Instrumental Methodology for Evaluation of Pharmaceutical Properties of Herbal Drugs and Formulations

ABBREVIATIONS

Abbreviation	Description
TGA	Thermogravimetric Analysis
LOD	Loss on Drying
SD	Solid Dispersion
ISO	International Organization for Standardization
He-Ne	Helium-Neon
GC-MS	Gas Chromatography-Mass Spectrometry
DoE	Design of Experiments
SLNs	Solid Lipid Nanoparticles
NLCs	Nanostructured Lipid Carriers
CUR	Curcumin
PMs	Physical Mixtures
EuD	Eudragit
o/w	Oil-in-Water
G′	Storage Modulus
G″	Loss Modulus
tan δ	Damping Factor
θ	Contact Angle
γLV	Liquid–Vapor Surface Tension
γSV	Solid–Vapor Surface Tension
γSL	Solid–Liquid Interfacial Tension

7.1 INTRODUCTION

Pharmaceutical possessions include the physical, chemical, and functional attributes of a drug substance or formulation that influence its quality, safety, efficacy, and overall performance. These properties – ranging from solubility and stability to rheology and particle size distribution – are particularly important in herbal formulations, where complex, multi-component compositions and variability in raw materials can pose challenges for standardization and quality control. Accurate evaluation of these possessions is essential for ensuring therapeutic effectiveness, patient safety, and compliance with regulatory requirements.

Instrumental methodologies provide a scientific and reproducible framework for assessing pharmaceutical properties, enabling precise characterization and monitoring of formulation parameters.

DOI: 10.1201/9781003540823-7

Modern analytical tools allow for the systematic evaluation of factors that directly impact product performance and stability. Techniques, as shown in Figure 7.1, such as contact angle measurement, assess surface wettability, influencing solubility and bioavailability; rheometers determine flow and deformation behavior critical to product consistency; moisture content analyzers monitor water levels that affect microbial stability and shelf life; particle size analysis optimizes dispersion and absorption; and zeta potential measurements evaluate electrostatic stability in colloidal systems.

FIGURE 7.1 Instruments for evaluation of pharmaceutical properties of herbal drugs and formulations

Additional techniques, including turbidity analysis, refractive index determination, and pH measurement, further ensure batch-to-batch uniformity, stability, and safety.

By integrating these instrumental approaches, formulators can obtain a comprehensive profile of pharmaceutical possessions, guiding the development, optimization, and quality assurance of both conventional and herbal pharmaceutical products. This chapter explores these methodologies in detail, highlighting their principles, applications, and significance in maintaining product reliability and therapeutic performance.

7.2 CONTACT ANGLE

The contact angle (θ) is a quantitative measure of how well a liquid wets a solid surface. It is defined as the angle formed at the point where the solid, liquid, and gas (or another immiscible liquid) phases meet. This angle reflects the balance of interfacial tensions between the three phases and indicates the degree of spreading or adhesion of the liquid on the solid surface [1]. Contact angle measurement serves as a valuable technique for assessing the spreading and adhesion properties of a formulation on a surface. Specifically, when a droplet of liquid is placed on a solid surface, the contact angle is the angle formed at the interface where the solid, liquid, and vapor phases converge [2]. A lower contact angle between the water droplet and the solid film indicates better wettability. Enhanced wettability, in turn, contributes to improved solubility and faster dissolution of the material.

This method is widely used to evaluate the wettability of various pharmaceutical products, including topical and transdermal formulations such as creams, gels, ointments, lotions, and patches. It is also applicable to advanced nanoformulations like nanoemulsions, liposomes, solid lipid nanoparticles (SLNs), nanostructured lipid carriers (NLCs), and polymeric nanoparticles, as well as conventional oral solid dosage forms such as tablets, capsules, and granules.

Types of contact angle measurement are as follows:

1. **Static contact angle:** Static (sessile drop) contact angles are measured when a droplet is stationary on a surface. They are commonly used to assess wettability, especially on smooth, homogeneous surfaces, and are useful for calculating surface free energy. This method relies on Young's equation, which assumes thermodynamic equilibrium of interfacial forces [1].
2. **Dynamic contact angle, hysteresis, and roll-off angle**: Dynamic contact angles include advancing (droplet spreads) and receding (droplet retracts) angles. The difference between them is contact angle hysteresis, caused by surface irregularities or chemical heterogeneity. A smaller hysteresis usually means a lower roll-off angle – the tilt angle at which a droplet begins to move [1].
3. **Roughness-corrected contact angles:** Surface roughness affects contact angle measurements. Roughness-corrected contact angles remove this effect and are essential for accurately determining surface free energy [1].

7.2.1 PRINCIPLE

The principle of contact angle is based on the balance of interfacial tensions between the solid, liquid, and gas phases at their point of contact. When a liquid droplet is placed on a solid surface, it assumes a shape determined by the equilibrium of three interfacial tensions: solid–gas interfacial tension, solid–liquid interfacial tension, and liquid–gas interfacial tension [3]. At equilibrium, the contact angle (θ) is described by Young's equation [3, 4]:

$$\gamma LV \times \cos(\theta) = \gamma SV - \gamma SL \tag{7.1}$$

where γLV is the surface tension of the liquid–vapor interface, γSV is the surface tension of the solid–vapor interface, γSL is the interfacial tension between the solid and the liquid, and θ is the contact angle [4].

Contact angle measurement reflects the degree of wetting, or the interaction between the liquid and the solid. A low contact angle (<90°) indicates good wetting and higher surface energy of the solid, while a high contact angle (>90°) suggests poor wetting and lower surface energy. Figure 7.2 depicts three droplets resting on a surface, each displaying a different level of interaction with that surface. The droplet on the far left exhibits a high contact angle, indicating minimal spreading across the solid, while the droplet on the far right shows a low contact angle, signifying extensive spreading. This behavior, referred to as "wetting," describes how a droplet either spreads out (wets) or retracts (de-wets) upon contact with a surface [2].

7.2.2 Advantages of Contact Angle Measurement for Herbal Formulations

- Analysis of contact angle helps to evaluate the herbal formulation behavior (wetting vs. beading), which affects absorption and adhesion.
- A good wetting often correlates with better surface contact, enhancing drug absorption, especially for poorly soluble herbal actives. It helps optimize dispersion in biological fluids.
- Contact angle measures the impact of surfactants, polymers, or excipients added to improve herbal formulation stability or performance, confirming the effectiveness of coating or surface functionalization [3].
- Contact angle helps assess properties like spreadability and stickiness, important in topical herbal formulations like creams, gels, and sprays [3].
- Changes in contact angle over time can signal instability, aggregation, or phase separation, especially in emulsions or nano-suspensions.

7.2.3 Instrumentation

A goniometer is an optical instrument used to measure the contact angle (θ) as observed in Figure 7.3. Its instrumentation comprises various combined components that work together.

- **Sample stage:** It includes a flat, adjustable sample stage designed to securely hold the solid sample. This stage is typically equipped with micrometer screws or motorized controls for precise movement along the X, Y, and Z axes, along with a levelling system, such as a bubble level or digital inclinometer, to ensure the surface is properly aligned and completely horizontal.

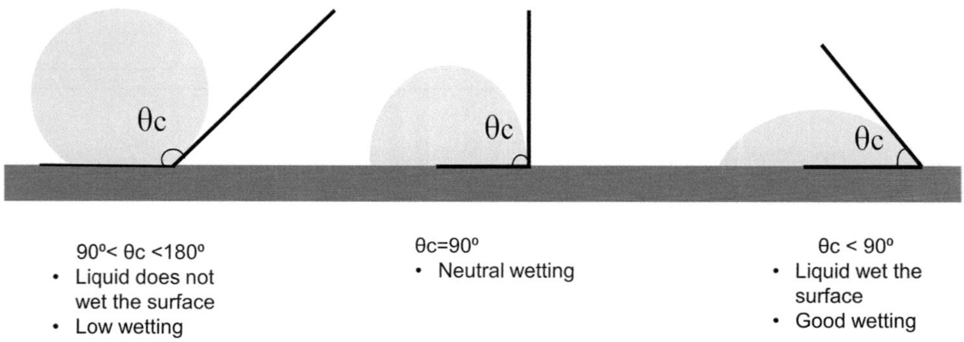

FIGURE 7.2 Contact angle variations representing different surface wettabilities

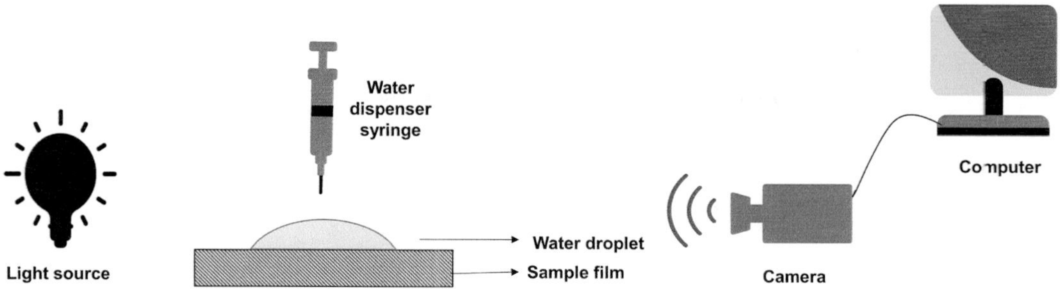

FIGURE 7.3 Schematic representation of contact angle measurement setup

- **Syringe:** The system includes a syringe unit, either manual or automated, mounted on an adjustable vertical arm to accurately dispense liquid droplets.
- **Light source:** Illumination is provided by an LED or halogen light source, often combined with a diffuser to produce even backlighting and enhance the contrast of the droplet's profile.
- **Optical devices:** The optical setup incorporates a high-resolution digital camera positioned at the same height as the droplet to capture clear, undistorted side-view images.
- **Recorder and integrator:** These images are processed using computer software that identifies the droplet's shape, the baseline, and the tangent at the contact point to calculate the contact angle, applying either geometric fitting methods or physical models like the Young-Laplace equation [5].

7.2.4 Procedure

7.2.4.1 Sample Preparation

Herbal formulations in solid or semi-solid form are first dissolved in a suitable solvent. The solution is then uniformly applied as a thin film over a defined area of the slide. Coated slides are placed in an oven and dried at a controlled temperature to allow complete solvent evaporation. This process ensures the formation of a smooth, uniform film suitable for contact angle analysis [2].

7.2.4.2 Method

Contact angle measurement involves the following steps [3, 6]:

- Microscope slides should be clean, smooth, and completely dry before use to ensure accurate contact angle measurement.
- The prepared and dried sample slide is placed on the stage of the contact angle goniometer or any other measurement device.
- Using a calibrated drop shape analyzer equipped with a micro-syringe, a small droplet (typically 2–5 µL) of distilled water or another test liquid is dispensed onto the surface of the dried film. The droplet should be placed gently to prevent splashing or excessive spreading, ensuring accurate and consistent contact angle measurement.
- The droplet shape is recorded using a high-resolution integrated camera.
- The software analyzes the droplet profile and automatically calculates the contact angle at the three-phase boundary (solid–liquid–gas interface). The measurement is determined based on the shape of the droplet and the point of contact between the droplet's edge and the sample surface.
- Experiments are conducted under controlled temperature and humidity conditions.

7.2.5 Challenges

- Herbal extracts tend to be a little sticky and thick, which makes forming a thin film surface difficult, making contact angle analysis in turn difficult.
- The dark-colored or opaque herbal formulations can make it problematic to detect the droplet edge clearly for the angle analysis.
- Since the herbal formulations comprise variability in surface composition, wetting behavior can differ across the sample, leading to inconsistent readings.

7.2.6 Application to Herbal Drugs

Contact angle measurement evaluates leaf surface wettability, essential for effective foliar application of agrochemicals on medicinal plants like *Ocimum sanctum*. It helps optimize the use of surfactants and wetting agents, ensuring better adhesion and absorption of treatments for disease and pest control [7].

It evaluates the interfacial wettability of andrographolide formulations, crucial for enhancing drug dissolution. In solid dispersions with Soluplus, reduced contact angles – especially at a 1:7 ratio – indicate improved wettability and support a carrier-controlled dissolution mechanism, aiding in formulation optimization for industrial development [8].

Contact angle analysis assesses the surface wettability of curcumin (CUR) and its solid dispersions (SDs). Pure CUR showed poor wettability, reflected by a high contact angle. While physical mixtures (PMs) with Eudragit (EuD) slightly reduced the contact angle, only solid dispersion processing significantly improved wettability by converting CUR to its amorphous form, thereby enhancing dissolution [9].

In a recent study, the contact angle was measured to investigate the wettability of the standardized extract and solid dispersion. The solid dispersion film showed a very low contact angle, nearly 4.18-fold less compared to the standardized *Piper longum* extract (Figure 7.4). The extract showed hydrophobicity, whereas the solid dispersion was found to be hydrophilic due to the amorphous modification [10].

7.3 RHEOLOGY

Rheology is the study of the deformation of solids and the flow of liquids [11]. A rheometer is an instrument that measures the material's mechanical response by evaluating important parameters such as shear stress (force per unit area), shear rate (rate of deformation), and deformation (change in shape or volume). These measurements help determine critical properties like viscosity, the material's resistance to flow, elasticity, and the ability to return to its original shape after deformation. It applies to materials with complex internal structures, such as polymers, emulsions, and various biological substances [12].

Newtonian law is represented as follows:

$$F = \eta G \tag{7.2}$$

where η is the viscosity, F is the shear stress, and G is the shear rate.

7.3.1 Principle

Rheology is governed by the interaction between shear stress (F) and shear rate (G), and includes important concepts such as yield stress, viscoelasticity, and flow behaviors like shear thinning and shear thickening. Newtonian fluids, such as water and alcohol, adhere to Newton's law and exhibit

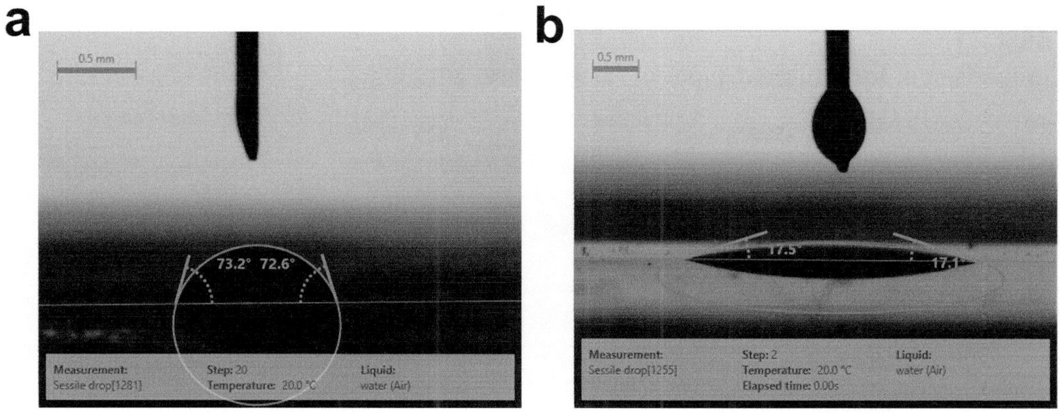

FIGURE 7.4 Contact angle measurements: (a) contact angle of a water droplet on standardized extract film, (b) contact angle of the water droplet on solid dispersion film

constant viscosity. However, non-Newtonian fluids exhibit viscosity that changes with applied stress and do not follow a linear flow relationship. Gels and other non-Newtonian fluids, which demonstrate complex rheological behavior, are frequently encountered in pharmaceutical formulation and manufacturing, as well as in the manipulation of various biological fluids within the body [13]. Many materials exhibit viscoelastic behavior, combining both viscous and elastic responses that vary with time, temperature, and applied force. This is characterized by parameters such as storage modulus (G'), loss modulus (G''), and damping factor (tan δ). Additionally, non-Newtonian fluids show variable viscosity depending on shear rate and time or require a minimum yield stress to begin flowing [12].

7.3.2 Advantages

- Rheology offers precise control of shear rate and stress.
- It is suitable for small sample volumes.
- It allows temperature-controlled measurements.
- It is possible assess both viscous and elastic components.
- It is suitable for characterizing materials like gels and creams, evaluating viscoelastic properties for product stability, and optimizing formulation consistency and spreadability.
- It offers good quality control and batch-to-batch comparison.
- It offers good predicting behavior under application or storage conditions.

7.3.3 Instrumentation

A comparison of various types of rheometers is shown in Table 7.1.

7.3.3.1 Rotational Rheometers

Rotational rheometers apply shear deformation to a material using rotating geometries, such as cones, plates, or cylinders. They measure the torque required to deform the sample and can operate in both steady and oscillatory modes. A schematic representation of rotational rheometer geometries and various types of flow models is shown in Figure 7.5.

TABLE 7.1
Comparison of the Different Types of Rheometers

Type	Deformation	Typical Use	Sample Size	Best For
Rotational	Shear	Pastes, gels, slurries, creams	Small to medium	Broad viscoelastic characterization
Capillary	Shear (flow)	Polymer melts, resins	Medium	High-shear processing conditions
Extensional	Stretching	Films, fibers, and adhesives	Small to medium	Extensional viscosity, strain hardening
Microfluidic	Shear (micro)	Biological and pharmaceutical fluids	Very small	Micro-volume, high-throughput analysis

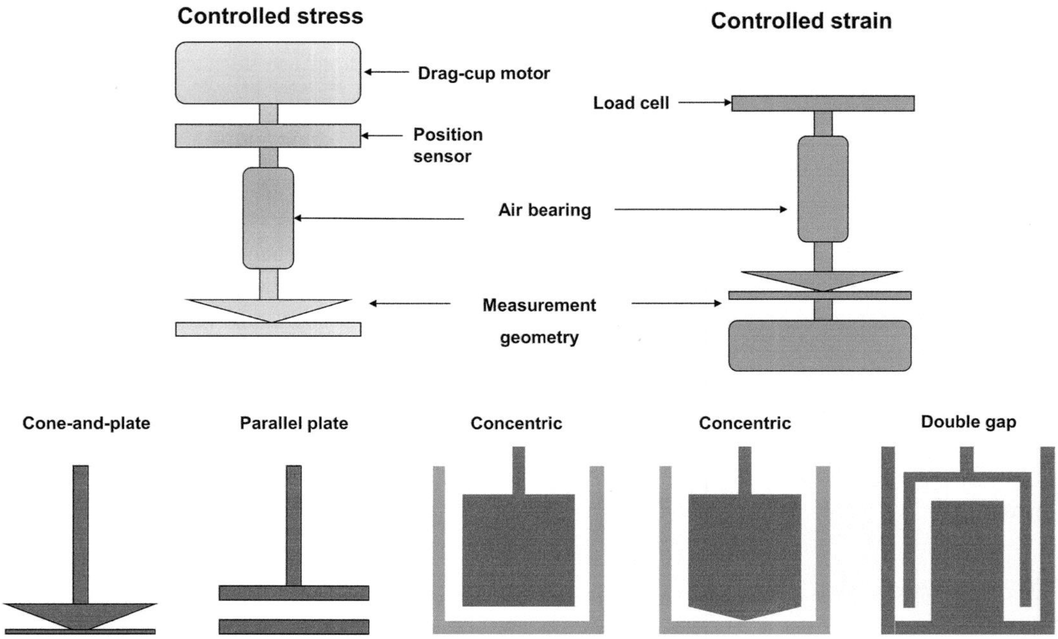

FIGURE 7.5 Schematic representation of rotational rheometer geometries and various types of flow models

Types of Rotational Rheometers

- **Cone-and-plate geometry**: A shallow cone rotates on a flat plate; provides a uniform shear rate; ideal for low-viscosity fluids with small sample volumes.
- **Parallel plate geometry**: Two flat plates, where one rotates or oscillates; suitable for soft solids, pastes, and structured fluids; allows control over gap size for thick or particulate samples.
- **Couette (cup-and-bob) geometry**: A rotating cylinder (bob) within a stationary cup; excellent for low-viscosity or suspending fluids; reduces edge effects and sample slippage.

7.3.3.2 Capillary Rheometers

Capillary rheometers are instruments used to measure the flow behavior of materials through a narrow tube under high pressure, with viscosity determined by the pressure drop and flow rate. They operate based on the Hagen-Poiseuille flow principle for Newtonian fluids, with extensions for non-Newtonian materials. These rheometers are particularly suited for evaluating shear viscosity at high shear rates, entrance pressure losses, and melt flow index, making them ideal for analyzing polymer melts, thermoplastics, elastomers, and other high-viscosity fluids. Commonly used in industrial process simulations, they offer the advantage of replicating real processing conditions, making them valuable for extrusion analysis and performance under extreme environments [12].

7.3.3.3 Extensional Rheometers

Extensional rheometers are designed to apply tensile (stretching) deformation rather than shear, assessing how materials behave when pulled apart. Common types include filament stretching rheometers, capillary break-up extensional rheometers (ex., CaBER), and the Sentmanat extensional fixture used for polymer films. These instruments measure extensional viscosity, strain hardening or softening, and the breakup dynamics of liquid threads. They are particularly useful in applications such as fiber spinning, film blowing, adhesives, inks, and biological fluids like saliva and mucus. Extensional properties are essential in processes where materials experience stretching rather than sliding, such as in printing or fiber production [12].

7.3.3.4 Microfluidic (Lab-on-Chip) Rheometers

Microfluidic rheometers are miniaturized devices that use micro-channels to evaluate the rheological properties of very small fluid samples, often incorporating optical sensors and real-time imaging for enhanced analysis. They measure shear viscosity at the microscale and generate flow curves in confined environments. These instruments are ideal for studying biological fluids such as blood and plasma, drug formulations, nano-dispersions, and for use in high-throughput screening. Their key advantages include requiring only microliter-scale sample volumes, offering high sensitivity with real-time feedback, and being well-suited for analyzing delicate or limited-quantity samples.

7.3.4 Procedure

- A small amount of the prepared sample is applied between the parallel plates of the rheometer, and the plate gap is adjusted, typically between 0.5 and 1 mm.
- Then the sample is allowed to reach thermal equilibrium at the desired temperature using the instrument's temperature control system.
- An appropriate rheological test method is selected, such as a flow curve, oscillatory test, or amplitude sweep.
- Then, the analysis is started. The upper plate will rotate or oscillate to apply shear to the sample, while the rheometer records the resulting torque and deformation.
- Rheological parameters to be recorded include viscosity, storage modulus (G'), loss modulus (G''), and yield stress, which are then calculated using the rheometer's software.

7.3.5 Challenges and Limitations

- Herbal formulations often contain plant extracts, oils, polysaccharides, and particulate matter, each contributing differently to the overall rheological profile. This heterogeneity can produce non-uniform measurements and complicate data interpretation.
- Temperature, humidity, and light can influence the viscosity and viscoelastic properties of herbal gels, emulsions, or suspensions, potentially affecting the reproducibility of results.
- Some herbal preparations may undergo phase separation, precipitation, or microbial growth during analysis, leading to inaccurate readings.
- Rheometers require careful calibration and may struggle with highly heterogeneous or fibrous samples, which can clog measuring geometries or produce inconsistent shear fields.
- Complex rheological curves from multi-phase herbal systems may require advanced modelling to extract meaningful parameters, limiting routine use in some quality control settings.

7.3.6 Applications to Herbal Drugs

Recently, a study was conducted to develop and evaluate oleogel and hydrogel formulations containing ethanolic and hydroalcoholic extracts of *Boswellia serrata*, *Ocimum basilicum*, *Sambucus nigra*, and *Galium verum* for burn wound healing. Second-degree burns were induced in Wistar rats, and formulations were applied for 21 days. Physicochemical properties (pH, viscosity, spreadability), wound pH modulation, and healing efficacy were assessed. Oleogels, particularly those with *B. serrata* and *O. basilicum*, showed superior healing, reduced inflammation, and improved epithelialization, correlated with lower pH (<6), higher viscosity, and thixotropic behavior [14].

In another study regarding transferosomal gel (TFG) of standardized *Piper longum* extract for melanoma therapy, rheology was used to investigate the formulations' consistency. All the gel formulations showed non-Newtonian behavior and exhibited a shear-thinning pseudoplastic type of flow, reflecting a decrease in viscosity with an increase in the shear rate. The rheogram of TFG was found to be superimposable with that of the reference marketed gel, representing optimal consistency for topical application (Figure 7.6) [15].

In a recent study, herbal oil-in-water (o/w) cream formulations for the treatment of acne vulgaris were evaluated by rheology. The formulations were prepared using natural fixed oils (ostrich and sesame oils) and essential oils from *Mentha piperita*, *Origanum vulgare*, and *Lavandula officinalis* as active antimicrobial agents. Rheological evaluation showed shear-thinning, thixotropic behavior, which was best modelled by the Ostwald-de Waele equation. Texture analysis showed that both aqueous and oil-soluble thickeners significantly impacted mechanical characteristics, including hardness, adhesiveness, and spreadability. The optimized formulation was further standardized using GC-MS for chemical integrity [16].

In another study, a natural anti-ageing emulsion serum *incorporating Acmella oleracea extract* and a glycolipid-based emulsifier was developed using a cold-processing method; the formulation

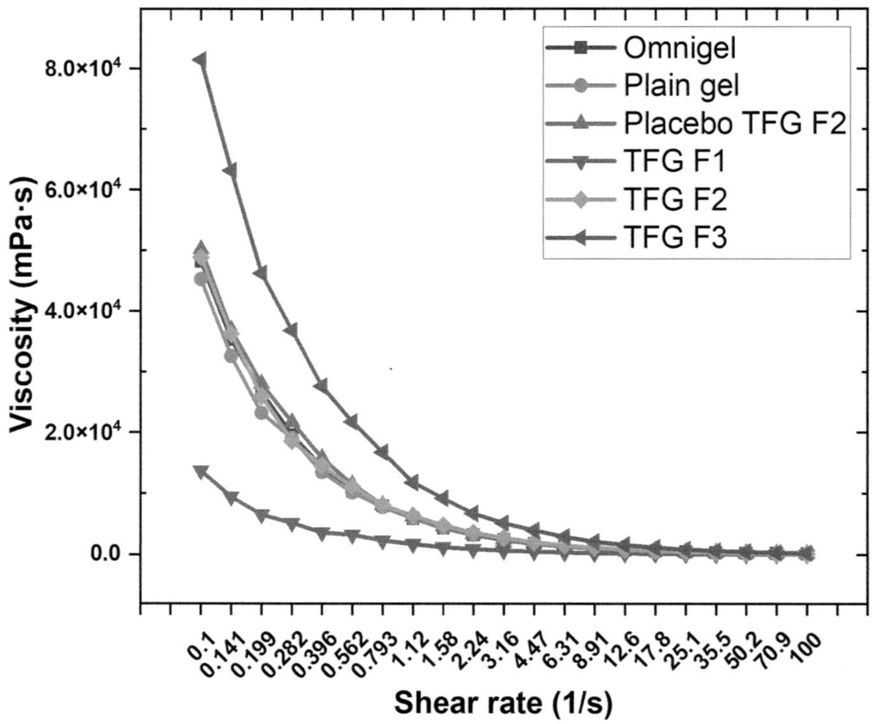

FIGURE 7.6 Rheogram of various transferosomal gel, plain gel, and marketed gel

exhibited stable shear-thinning and elastic rheological behavior, along with good physical stability. Using a design of experiments (DoE) approach, an optimized formulation containing 15% emollients and 1% emulsifier was identified. *In vivo* evaluations confirmed excellent skin compatibility and demonstrated significant wrinkle reduction within two weeks. These findings support the serum's potential as a safe, effective, and environmentally friendly anti-wrinkle cosmetic product [17].

Another study investigated the influence of different types and concentrations of oils on the mechanical and rheological properties of cosmetic creams. Using a hot homogenization method, both blank formulations and anti-age variants were developed. The latter is enriched with *Jasminum sambac* extract and Argireline. The study aimed to determine how oil composition impacts texture and application performance. Among the tested formulations, the cream containing 6% babassu oil and 2% *Camelina sativa* oil demonstrated the most favorable rheological and textural characteristics, supporting effective skin application and prolonged contact, which are critical for enhanced cosmetic performance [18].

7.4 MOISTURE CONTENT ANALYZER

A moisture content analyzer is an instrument used to measure and analyze moisture content in various samples. This instrument is a substitute for traditional hot air oven testing methods, employing halogen bulbs to effectively heat and dry samples. The instrument measures the weight loss over time, which corresponds to the amount of moisture evaporated. The result is calculated as a percentage of the original sample weight. This technique is important for assessing the stability, shelf life, and quality of herbal extracts, granules, tablets, and solid nano-formulations like solid dispersion and other dosage forms. Figure 7.7 illustrates different methods of analyzing moisture content.

Determining moisture content in herbal formulations is essential for maintaining product quality, stability, and performance. Moisture levels influence the flow properties, compressibility, and blending consistency of herbal powders and extracts, which are critical during mixing, granulation, and tableting. Excess moisture can cause microbial growth, chemical degradation, caking, and alterations in the crystalline structure of active constituents, while insufficient moisture may lead to poor binding and tablet friability. Accurate moisture analysis ensures product uniformity, prolongs shelf life, and preserves therapeutic effectiveness. Furthermore, monitoring moisture content aids in selecting suitable packaging to prevent clumping, phase separation, or potency loss during storage and distribution [19, 20].

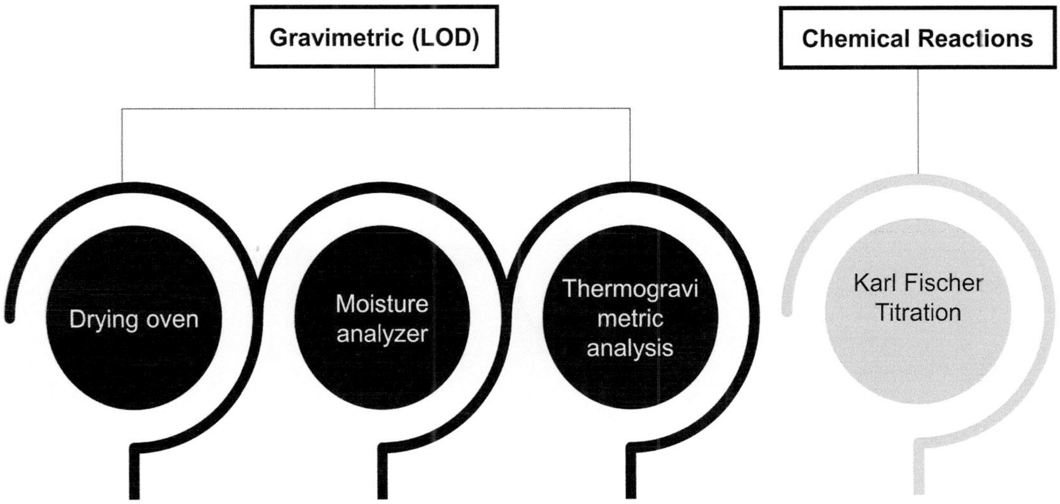

FIGURE 7.7 Different methods of analyzing moisture content

7.4.1 Principle

Moisture analyzers operate based on the principle of thermogravimetric analysis (TGA), where a sample is heated and its weight is continuously monitored to determine moisture loss. Most of the analyzers use the loss on drying (LOD) method [20], which involves heating the sample to evaporate moisture and measuring the resulting weight change [19]. A precision balance records the initial weight, and a heating element-commonly a halogen lamp raises the sample temperature. As the moisture evaporates, the instrument tracks the decrease in weight until a constant mass is achieved, indicating complete drying. The moisture content is then calculated from the difference between the initial and the final weights, providing accurate and reliable results [19].

7.4.2 Advantages of Moisture Analyzers

- Moisture analyzers provide faster results compared to a traditional oven, enhancing productivity and reducing wait times.
- They use advanced sensors and controlled conditions to deliver consistent, reliable, and repeatable measurements.
- The introduction of automatic drying and measurement processes eliminates manual handling errors and ensures consistent operation.
- Moisture analyzers operate in a closed system, protecting the sample from external contamination during analysis.
- They are designed to meet industry standards, providing traceable and accurate moisture data for quality assurance.
- Compact design requires less laboratory space compared to conventional oven drying setups.

7.4.3 Instrumentation

Moisture analyzers consist of a heating unit, a precision weighing balance, a temperature control system, and a moisture sensor (Figure 7.8). The moisture sensor can be based on various technologies, such as capacitance, halogen, or infrared, depending on the specific instrument. These components work together to ensure precise measurements and consistent performance [20].

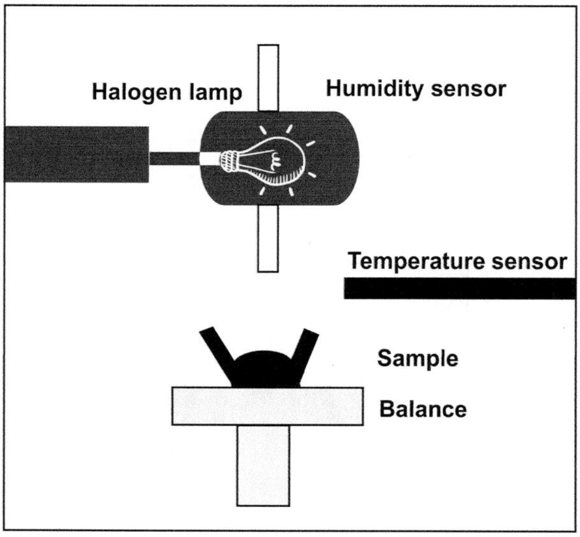

FIGURE 7.8 Instrumentation of the halogen moisture analyzer

There are two main types of moisture analyzers:

1. **Standard moisture analyzers:** These analyzers use a ceramic heating element to dry the sample. They are less expensive than halogen moisture analyzers, but they are also slower and less accurate.
2. **Halogen moisture analyzers:** These analyzers use a halogen heating lamp to dry the sample. They are more expensive than standard moisture analyzers, but they are also faster and more accurate [20].

7.4.4 Procedure

- The herbal formulations and extracts are prepared by thoroughly drying the samples, lyophilized if necessary, and finely powdering them to ensure uniformity.
- A clean, dry pan is used and placed on the analyzer's weighing platform and tared to zero before adding the sample. Then 1 to 5 grams of the dried sample is added and spread evenly in a thin layer to ensure consistent heating during analysis.
- Then, the analyzer lid is closed to minimize heat loss and appropriate drying parameters are set based on the sample type. The drying temperature, drying mode, and the shutoff criteria are set.
- The drying cycle is initiated and the initial weight is recorded; heat is applied via a halogen lamp and real-time weight loss is monitored until the sample reaches a constant weight.
- Once the endpoint is reached, the analyzer displays the moisture content (%) of the sample. The pan is allowed to cool before removal, and cleaned thoroughly before reuse.
- The heating chamber is cleaned to avoid contamination from previous samples. Overloading or using samples with high volatile solvent content is avoided, as they may distort results.
- Herbal samples may contain fibrous or resinous components, requiring optimized drying conditions to prevent degradation or loss of volatile oils. Using a temperature that is too high can cause degradation or loss of other sample components, leading to inaccurate moisture measurements due to changes in the sample's overall mass.
- Overloading the moisture analyzer or failing to clean it properly after each use can compromise the accuracy of results. Residues from previous samples or excessive sample quantity may interfere with the weighing mechanism.
- Moisture analyzers are high-precision instruments and require routine calibration to maintain measurement accuracy. Failure to calibrate the device regularly can result in unreliable or inconsistent data.

7.4.5 Challenges with Herbal Formulations

- The essential oils and residual solvents tend to evaporate faster, which may falsely increase moisture readings.
- Thermal degradation of sensitive compounds can mimic moisture loss during analysis.
- Variability in particle size and composition results in uneven or unreliable measurements.
- No universal method fits all herbal formulations, affecting reproducibility.
- Leftover solvents in extracts are misread as moisture, skewing results.
- Herbal samples can absorb ambient moisture, altering results during preparation.

7.4.6 Applications to Herbal Drugs

Moisture content analysis plays an important role in the quality control and development of herbal formulations. It ensures the quality of raw materials by detecting remaining moisture that could lead to microbial growth, mold, or degradation of active constituents. During the drying process of the herbal plants, moisture analysis helps determine the optimal endpoint, preventing over- or under-drying that can affect potency and stability.

In standardized extracts, controlling moisture content is essential for maintaining batch-to-batch consistency and the concentration of bioactive compounds. Moisture analysis also aids in shelf life prediction, as higher moisture levels can accelerate oxidation and microbial degradation. Additionally, it ensures compliance with pharmacopeial standards, which define acceptable moisture limits for various herbal materials. In formulation development, moisture control is particularly important for moisture-sensitive dosage forms, where excess moisture can compromise product stability and performance.

A recent study focused on the development of a solid dispersion (SD) formulation of *Piper longum* extract to enhance its bioavailability for melanoma treatment. The SD was prepared using the solvent evaporation method with appropriate polymers and surfactants to improve solubility and dissolution. Moisture content was assessed using a halogen moisture analyzer at 70°C. The measured values were consistent with those obtained from TGA, confirming the reliability of the results. The low moisture content observed is likely due to limited atmospheric absorption during sample handling. Maintaining low moisture levels is crucial for ensuring the physical stability of the SD, as excess moisture can plasticize the matrix and lower its glass transition temperature, potentially affecting performance [10].

A recent study was conducted to evaluate the impact of storage temperature on the moisture content of encapsulated *Orthosiphon stamineus* spray-dried powder over a four-week time period at 10°C and 25°C. Moisture content was measured using a halogen moisture analyzer. Results indicated that higher storage temperatures led to an increase in moisture content. However, the type of packaging – whether vacuum or non-vacuum – did not significantly influence the moisture content at either temperature. A slight increase in moisture content was observed across all samples over the storage period. To extend the shelf life of encapsulated herbal products, storage at lower temperatures (around 10°C) is recommended. Higher temperatures may affect the physical integrity of the capsules by altering their permeability, potentially compromising product stability [21].

Another study investigated the spray-drying performance and product quality of aqueous extracts from seven medicinal plants: *Rhamnus purshiana*, *Peumus boldus*, *Centella asiatica*, *Valeriana officinalis*, *Hypericum perforatum*, *Hamamelis virginiana*, and *Cynara scolymus*. These herbs were selected based on their therapeutic significance and diverse phytochemical profiles. The study focused on important quality parameters, including moisture content, hygroscopicity, and flow properties, to evaluate the suitability of spray drying as a method for producing stable, free-flowing herbal powders. The results provide insight into optimizing formulation and drying conditions for the development of high-quality herbal products [22].

7.5 PARTICLE SIZE ANALYSIS

Particle size analysis determines the size and distribution of particles within a formulation. This is essential for optimizing product development and ensuring the quality, consistency, and performance of final dosage forms and drug delivery systems. It is an important characterization technique used to evaluate the particle size distribution in various sample types, including solid materials, suspensions, emulsions, and aerosols [23].

7.5.1 Laser Diffraction for Particle Sizing

Laser diffraction is a particle-sizing technique that measures the entire sample collectively, unlike analyzing individual particles. It is widely used for analyzing particles ranging from 0.5 to 1000 microns and applies to dry powders, suspensions, emulsions, and aerosols. Its popularity is due to its speed, reproducibility, and broad measurement range [12].

7.5.2 Principle of Laser Diffraction

Laser diffraction operates on the principle that particles scatter light at angles inversely related to their size; when a laser beam passes through a particle sample, the particles scatter the light at various angles and intensities. the smaller particles scatter light at wider angles, while the larger particles scatter it at narrower angles. An array of detectors captures this scattering pattern [12]. To determine the particle size distribution, the scattering data is analyzed using mathematical models through a process called deconvolution. The two most commonly used models are the Mie theory and the Fraunhofer approximation – the latter being a simplified version of Mie theory. As specified in ISO 13320:2009, Mie theory is generally preferred, especially for wide size ranges and finer particles, due to its higher accuracy. While both models produce similar results for larger particles, Fraunhofer can result in unpredictable errors for small particles. It's important to note that both models assume spherical particles; therefore, in the case of irregularly shaped particles, the reported sizes are based on an equivalent spherical diameter [12].

7.5.3 Advantages of Laser Diffraction

- It can analyze a wide range of particle sizes, ranging from 0.5 μm to several mm.
- The method is fast and offers real-time analysis.
- High accuracy and reproducibility.
- Various sample types, like powders, suspensions, emulsions, and aerosols, can be analyzed.
- Minimal sample preparation is required.
- It is non-destructive.
- It provides full particle size distribution.

7.5.4 Instrumentation of the Laser Diffraction Setup

Instrumentation of laser diffraction for measuring particle size is shown in Figure 7.9. The major components of Laser diffraction include the following:

- **Light source**: A He-Ne or diode laser emits a monochromatic, coherent, and collimated beam. This beam is expanded and filtered to ensure uniform intensity, typically reaching a diameter of a few millimeters [24].
- **Sample dispersion unit**: Particles are dispersed in a liquid or air stream and passed through the laser beam at optimal concentration. The sample is placed in a measurement cell located in the focal region to avoid agglomeration and ensure accurate scattering [24].
- **Fourier lens:** A positive lens or lens system focuses the scattered light onto the detector, converting angular scattering into spatial data [25].
- **Detector:** A multi-element detector, arranged in rings or linear arrays, captures the intensity of scattered light at various angles, including the unscattered beam, across a wide angular range [25].
- **Computer and software interface:** The detector send data to a computer, where measurements are averaged to enhance precision. Software models like Mie or Fraunhofer theory and deconvolution algorithms are used to calculate the particle size distribution, even for complex or multimodal samples [25].

7.6 DYNAMIC LIGHT SCATTERING

The particle size reported is the diameter of a sphere that diffuses at the same rate as the measured particle. Several commercial dynamic light scattering (DLS) instruments are available for sizing

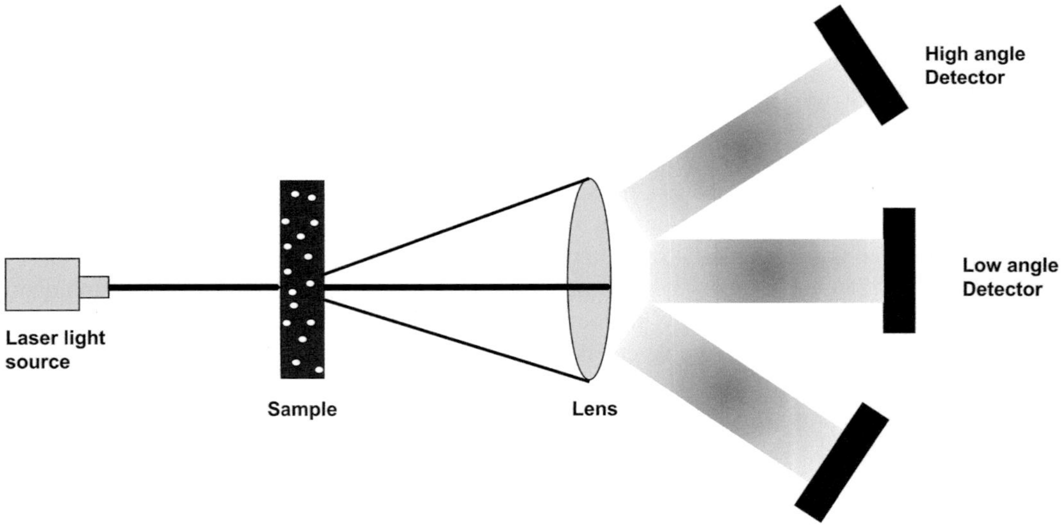

FIGURE 7.9 Instrumentation of laser diffraction for measuring particle size

particles and polymer molecules in the submicron range. Due to its reliance on diffusion, DLS is generally effective for particles up to approximately 3 micrometers in size [24].

7.6.1 Principle

DLS is based on the principle that particles suspended in a gas or a liquid undergo random movement, known as Brownian motion. These particles move randomly in all directions as a result of constant energy transfer from solvent molecule collisions. Because this transferred energy remains relatively consistent, smaller particles, having less mass, are more significantly affected and therefore move faster than larger ones. This motion results from continuous collisions with surrounding solvent molecules [26]. In DLS, nanoparticles scatter incident light with an intensity proportional to the sixth power of their radius. When particle size is smaller than one-tenth of the light's wavelength (λ/10), scattering is elastic and angle-independent, known as Rayleigh scattering. For particles larger than this threshold, scattering becomes angle dependent and inelastic, referred to as Mie scattering (Figure 7.10). In this case, scattered light is strongest in the forward direction. The $\lambda/10$ size threshold reflects how electromagnetic waves interact with particles, though detailed explanations lie beyond the scope of this chapter [26]. Depending on the sample's turbidity, either side scattering at 90° or back scattering at 175° may be more appropriate. Additionally, a forward angle of 15° is useful for monitoring particle aggregation. By measuring the rate of particle movement and factoring in variables such as temperature and viscosity, the hydrodynamic diameter of the particles can be accurately determined. The diffusion behavior of these particles is described by the Stokes-Einstein equation [27]:

$$D = \frac{kBT}{6\pi\eta r} \tag{7.3}$$

where,
D = diffusion coefficient, m²/s
kB = Boltzmann constant
T = absolute temperature, Kelvin
η = dynamic viscosity of the fluid, Pa s or kg/(m s)
r = radius of the spherical particle, meters

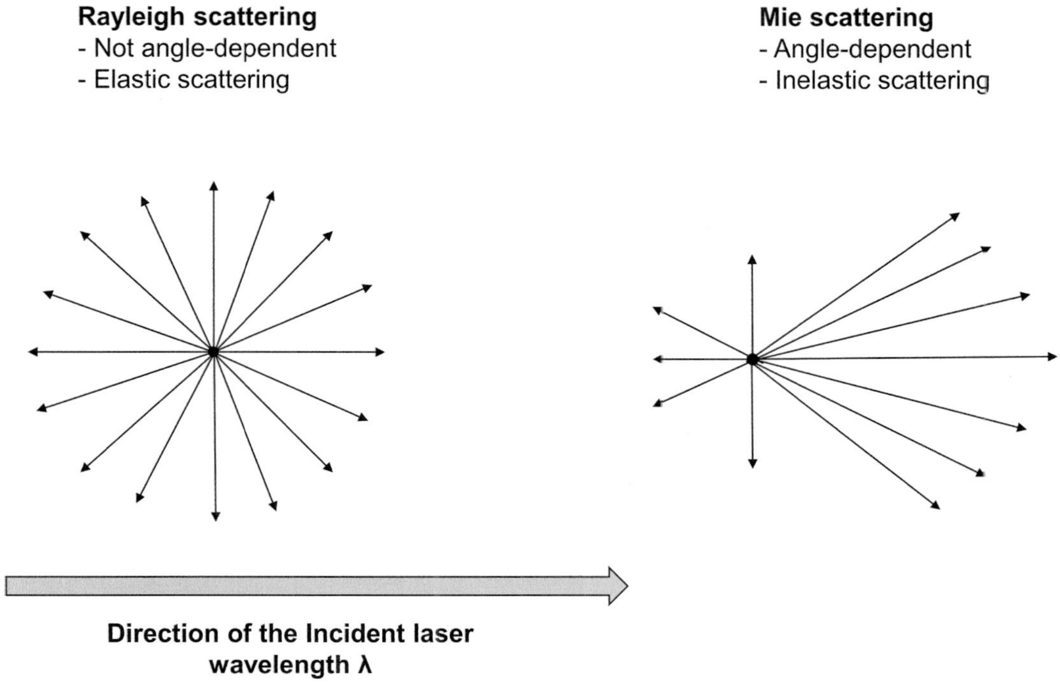

FIGURE 7.10 Particle size-dependent transition from Rayleigh to Mie scattering

7.6.2 Instrumentation

A schematic representation of the DLS instrument is shown in Figure 7.11.

- **Laser light source**: The components of a DLS device include a stable laser light source, commonly a helium-neon (He-Ne) laser, which is used to illuminate the sample.
- **Sample holder/cuvette**: The sample is kept in a cuvette, which may be made of glass or quartz, depending on the optical requirements.
- **Detector:** As the laser light interacts with the particles in the sample, it scatters, and the resulting fluctuations in scattered light intensity are detected by a sensitive detector such as a photomultiplier or an avalanche photodiode.
- **Correlator:** Fluctuations in intensity are analyzed by a digital signal processor known as a correlator, which calculates the diffusion coefficient and particle size. The scattering angle at which the light is detected plays a critical role in the accuracy of the analysis, with common detection angles being 90° degrees (right angle) and 173° (back angle) [26].

7.6.3 Procedure

- The formulation is diluted with deionized water or a suitable buffer to an appropriate concentration. The sample, if needed, is filtered using a 0.22 μm syringe filter to remove dust and large particles. If required, the sample is sonicated quickly to disperse any aggregates.
- Then, the sample is filled into a clean quartz or glass cuvette, avoiding air bubbles, and placed into the sample holder of the DLS instrument.
- The measurement parameters, including temperature (typically 25°C) and scattering angle (commonly 90° or 173°), are set.
- The measurement is initiated by illuminating the sample with a laser. The instrument records the intensity fluctuations of the scattered light.

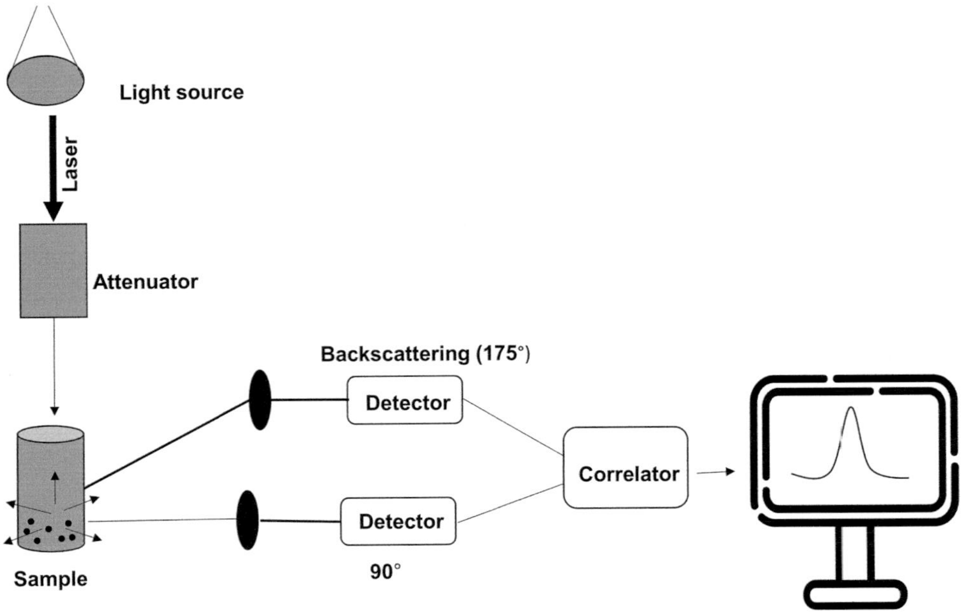

FIGURE 7.11 Dynamic light scattering instrumentation

- The data is processed by the correlator to calculate the diffusion coefficient and determine particle size using the Stokes-Einstein equation. The measurements are repeated to ensure accuracy and reproducibility. The sample concentration is optimized to prevent multiple or weak scattering.
- As DLS cannot resolve particles of similar sizes, polydisperse samples may show broad, overlapping peaks. Particle size should primarily be reported from the intensity distribution, and accurate refractive index and viscosity values for both dispersant and particles – especially for sizes above 100 nm are essential for reliable results [26].

7.6.4 Challenges

- Herbal powders or extracts often contain sticky, fibrous, or oily materials that resist uniform dispersion in the selected medium.
- Some plant-based formulations can form agglomerates, which are hard to disperse in the suitable solvent medium, even after vortexing or sonication, leading to inaccurate size measurements.
- Dense or poorly dispersed herbal particles tend to settle quickly, reducing homogeneity and affecting analysis.
- The presence of saponin compounds in herbal formulations can cause foaming during mixing or sonication, which introduces air bubbles and affects light-based measurements.
- Filtering herbal suspensions to remove large particles or undissolved residues can lead to loss of active components or fine particles critical for accurate analysis.
- Since there is variability in plant material phytoconstituents, it may lead to differences in particle behavior, dispersion, and appearance across different preparations.

7.6.5 Application to Herbal Drugs

In a study, the particle size analysis of herbal extracts, that is, *Hippophae rhamnoides* L. (antiviral), *Arnica montana* L. (anti-inflammatory), and *Malva sylvestris* L. (expectorant), using Mastersizer

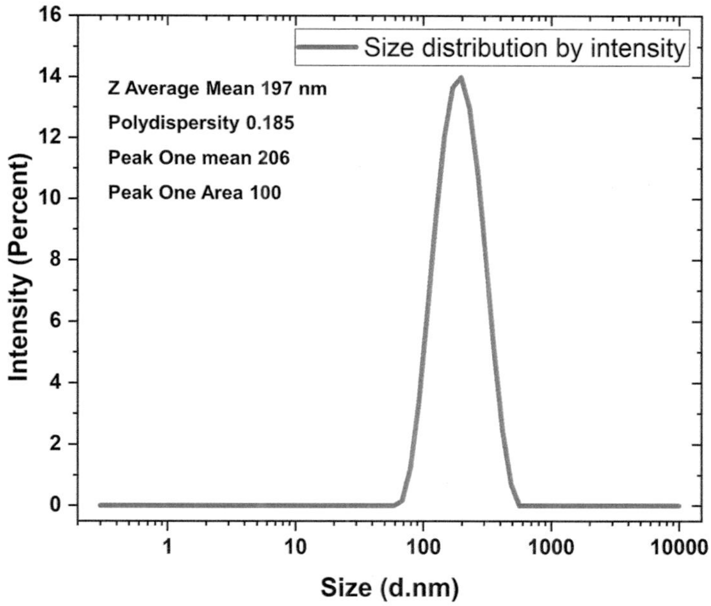

FIGURE 7.12 Vesicle size (Z_{avg}) and polydispersity index (PDI) of transferosome

and Zetasizer revealed extract-specific and irregular distributions. Upon dissolution, only the finer particles of *Arnica foliosa* remained, while *Malva sylvestris* showed a highly polydisperse profile. In the case of *H. rhamnoides*, incorporation into a plaster matrix simplified the system to a more uniform particle population, improving distribution – an important factor for consistent biological activity and dosage form performance [28].

Recently, a transferosome of *Piper longum* standardized extract was prepared and further converted to a transferosomal gel for melanoma therapy. The hydrodynamic vesicle size and polydispersity index (PDI) were evaluated by a particle size analyzer using DLS. The transferosomes were found to be within 200 nm with a PDI value of <0.3, representing monodispersity and narrow size distribution of the vesicles (Figure 7.12) [15].

In a recent study, the influence of particle size and extraction time on the aqueous recovery of antioxidant compounds from *Agrimonia eupatoria* (agrimony), *Salvia officinalis* (sage), and *Satureja hortensis* (savoury) was evaluated. Results indicated that smaller particle sizes significantly enhanced extraction efficiency, with an optimal size around 0.2 mm. While extraction time effects varied by plant species, a minimum of 5 minutes was sufficient for acceptable antioxidant recovery [29].

DLS was reported to characterize the particle size of green synthesized silver nanoparticles (AgNPs) using *Swertia paniculata* extract, which acted as both a reducing and a stabilizing agent. The resulting nanoparticles were found to be spherical with a particle size ranging from 31 to 44 nm. Synthesis conditions, such as temperature and reaction duration, played a significant role in determining particle characteristics. The AgNPs remained stable for up to 90 days and exhibited strong antimicrobial activity, particularly against *Pseudomonas aeruginosa* and *Klebsiella pneumoniae* [30].

7.7 ZETA POTENTIAL

Zeta potential, also known as electrokinetic potential, refers to the electrical charge that forms at the interface between a solid particle and the surrounding liquid. It represents the potential difference between the surface of the particle and the bulk of the fluid, caused by the accumulation of charged species at the particle – liquid boundary. This potential, typically ranging from +100 mV to −100 mV, is influenced by factors such as the particle's dielectric properties and the ionic strength of the surrounding medium. Zeta potential is measured at the slipping plane within the electrical double layer and indicates the degree of electrostatic repulsion or attraction between particles. A high absolute zeta potential (positive or negative) suggests strong repulsion, resulting in stable dispersions, while a low zeta potential often leads to particle aggregation or sedimentation. Understanding zeta potential is crucial for controlling stability in colloidal systems and plays a significant role in processes like coagulation, flocculation, and flotation [31].

7.7.1 Principle

Electrophoretic light scattering (ELS) is based on the principle of electrophoresis – the movement of charged particles in a fluid under an applied electric field. The zeta potential (ZP), or electrokinetic potential, refers to the electric potential at the slipping (shear) plane, which separates the tightly bound ionic layer around a particle from the surrounding dispersant. It reflects the potential difference between the electric double layer (EDL) of the particle and the dispersing medium at this boundary [26]. When an electric field is applied, particles move at a velocity known as electrophoretic mobility, which is measured using the Doppler shift in scattered laser light. The resulting frequency shift, caused by the Doppler effect, is used to calculate the electrophoretic mobility (μe), from which the zeta potential (ζ) is determined using the Henry equation [27].

$$\mu = \frac{2}{3} \times \left(\varepsilon \times \frac{\zeta}{\eta} \right) \times f(\kappa a) \quad (7.4)$$

where
μ = electrophoretic mobility
ε = dielectric constant of the medium
ζ = zeta potential
$f(\kappa a)$ = Henry's function (depends on particle size and ionic strength)
η = viscosity of the medium

ZP is a key indicator of colloidal stability, with values of ±0–10 mV indicating high instability, ±10–20 mV indicating relative stability, ±20–30 mV indicating moderate stability, and above ±30 mV indicating high stability. According to DLVO theory, stability results from the balance between van der Waals attraction and electrostatic repulsion from the EDL. Since ZP reflects only repulsion, it cannot fully predict behavior, and factors such as PEGylation or water-in-oil emulsions may enhance stability regardless of ZP values. ZP is also used to estimate nanoparticle surface charge, though it measures surface potential rather than charge density. While most natural surfaces are negatively charged and cationic surfaces are often synthetic, the notion of "neutral" nanoparticles is misleading, as all particles in dispersion acquire some degree of surface charge [26].

Accurate ZP measurement requires knowledge of fluid properties such as viscosity and dielectric constant, and consideration of sample turbidity to avoid multiple scattering interference. Unlike DLS, standard references are limited, with NIST recommending freshly prepared goethite [α-FeO (OH), ~+32.5 mV] as a reference. Samples should not be reused, and electrodes must be regularly inspected, as they may react with metallic ions like Fe^{3+}, potentially causing damage and affecting results.

7.7.2 Instrumentation

Zeta potential measurement systems consist of different components. A coherent light source, mostly used is a He-Ne or diode laser (633–830 nm), which directs a beam at the sample. The sample is loaded in a transparent electrophoretic cell, often made of quartz or polystyrene, which includes electrodes connected to a power supply to generate a uniform and adjustable electric field. As charged particles move under this field, they scatter the incident light. This scattered light is collected at fixed angles (commonly 17°, 90°, or 173°) and detected by a photodetector such as a photomultiplier tube or avalanche photodiode. The Doppler-shifted signal is processed by a correlator and signal processor using the Fourier Transform or heterodyne detection to determine electrophoretic mobility. Software then calculates zeta potential and provides data visualization. Accurate measurements also require temperature control and consistent pH and ionic strength.

7.7.3 Advantages of Electrophoretic Light Scattering

- Highly sensitive to small changes in particle surface charge.
- Applicable to nano- and sub-micron particles.
- Non-destructive and rapid.
- Minimal sample preparation.

7.7.4 Procedure

- The sample is diluted in a suitable solvent, usually water, to avoid multiple scattering and filtered to remove aggregates, and if necessary, sonicated for a few minutes to further reduce the size of the particles.
- To start with the analysis, the laser is allowed to warm up. Parameters, such as temperature, angle, and measurement duration, are set correctly.
- Then the sample is loaded in the cuvette or cell, and the measurement is started.
- The instrument shines a laser beam through the sample. The photodetector records the intensity fluctuations of the scattered light over time.
- Using the Stokes-Einstein equation, the particle's hydrodynamic diameter (d_h) is calculated.
- The result is then interpreted as Z-average diameter, polydispersity index (PDI), which indicates homogeneity of size distribution (by intensity, volume, or number). The zeta potential (ζ) is determined using the Henry equation.
- The measurement can be taken in repetition for accuracy, and validation can be done.
- Factors such as pH, ionic strength, and concentration affect ZP. So, these should be considered during analysis. ZP becomes more positive at acidic pH and more negative at basic pH. Higher ionic strength compresses the EDL, leading to a decrease in ZP magnitude. Ions with higher valency (e.g., Ca^{2+}, Al^{3+}) compress the EDL more than monovalent ions (e.g., Na^{+}), resulting in lower ZP values [26]. At low concentrations, surface adsorption dominates, and ZP generally increases. At higher concentrations, EDL compression becomes more significant, often reducing ZP and decreasing dispersion stability [26].

7.7.5 Challenges

- Due to the complex mixtures of components in herbal extract formulations, attaining a homogenous dispersion is difficult.
- Often, the herbal extracts are usually darker or turbid, which may lead to multiple scattering, resulting in inaccurate measurements.
- Natural compounds in the formulation may rapidly aggregate or sediment, compromising the stability of the dispersion and the reproducibility of the measurement results.
- Some herbal components may degrade or change properties with temperature, requiring careful control during measurement.

7.7.6 Application

In the characterization of the *Curcuma longa* L. nanosuspension, zeta potential analysis revealed a value of –33.3 mV for the formulation stabilized with sodium lauryl sulfate (SLS). This highly

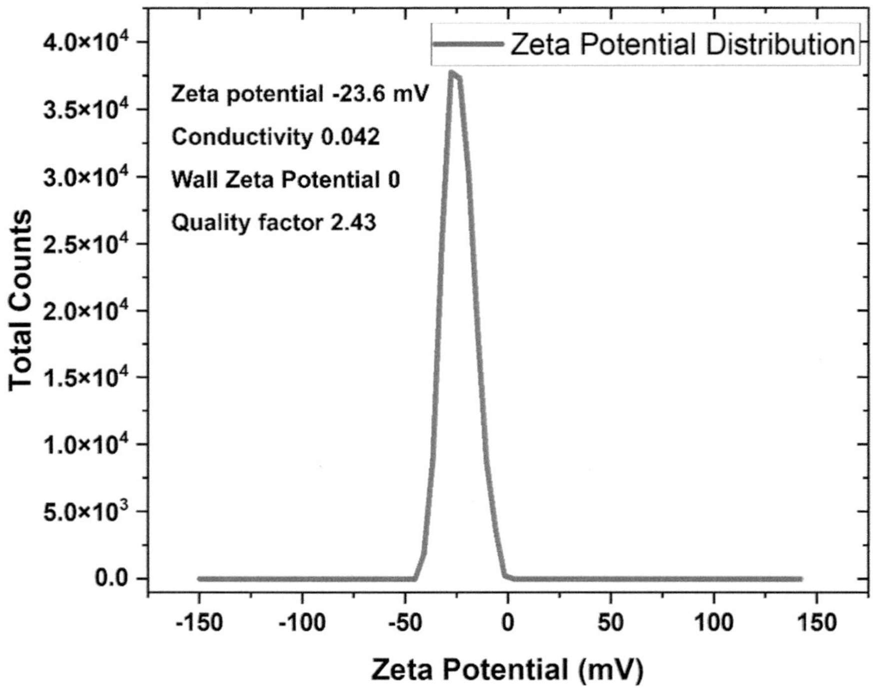

FIGURE 7.13 Zeta potential of standardized extract-loaded transferosome

negative zeta potential indicates strong electrostatic repulsion between particles, contributing to the overall stability of the nanosuspension [32].

Recently, a zeta potential measurement was carried out to study the colloidal stability of the transferosome of *Piper longum* standardized extract. The zeta potential of the optimized transferosome was found to be below ±30 mV, representing a moderately stable nature (Figure 7.13) [15].

Kim et al. (2019) evaluated AGN/RSV nanoparticles and highlighted the role of zeta potential, measured using laser Doppler electrophoresis, in assessing colloidal stability. The nanoparticles exhibited a consistently negative zeta potential, indicating strong electrostatic repulsion that prevented aggregation across media such as distilled water, PBS, and 50% FBS. This stability supports their potential for intravenous use and effective tumor targeting via the EPR effect [33].

Chittineedi et al. (2023) developed a polyherbal formulation (PHF) conjugated to gold nanoparticles (AuNPs) to enhance PHF bioavailability and reduce doxorubicin dosage in drug-resistant breast cancer stem cells. The zeta potential analysis revealed a high negative charge, indicating robust colloidal stability critical for effective drug delivery. Functionally, the AuNP-PHF induced ferroptosis by promoting ferritin degradation, and when combined with low-dose AuNP-Dox, showed enhanced cytotoxicity and inhibition of cell migration and proliferation via autophagy/necroptosis pathways [34].

7.8 TURBIDITY

Turbidity refers to the cloudiness or haziness of a liquid resulting from suspended particles that scatter or absorb light. Nano herbal formulations, measuring turbidity is crucial for assessing particle dispersion, uniformity, and colloidal stability. It serves as an indicator of potential aggregation, precipitation, or instability over time [35]. Similarly, in microemulsions – which are thermodynamically stable, optically isotropic systems composed of water, oil, and surfactant/co-surfactant mixtures – turbidity measurements offer valuable insights into the system's stability and overall characteristics.

7.8.1 Principle

Turbidity measurement is based on the interaction of light with suspended particles in a liquid, where the degree of light scattering or absorption is directly related to the concentration and size of the particles. This principle enables the quantification of turbidity using optical methods, primarily nephelometry, backscattering, and turbidimetry [36].

7.8.2 Instrumentation and Methods

A colorimeter or spectrophotometer can be used as a turbidimeter. The major components of a turbidimeter include the following (Figure 7.14):

- **Light source:** Provides incident radiation for measurement. Tungsten lamp (visible region) and mercury vapor lamp (UV region) are commonly used as light sources.
- **Filters and monochromators:** They ensure that only the desired wavelength reaches the sample. When monochromatic light is required, a filter or monochromator is used.
- **Sample cells/cuvette:** It holds the turbid sample, usually made of glass or quartz, depending on the UV and visible wavelength range. Various sample cells are used in turbidometry. They may be cylindrical and rectangular with a path length of 1 cm.
- **Photoresistor/photodetector:** It senses the intensity of light that has passed through the turbid sample. In turbidometry, it measures the decrease in transmitted light due to scattering by suspended particles. Photovoltaic cells or phototubes are used as detectors.
- **Amplifier:** It amplifies weak electrical signals from the detector and improves sensitivity and signal-to-noise ratio.
- **Readout system:** It converts amplified signal into a readable output and often displays turbidity in percentage transmittance or absorbance units.

There are different methods used for turbidity measurement, which include the following: turbidimetry, turbidity spectroscopy (TUS), nephelometry, backscattering, and wavelength exponent method.

- **Turbidimetry** measures the decrease in light intensity as it passes directly through the sample. It is suitable for samples with higher turbidity levels and offers a broad dynamic range. However, it can be influenced by variations in particle size, shape, and refractive index [36]. Turbidimetry is a quantitative technique that measures the transmission of light by a substance as a function of wavelength. It assesses the intensity of light across different wavelengths and provides a simple, cost-effective method for evaluating formulation

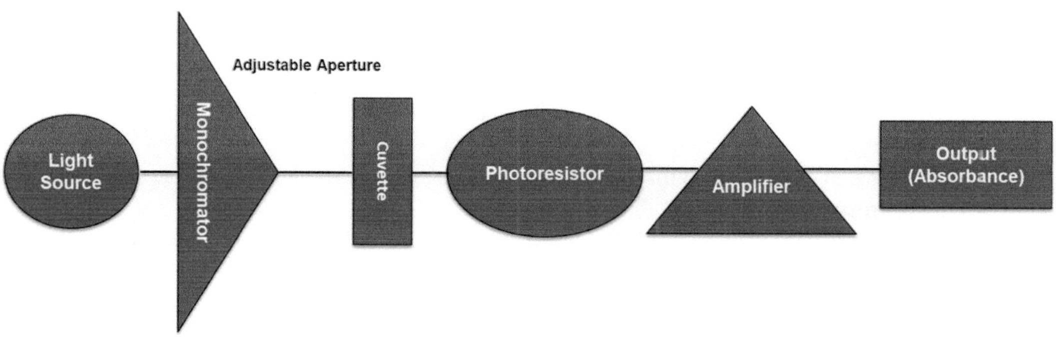

FIGURE 7.14 Schematic representation of turbidimeter instrumentation

stability. By linking turbidity to particle size distribution, turbidimetry serves as an indirect yet effective approach to analyzing colloidal systems [35].

- **Nephelometry** detects light scattered at a 90° angle to the incident beam. It is highly sensitive and effective for measuring low turbidity levels. The accuracy of this method depends on both the size and the number of suspended particles [36]. The intensity of scattered light is proportional to concentration.
- **Backscattering** involves the detection of light scattered in the reverse direction (180°). This technique is particularly useful for analyzing samples with high turbidity or those that are more concentrated, where forward or side scattering may be less effective [36].
- **Turbidity spectroscopy** measures the attenuation of transmitted light across a range of wavelengths. This method enables the determination of particle size and distribution within a dispersed system, providing detailed information about the formulation [36]. The intensity of transmitted light is inversely proportional to the concentration of the sample.
- **Wavelength exponent method** analyzes how turbidity changes with varying wavelengths of light. By examining these changes, it is possible to estimate particle size, making this method valuable for characterizing complex colloidal systems [36].

7.8.3 Procedure

The turbidimeter was first calibrated using standard turbidity solutions, such as formazin standards, to ensure measurement accuracy. Freshly prepared herbal micro- and nanoemulsions were then transferred into clean, dry turbidity cuvettes. The outer surface of each cuvette was carefully wiped to remove any fingerprints or moisture before being placed into the instrument. Turbidity values were recorded for each sample, with measurements repeated at least three times to ensure reproducibility. The average value from these replicates was reported. For time-based stability analysis, the emulsion samples were drawn into 5 mL plastic syringes, sealed with caps, and positioned upright in a rack to allow creaming over intervals ranging from 0 to 300 minutes. The emulsification process was repeated at different time points to capture temporal changes. For each sampling time, approximately 0.5–1 mL of emulsion was carefully extracted from the syringe, diluted with a suitable volume of filtered water, and immediately transferred into a UV cuvette. Absorbance measurements were carried out using a UV-visible spectrophotometer at the wavelength range of 350–800 nm [35].

During experimentation, homogeneous and bubble-free micro- or nano-emulsion herbal formulation must be ensured. If needed, dilution must be performed using an appropriate solvent to fit the instrument's range, and filtration must be done only to remove large particulates without losing nano-sized components. Gentle mixing must be done before measurement to maintain uniformity.

7.8.4 Challenges

- Foaming or bubbles can interfere with light scattering.
- High turbidity may require dilution for accurate readings.
- Light-absorbing herbal components may affect accuracy.
- Aggregation of nanoparticles may cause fluctuating results, requiring proper dispersion before testing.
- Dilution is often necessary to maintain single scattering conditions, as multiple scattering can distort results.
- Turbidity measurements are less reliable for very small (<0.1 μm) or very large (>10 μm) particles.
- Colored components can absorb light and affect accuracy, particularly with white light sources; using near-infrared light helps reduce this issue.
- Variations in turbidimeter design (light source, detector angle, path length) impact results; adherence to standardized methods (e.g., EPA, ISO) ensures consistency.

7.8.5 Application

In a recent study on peppermint and myrtle nanoemulsions, turbidity was evaluated using a UV-visible spectrophotometer. The freshly prepared nanoemulsions appeared nearly translucent, particularly those formulated with peppermint essential oil. However, statistically significant differences in turbidity ($p > 0.05$) were observed among the formulations, largely influenced by the nature of the essential oil. Myrtle nanoemulsion exhibited the lowest turbidity, whereas peppermint nanoemulsions at 3% and 5% concentrations displayed the highest turbidity values. These findings are consistent with previous reports that noted that turbidity in nanoemulsions is affected by factors such as essential oil composition and droplet size. Thus, nanoemulsions with droplet sizes smaller than the wavelength of visible light exhibit minimal light scattering and appear more transparent, supporting the observed relationship between particle size and turbidity in the peppermint and myrtle formulations.

In another study, a nanoemulsion formulation was prepared using plant essential oil to check its larvicidal activity. The turbidity of the nanoemulsion was evaluated by measuring the absorbance at 600 nm using a spectrophotometer. The coarse emulsion, before sonication, appeared milky and turbid due to the presence of micrometer-sized droplets. After sonication, the emulsion became optically transparent, which was attributed to a significant reduction in droplet size, resulting in weaker light scattering [37].

7.9 REFRACTIVE INDEX

The refractive index (RI) is a physical property that defines how light propagates through a substance. It is expressed as the ratio of the speed of light in a vacuum to the speed of light in the material. The refractive index is measured to determine the nature of sample. Refractive index is determined by using an instrument called a refractometer, namely the Abbe refractometer, immersion refractometer, or Pulfrich refractometer.

7.9.1 Principle of the Abbe Refractometer

The Abbe refractometer determines the refractive index (n) of liquids or solids based on the principle of total internal reflection. The sample is placed between two prisms, and light passing through the system strikes the interface between the prism and the sample at various angles. When the angle of incidence is below the critical angle [$\delta = \arcsin(n/N)$, where N is the refractive index of the prism material], light is transmitted through the sample, creating a bright region in the viewing field. When the angle exceeds this value, total internal reflection occurs, forming a dark region [38]. The sharp boundary between these bright and dark regions marks the critical angle, which is used to calculate the refractive index of the sample. This boundary is aligned with a reticle or crosshair, and the corresponding uncorrected refractive index (n_0) is read from a scale. Since the refractive index can vary with wavelength due to material dispersion, manufacturers provide correction tables to adjust n_0 and obtain the accurate refractive index (n), typically with an accuracy of $\pm 10^4$ [38, 39]. The refractive index is read from a fixed scale, and temperature is controlled via water circulation around the prisms [40]. Samples with different refractive indices shift this boundary. By calibrating the scale, the position of the borderline indicates the sample's refractive index. The refractometer can be calibrated using standard liquids (Table 7.2).

7.9.2 Advantages of the Abbe Refractometer

- Convenient, reliable and simple instrument.
- Small sample size requirement.
- An ordinary light source can be used.
- Easy maintenance and economy.
- Easy determinations of RI.

TABLE 7.2
Calibrating Standards for Refractive Index Instrument

Calibrating Liquid	Refractive Index (n)
Water	1.3325
Carbon tetrachloride	1.4969
Toluene	1.4969
α-Methylnaphthalene	1.6176

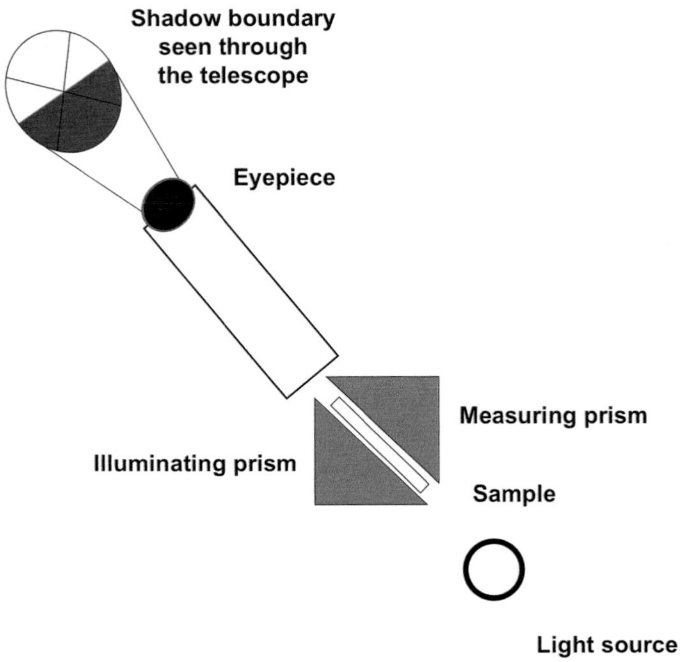

FIGURE 7.15 Instrumentation of the Abbe refractometer

7.9.3 Instrumentation of the Abbe Refractometer

The instrumentation of the Abbe refractometer is shown in Figure 7.15.

- **Light source:** The instrument is equipped with a light source commonly used, the D-line of sodium or a stable LED, which directs light through the system.
- **Mirror and prism:** This light is guided by a light reflection mirror and enters the prism assembly, which is composed of two prisms: the illuminating prism and the measuring prism. The sample is applied between these two prisms, allowing light to pass through it.
- **Optical system**: The optical system, which includes Amici prisms within the telescope, enables the user to observe the refracted light, either visually through an eyepiece or via a digital display in modern versions.
- **Temperature controller:** Additionally, the instrument is often equipped with temperature control mechanisms, such as a built-in thermostat or water bath connection, to maintain a constant temperature, since refractive index measurements are sensitive to temperature variations [40].

7.9.4 Procedure

- The liquid herbal formulation is measured as a pure sample form. In the case of the solid formulation, a small amount of the sample is dissolved in a liquid with a known refractive index.
- To begin with the analysis, the prism is first cleaned using a soft cloth or tissue with distilled water or alcohol to gently clean the measuring prism and cover prism.
- Then 1–2 drops of the liquid sample are applied to the lower prism. The upper prism is carefully closed to spread the sample evenly between the prisms, ensuring no air bubbles are trapped.
- The light source is turned on into the mirror, observed through the eyepiece and adjusted until the boundary between the light and dark areas is visible.
- The refractometer dial is rotated until the boundary line aligns with the crosshairs. Then, the refractive index value is recorded directly from the scale. A minimum of 3 readings are recorded, and the average is reported.
- The temperature is recorded during the measurement.
- After taking the measurement, the upper and lower prisms are cleaned with distilled water or alcohol to avoid residue build-up or contamination.

7.9.5 Application to Herbal Drugs

Refractive index values help in identifying the molecular weights and structures of organic compounds using their molar refraction, commonly used to measure the concentration of solutes in aqueous solutions. Refractive index is used to assess the purity and consistency of herbal extracts and formulations by monitoring batch-to-batch variations. It helps in detecting adulteration or substitution in herbal products by comparing the refractive index with standard values. It assists in standardizing herbal formulations by providing a quick and non-destructive method to evaluate composition and concentration. By monitoring changes in refractive index over time, the stability and shelf life of herbal preparations are evaluated.

In research carried out on essential oils of *Lippia origanoides*, it was found that the refractive index closely correlates with thymol content. Since thymol (RI = 1.5227) is the major component in many chemotypes, oils with higher thymol concentrations showed higher RI values. Other constituents like ρ-cymene, α-terpinene, and γ-terpinene exhibited lower RI values, leading to a decrease in overall RI when thymol content declined. The Patía chemotype showed consistent RI and thymol levels, while in Cítrica and Típica chemotypes, variability was observed, yet still, higher RI values consistently matched higher thymol concentrations. The relationship followed a linear model ($r = 0.9439$), confirming that RI can serve as a rapid and reliable indicator of thymol content in these essential oils [41].

In a study, the RI of dual fluorescent carbon nanodots synthesized from leaves of a medicinal plant was measured using a digital Abbe-type refractometer for a stability study. The measurements were carried out intermittently at various time points and at the end of the stability study. The RI of carbon dots was found to be uniform throughout the liquid sample, representing its isotropic and homogeneous nature. No significant change in the refractive index was observed during the stability study, which reflected sufficient physical stability of synthesized carbon dots. The unaltered refractive index of the synthesized nanodots signified the retention of optical clarity and homogeneity [42].

7.10 pH

pH is the quantitative measure of the acidity or basicity of aqueous solution or other liquid solutions. A pH meter is used to determine the acidity or alkalinity of the solution. pH is the concentration of hydrogen ions in the solution. A solution containing more H^+ ions remains acidic, while a solution containing more OH^- ions remains alkaline. pH value of solutions ranges from 1 to 14. where a pH less than 7 denotes an acidic solution, a pH of 7 indicates neutrality, and a pH greater than 7 represents a basic (alkaline) solution. It is an essential parameter in chemistry, biology, environmental science, and many industrial processes [43].

7.10.1 Ways of Measuring pH

1) **Using pH Test Strips:** A pH test strip, also known as pH paper, is coated with a pH-sensitive dye and dipped into the solution being tested. Upon contact, the strip changes color based on the pH of the solution. The strip is then removed, and the resulting color is compared to a reference chart to determine the corresponding pH value.
2) **Using pH indicator solutions:** A pH indicator solution can be used to determine the pH of an unknown liquid by mixing the sample with the indicator. The indicator changes color depending on the acidity or alkalinity of the solution. This color change is then compared to a standard pH color chart or wheel to estimate the solution's pH value.
3) **Using a pH sensor (electronic method):** A pH sensor allows for direct measurement by inserting the probe into the solution, where it instantly provides a digital pH reading on an electronic meter. This method is quick, is convenient, and offers high accuracy.

7.10.2 Principle

The principle of pH measurement is based on the electrochemical potential difference generated by hydrogen ions in a solution. A pH meter measures this potential using a combination of electrodes – typically a glass electrode and a reference electrode – immersed in the test solution. The measured voltage is converted to pH units based on the Nernst equation. A pH meter basically works on the fact that an interface of two liquids produces an electric potential, which can be measured. In other words, when a liquid inside an enclosure made of glass is placed inside a solution other than that liquid, there exists an electrochemical potential between the two liquids [44].

7.10.3 Advantages

- It is a more accurate method than the pH strip.
- Provides rapid and accurate measurements.
- Useful for both laboratory and field applications.
- Requires only a small sample volume.
- Non-destructive to the sample.
- Applicable to a wide range of industries and research fields.
- They are easy and convenient to use.

7.10.4 Instrumentation

A typical pH measurement setup includes a pH meter, a glass electrode, and sometimes a temperature sensor for thermal compensation (Figure 7.16). The glass electrode houses two components: a measuring (sensor) electrode and a reference electrode, both enclosed in glass tubes. The measuring electrode contains a pH 7 buffer solution and a silver wire coated with silver chloride, while the reference electrode is filled with a saturated potassium chloride solution and also contains a silver/silver chloride wire. The bulb of the measuring electrode is made of porous glass coated with silica and metal salts, allowing it to interact with hydrogen ions in the test solution. When the probe is immersed in the solution, hydrogen ions replace metal ions on the bulb surface, generating a voltage captured by the silver wire. The reference electrode provides a stable zero-voltage point to complete the circuit. The pH meter then measures the voltage difference between the two electrodes and converts it into a corresponding pH value.

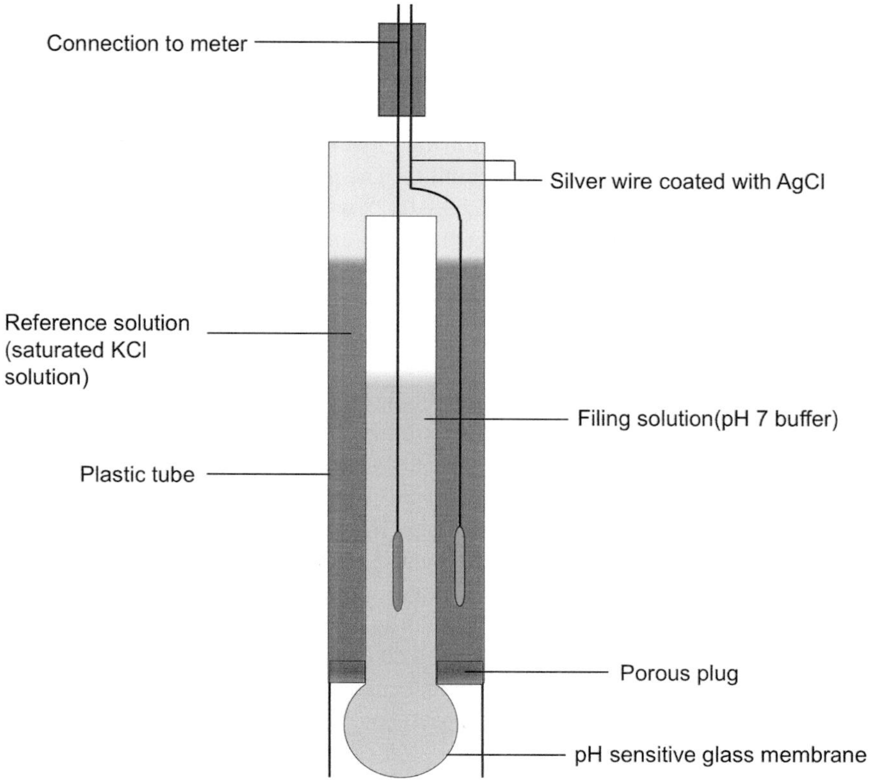

FIGURE 7.16 Instrumentation of an electronic pH meter

7.10.5 Procedure

- The pH meter is first calibrated using standard buffer solutions (usually pH 4.0, 7.0, and 10.0) before any measurement.
- The electrode is rinsed with distilled water and dried with tissue paper.
- Then, the electrode is immersed in the sample solution, and reading is observed until a stable value.
- Then, the electrode is cleaned after each measurement and stored in an appropriate storage solution.
- For accurate pH measurement, the stability of sample temperature is ensured, as fluctuations can affect readings. The high resistance of the glass electrode may cause voltage drops and inaccuracies, so a high-impedance meter or null-balance setup is recommended.

7.10.6 Applications to Herbal Drugs

The pH of topical herbal formulations is a critical parameter in ensuring skin compatibility and formulation stability. In one study, the pH of a herbal formulation was measured using a calibrated digital pH meter with standard buffer solutions at pH 4.0, 7.0, and 9.0. The pH of the simple ointment base and the topical herbal formulations ranged from 6.5 ± 0.03 to 7.1 ± 0.08, aligning with the normal pH range of human skin and indicating a low risk of irritation. Furthermore, no significant pH variation was observed over time, suggesting good formulation stability [45].

In a recent study of a formulation of *Andrographis paniculata* aqueous herbal cream, pH determination was carried out, and the values of different cream batches were found to range from 5.54

to 6.84, which corresponds to the natural acidic pH of the stratum corneum, the outermost layer of the skin. This range confirms the formulation's non-irritant nature and indicates that the pH is sufficiently low to help inhibit the growth of fungi and bacteria, contributing to the microbiological safety of the product [46].

In another study, *Lawsonia inermis* extract-loaded hydrogel for wound dressing was evaluated by pH measurement for evaluating skin compatibility. The pH of all the prepared formulations was found to be toward neutrality (6.53 ± 0.067 to 6.8 ± 0.082), which demonstrated a very low chance of skin irritation. The skin irritancy results also represented excellent skin compatibility and the non-irritancy nature of the prepared hydrogel for wound dressing application [47].

7.11 CONCLUSION

The application of advanced instrumental methodologies is essential in modern pharmaceutical research and development. Every technique discussed in this chapter, starting from contact angle analysis and rheometry to particle size measurement, zeta potential determination, and pH analysis, offers critical insights into the physical and chemical properties of pharmaceutical formulations. These techniques not only aid in characterizing herbal drugs but also play a pivotal role in optimizing formulation design, enhancing product stability and ensuring therapeutic efficacy. By enabling accurate, reproducible, and non-destructive evaluations, these instruments support stringent quality control and regulatory compliance throughout the drug development lifecycle. As the research in herbal formulation, phytoformulations, and targeted delivery of herbal formulation is continuously evolving, the integration of such precise analytical techniques will remain fundamental to the development of safe, effective, and high-quality therapeutic products.

REFERENCES

1. Sahu, A.N. and D. Mohapatra, *Herbal Drug Formulation and Standardization*. 2021, New Delhi: Ane Books Pvt. Ltd. ISBN-10: 9390658365, ISBN-13: 978–9390658367.
2. Erbil, H. Yildirim, Dependency of contact angles on three-phase contact line: A review. *Colloids and Interfaces*, 2021. 5(1): p. 8.
3. Kwok, D.Y. and A.W. Neumann, Contact angle measurement and contact angle interpretation. *Advances in Colloid and Interface Science*, 1999. 81(3): p. 167–249.
4. Adam, N., Use of the term 'Young's Equation' for contact angles. *Nature*, 1957. 180(4590): p. 809–810.
5. Fiorentin, P. and A. Scroccaro, Analysis of the performance of a goniometer for studying surface reflection. *IEEE Transactions on Instrumentation and Measurement*, 2008. 57(11): p. 2522–2527.
6. Krishnan, A., et al., An evaluation of methods for contact angle measurement. *Colloids and Surfaces B: Biointerfaces*, 2005. 43(2): p. 95–98.
7. Shah, P., et al., Wetting parameters, surface free energy, and wettability enhancement of Ocimum sanctum leaves through adjuvants. *Chemical Papers*, 2025: p. 1–15.
8. Song, B., et al., Andrographolide solid dispersions formulated by Soluplus to enhance interface wetting, dissolution, and absorption. *Journal of Applied Polymer Science*, 2020. 137(6): p. 48354.
9. Gangurde, A.B., et al., Enhanced solubility and dissolution of curcumin by a hydrophilic polymer solid dispersion and its insilico molecular modeling studies. *Journal of Drug Delivery Science and Technology*, 2015. 29: p. 226–237.
10. Mohapatra, D., et al., Quality by design – based development and optimization of fourth-generation ternary solid dispersion of standardized Piper longum extract for melanoma therapy. *Drug Delivery and Translational Research*, 2023. 13(12): p. 3094–3131.
11. Burey, P., et al., Confectionery gels: A review on formulation, rheological and structural aspects. *International Journal of Food Properties*, 2009. 12(1): p. 176–210.
12. Virden, A., Method development for laser-diffraction particle-size analysis. *Pharmaceutical Technology*, 2010. 34(11): p. 1–7.
13. Aho, J., S. Hvidt, and S. Baldursdottir, Rheology in pharmaceutical sciences, in *Analytical Techniques in the Pharmaceutical Sciences*. 2016, Springer. p. 719–750.
14. Roşca, O.-J., et al., Optimizing burn wound healing: The critical role of pH and rheological behavior in plant-derived topical formulations. *Pharmaceutics*, 2025. 17(7): p. 853.

15. Mohapatra, D., et al., Quality-by-design-based development of ultradeformable nanovesicular transgelosome of standardized Piper longum extract for melanoma. *Nanomedicine*, 2023. 18(14): p. 963–985.
16. Gandomkar, M., A. Pasdaran, and G. Yousefi, Development of an anti-acne cream based on natural oils: Investigation of the effect of ingredients on rheology, texture properties, and physical stability. *Trends in Pharmaceutical Sciences and Technologies*, 2023. 9(2): p. 93–104.
17. Savic, S.M., et al., 'All-natural'anti-wrinkle emulsion serum with Acmella oleracea extract: A design of experiments (DoE) formulation approach, rheology and in vivo skin performance/efficacy evaluation. *International Journal of Cosmetic Science*, 2021. 43(5): p. 530–546.
18. Andonova, A., et al., Influence of natural oils on the textural and rheological properties of cosmetic creams. *Pharmacia*, 2024. 71: p. 1–10.
19. Lad, S., et al., Review on Moisture Content: A stability problem in pharmaceuticals. *EPRA International Journal of Research & Development*, 2022. 7: p. 27–33.
20. Razvi, S.Z.A., et al., Loss on drying using halogen moisture analyzer: An orthogonal technique for monitoring volatile content for in-process control samples during pharmaceutical manufacturing. *Organic Process Research & Development*, 2021. 25(2): p. 300–307.
21. Razak, N., N. Hamid, and A. Shaari. Effect of storage temperature on moisture content of encapsulated Orthosiphon stamineus spray-dried powder, in *AIP Conference Proceedings*. 2018. AIP Publishing LLC.
22. Gallo, L., et al., A comparative study of spray-dried medicinal plant aqueous extracts. Drying performance and product quality. *Chemical Engineering Research and Design*, 2015. 104: p. 681–694.
23. Shekunov, B.Y., et al., Particle size analysis in pharmaceutics: Principles, methods and applications. *Pharmaceutical Research*, 2007. 24: p. 203–227.
24. Wang, L. and H. Cheng, Principles and application of laser light scattering (LLS) in characterization of the spatial configuration of microbial products in aqueous solution. *Trends in Environmental Analytical Chemistry*, 2015. 8: p. 12–19.
25. Bittelli, M., et al., Experimental evidence of laser diffraction accuracy for particle size analysis. *Geoderma*, 2022. 409: p. 115627.
26. Bhattacharjee, S., DLS and zeta potential – what they are and what they are not? *Journal of Controlled Release*, 2016. 235: p. 337–351.
27. Sandhu, R., et al., Dynamic light scattering (DLS) technique, principle, theoretical considerations and applications. *Nano-Technological and Biochemical Techniques for Assessing the Quality and Safety of Milk and Milk Products*, 2018. 8: p. 135–7.
28. Ulyantsev, A., et al., Particulate characterization of herbal extracts for its standardization and quality control. *Chemical Engineering Transactions*, 2011. 24: p. 691–696.
29. Gião, M.S., et al., Effect of particle size upon the extent of extraction of antioxidant power from the plants Agrimonia eupatoria, Salvia sp. and Satureja montana. *Food Chemistry*, 2009. 117(3): p. 412–416.
30. Ahluwalia, V., et al., Nano silver particle synthesis using Swertia paniculata herbal extract and its antimicrobial activity. *Microbial Pathogenesis*, 2018. 114: p. 402–408.
31. Salopek, B., D. Krasic, and S. Filipovic, Measurement and application of zeta-potential. *Rudarsko-geolosko-naftni zbornik*, 1992. 4(1): p. 147.
32. Hussain, A., et al., Nanoformulation of Curcuma longa root extract and evaluation of its dissolution potential. *ACS Omega*, 2022. 8(1): p. 1088–1096.
33. Nam, S., et al., Development of resveratrol-loaded herbal extract-based nanocomposites and their application to the therapy of ovarian cancer. *Nanomaterials*, 2018. 8(6): p. 384.
34. Chittineedi, P., et al., Polyherbal formulation conjugated to gold nanoparticles induced ferroptosis in drug-resistant breast cancer stem cells through ferritin degradation. *Frontiers in Pharmacology*, 2023. 14: p. 1134758.
35. Alade, O.S., et al., Rapid determination of emulsion stability using turbidity measurement incorporating artificial neural network (ANN): Experimental validation using video/optical microscopy and kinetic modeling. *ACS Omega*, 2021. 6(8): p. 5910–5920.
36. Matos, T., et al., A review of methods and instruments to monitor turbidity and suspended sediment concentration. *Journal of Water Process Engineering*, 2024. 64: p. 105624.
37. Ghosh, V., A. Mukherjee, and N. Chandrasekaran, Formulation and characterization of plant essential oil based nanoemulsion: Evaluation of its larvicidal activity against Aedes aegypti. *Asian Journal of Chemistry*, 2013. 25(Supplementary Issue): p. S321.
38. Rheims, J., J. Köser, and T. Wriedt, Refractive-index measurements in the near-IR using an Abbe refractometer. *Measurement Science and Technology*, 1997. 8(6): p. 601.
39. Tilton, L.W., Testing and accurate use of Abbe-type refractometers. *Journal of the Optical Society of America*, 1942. 32(7): p. 371–381.

40. Gooch, J.W., Abbe' Refractometer, in *Encyclopedic Dictionary of Polymers*. 2011, Springer. p. 1–2.
41. De Refracción, R.E.E.Í. and Y. La Concentración, Scientific note relationship between refractive index and thymol concentration in essential oils of lippia origanoides Kunth. *Chilean Journal of Agricultural & Animal Sciences*, 2016. 32(2): p. 127–133.
42. Mohapatra, D., et al., Bioengineered dual fluorescent carbon nano dots from Indian long pepper leaves for multifaceted environmental and health utilities. *Environmental Science and Pollution Research*, 2023. 30(18): p. 52182–52208.
43. Persat, A., R.D. Chambers, and J.G. Santiago, Basic principles of electrolyte chemistry for microfluidic electrokinetics. Part I: Acid – base equilibria and pH buffers. *Lab on a Chip*, 2009. 9(17): p. 2437–2453.
44. Karastogianni, S., S. Girousi, and S. Sotiropoulos, pH: principles and measurement. *Encyclopedia of Food and Health*, 2016. 4: p. 333–338.
45. Toppo, F.A. and R.S. Pawar, Development, optimization and evaluation of different herbal formulations for wound healing. *International Journal of Pharmacy and Pharmaceutical Sciences*, 2015. 7: p. 447–52.
46. Afokoghene, A.J., O.S. Eziuzo, and N.C. Lynda, Antimicrobial investigation, formulation and evaluation of Andrographis paniculata aqueous herbal cream for topical application. *Research Journal of Pharmacy and Technology*, 2022. 15(8): p. 3553–3558.
47. Dixit, K., et al., Formulation development and evaluation of Lawsonia inermis extract loaded hydrogel for wound dressing application. *Indian Journal of Pharmacy*, 2022. 84(4): p. 848–862.

8 Instrumental Methodology for Thermal Analysis of Herbal Drugs and Formulations

ABBREVIATIONS

Abbreviation	Description
CMC	Carboxymethyl Cellulose
DMA	Dynamic Mechanical Analysis
DMTA	Dynamic Mechanical Thermal Analysis
DSC	Differential Scanning Calorimetry
DTA	Differential Thermal Analysis
EGA	Evolved Gas Analysis
HIDRO	Hydroxyethyl Cellulose
HPMC	Hydroxypropyl Methylcellulose
SLN	Solid Lipid Nanoparticles
TGA	Thermogravimetric Analysis
TG-SPI-TOFMS	Thermogravimetry Coupled to Single Photon Ionization Time-Of-Flight Mass Spectrometry
TMA	Thermomechanical Analysis

8.1 INTRODUCTION

Thermal analysis involves the investigation of properties of materials that change with temperature. It involves use of various analytical instruments to study the changes in physicochemical properties of substances when they are subjected to various temperatures (heated, cooled, or kept at a specific temperature) [1, 2]. The controlled temperature program is divided into two categories: static measurement and dynamic measurement. Under static conditions, the temperature remains constant throughout the study, whereas under dynamic measurement, heating is provided at a defined rate [2]. A wide variety of samples, including drugs, composites, polymers, ceramics, thin films, and metallic glasses, can be investigated by thermal analysis techniques. Such techniques are extensively used in the evaluation of the thermal properties of herbal drugs and their formulations [1]. The information on thermal properties provides a comprehensive understanding of the physicochemical properties of herbal drugs, aids in their authentication, standardization, stability, shelf life prediction, drug-excipient compatibility in herbal formulations, and selection of appropriate storage conditions of herbal formulations. Such techniques ensure the preservation of safety, quality, and therapeutic activity of herbal drugs and formulations [2]. However, the evaluations of thermal properties of herbal drugs are relatively difficult compared to those of isolated phytoconstituents and synthetic drugs due to their chemical complexity [2]. The thermal properties of multicomponent-based herbal extracts are influenced by their chemical composition, the extraction method used, and the presence of residual solvents. The plant extracts, comprising various bioactive phytoconstituents like alkaloids, flavonoids, polyphenols, and essential oils, contribute to their unique thermal properties. Different phytoconstituents possess distinct thermal properties, resulting in varied thermal

DOI: 10.1201/9781003540823-8

responses during analysis [2]. The choice of extraction techniques also impacts the chemical composition of the herbal extract, thus affecting the thermal properties [2].

Various thermal analytical techniques (Figure 8.1), such as DSC, DTA, TGA, and hot stage microscopy, provide detailed insights into the thermal properties, phase transitions, and stability of phytoconstituents and finished herbal formulations. In addition to these, dynamic mechanical analysis (DMA), thermo-optical analysis, dilatometry, dielectric thermal analysis, evolved gas analysis (EGA), thermomechanical analysis (TMA), laser flash analysis, and derivatography are exploited for thermal analysis of herbal drugs [1]. DSC measures heat difference; DMA provides information on mechanical stiffness and damping; thermo-optical analysis measures the optical properties; dilatometry measures the volume; dielectric thermal analysis measures dielectric permittivity and loss factor; EGA measures gaseous decomposition products; DTA measures the temperature difference; TGA measures percentage weight loss, TMA measures dimension; laser flash analysis measures thermal diffusivity and thermal conductivity; and derivatography is the measurement of various basic and derived values in thermal analysis [1].

8.2 DIFFERENTIAL SCANNING CALORIMETRY

Differential scanning calorimetry (DSC) is an emerging thermal technique for investigating thermodynamic properties of herbal drugs and phytoconstituents. It investigates the heat flow associated with phase transitions in a substance as a function of temperature or time. It provides a comprehensive understanding of melting point, glass transition temperatures, specific heat capacity, stability, purity, quality, polymorphic transitions, crystallization, drug-excipient compatibility, and optimal storage conditions of phytoconstituents, herbal drugs, and finished herbal formulations [2]. The principles, advantages, instrumentation, procedure, challenges, and their application for examining the crystallinity and polymorphism of herbal drugs are described in Chapter 6. In this chapter, their applications for examining the thermal properties, stability, and drug-excipient compatibility are described.

In herbal drugs, DSC is used to study thermal properties of bioactives, herbal extracts, fractions, and excipients [2]. Impurities or mixed phases can alter thermal profiles, helping to assess the purity and consistency of herbal extracts or powders. In a recent study, DSC of crude peanut extract (CPE) and physical mixture (CPE and polymer) was carried out to understand the drug-excipient compatibility in CPE-loaded fast-disintegrating tablets. The DSC thermograms of CPE and the physical mixture showed pointed endothermic peaks, which were attributed to the melting of the CPE. The physical mixture showed a slight shifting of the endothermic peak due to the presence of polymer, representing no significant interaction among the CPE and excipients used [3].

The thermal properties of chitosan (CS)/agarose (AG) based bioscaffold with *Matricaria recutita* herbal extract (CH) were reported using DSC study. The interactions among the polymer chain and bioactive components of the herbal drug in the bioscaffolds during heat were studied from the DSC thermograms of chitosan/agarose and CS-AG-CH bioscaffold. The CS-AG-CH bioscaffold demonstrated a shift in the transition temperature (T_m), which was due to the thermodynamic factors, disintegration of intermolecular side chains, and interactions among the biopolymers and the chamomile extract [4].

In another study, the thermal behavior of 5-fluorouracil and andrographolide-loaded solid lipid nanoparticles (SLNs) for cancer therapy was evaluated by DSC. The physical state and melting point analysis of cetyl alcohol (CA), physical mixture/andrographolide-cetyl alcohol (AG-CA-PM), andrographolide SLNs (AG-SLNs), 5-FU-cetyl alcohol physical mix (5-FU-CA-PM), and 5-FU-SLNs were investigated by DSC analysis. The DSC thermograms of AG-CA-PM and 5-FU-CA-PM showed representative peaks for AG, 5-FU, and CA. However, after loading the drugs in SLNs, a single peak of lipid was observed, which revealed encapsulation of the drug in the lipid [5].

DSC is also used to investigate the thermal stability and fatty acid profile of herbal drugs. By analyzing these thermal events, the thermal stability of the herbal drugs can be estimated [2]. Oils

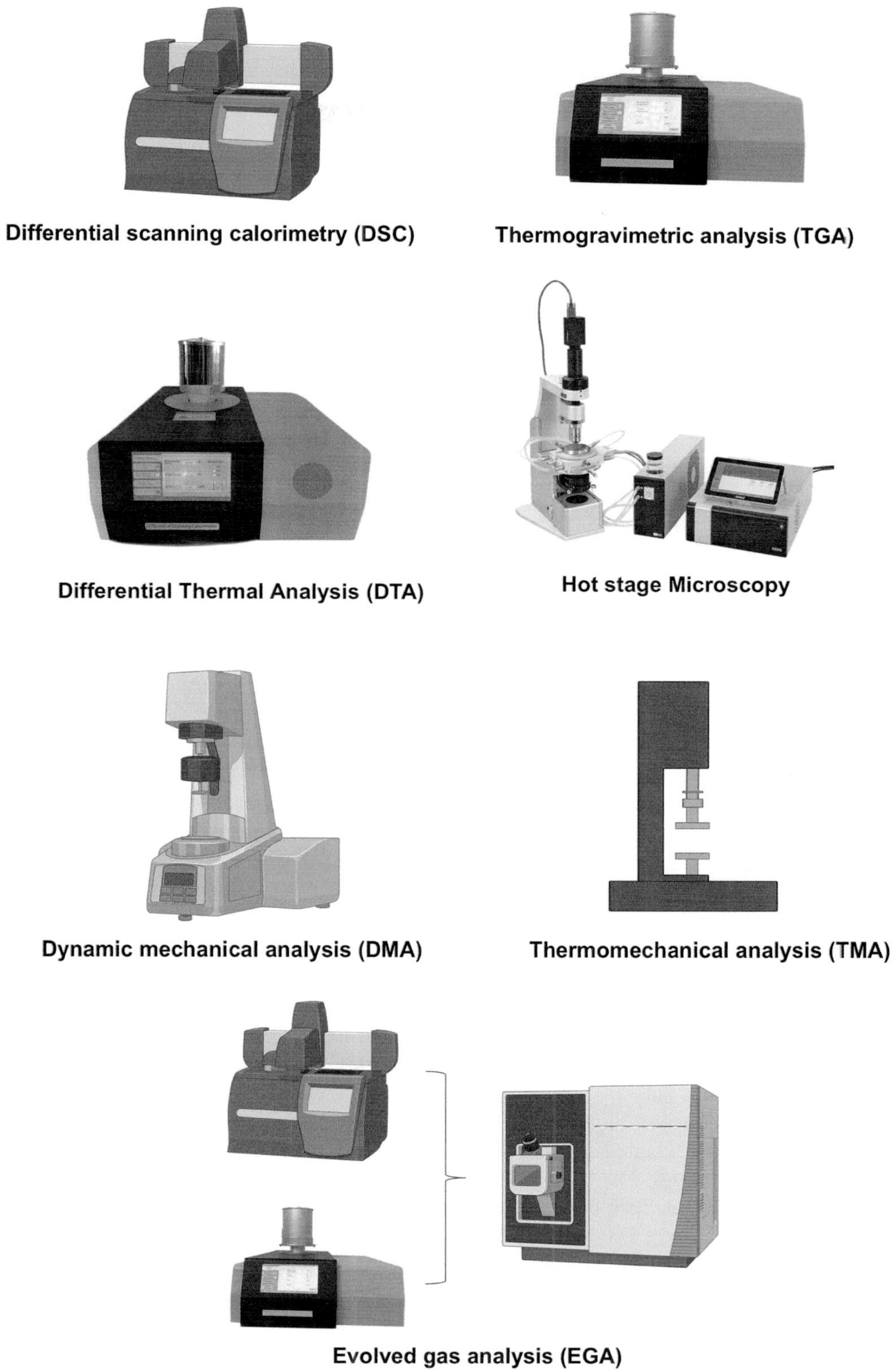

FIGURE 8.1 Instruments for thermal analysis of herbal drugs and formulations

with a high proportion of polyunsaturated fatty acid content show more susceptibility to oxidation with the appearance of DSC thermal events. The presence of natural antioxidants in oil offers high stability. Thus, DSC study can envisage the thermal properties, fatty acids, and antioxidant compositions [2].

Fat from *Astrocaryum vulgare* was explored as a pharmaceutical excipient for lipid nanoparticle production. The thermal properties were evaluated using DSC. The DSC thermogram of fat displayed an endothermic peak at 34.74°C, representing the melting point of the fat. The melting point of vegetable fats is highly influenced by fatty acid composition. The fat of *Astrocaryum vulgare* is reported to be composed of 85.15% saturated fatty acid (lauric acid) and the remaining unsaturated fatty acids (oleic acid). The triglyceride formed solely by lauric acid possesses a melting point of 44°C, and triglycerides formed only by oleic acid possess a melting point of 5°C. It is reported that the melting point of *Astrocaryum vulgare* fat is consistent with its fatty acid profile. Since fat is liquid at body temperature (37°C), various lipid mixtures are prepared with carnauba wax (10%–50%) to achieve thermal stability up to 40°C for use as an excipient in the preparation of nanostructured lipid carrier (NLC). The mixture with smaller endothermic peaks and widening ("shoulders") was considered to be best for the NLC excipient. Such thermal change is caused by a great disorganization of the samples and thus a lowering of the crystallinity of the system. All lipidic mixtures were found to disrupt the melting peak of carnauba wax as well as alter its melting point and so decrease its crystallinity. Nevertheless, a lipidic mixture with 50% carnauba wax was found to be suitable due to the absence of endothermic event within 40°C, and hence it was chosen for the formulation of NLCs [6].

In a recent study, DSC was used to study the thermal properties of *Malania oleifera* seed oil. The thermo-oxidative processes that occur in heated oil as a function of temperature and time were studied. The DSC curve of oil showed melting peaks at −30°C, −20°C, and 7°C due to the diversity of the composition of the triglycerides of oils. The oil showed complete melting at 7°C, representing good performance at the low temperature due to its high degree of unsaturation. The crystallization thermographs showed a crystallization peak of oil at −49°C, which further proves the good performance of oil at low temperature [7].

DSC is also useful for investigating the successful encapsulation of the herbal drugs in polymeric matrices or excipients during formulation and development. The DSC herbal drugs or phytoconstituents, and their formulations, are studied and compared. The absence of an endothermic DSC peak in the final formulation is an indication of successful encapsulation into the excipients [2]. Nanostructured lipid carriers (NLCs) of plant bioactives have been prepared with Compritol® 888 ATO as solid lipid and squalene as liquid lipid to overcome the solubility and permeability issues of quercetin and piperine. DSC was carried out to examine the thermal events of drugs in the formulation. The DSC thermogram of piperine and quercetin showed sharp endothermic peaks due to the melting of the drugs. However, the thermogram of NLCs showed the absence of sharp endothermic peaks, with the appearance of a DSC thermogram similar to that of the blank formulation. Such outcomes reflected successful encapsulation into the lipid matrix with amorphous modifications [8].

The thermal analysis of herbal drugs by DSC can also be used to investigate the purity of the oil extracted from herbal drugs. The oil purity can be examined by analyzing the heat of fusion and the melting point of the oil. Such information can be used to identify any impurities or adulterants in natural oils [2]. In a study, DSC was used to characterize the thermal properties of lemongrass, orange, and basil essential oils. The predominant substances in these oils were citral (66% of lemongrass oil), limonene (90% of orange oil), and methyl chavicol (84% of Basil oil). DSC was used to investigate the enthalpies of evaporation of essential oils and compare them with the estimated enthalpies of vaporization of their major components. The estimated and calculated vaporization enthalpies were found to be comparable. The closer the enthalpy of vaporization of the essential oil is to that of the pure substances, the more it indicates higher purity with respect to the major components [2].

The thermal behavior of plant extracts is highly influenced by composition and the method of extraction [2]. DSC is also useful for estimating the boiling point of herbal drugs. It is based on investigating the changes in the heat flow of the sample during phase transitions. The boiling point can be identified by observing the endothermic peak on the DSC thermogram, which corresponds to the vaporization of the sample. Such information is required to check the purity of the herbal drugs, as impurities may affect the boiling point of the sample. In addition to that, the boiling point can be used to predict the stability of herbal drugs at high temperatures [2].

DSC is also used to study the moisture content and impurity analysis of herbal drugs. The moisture content analysis of essential oils by DSC offers several advantages, including short analysis time and the use of small samples. This method allows the quantification of phytocomponents from other volatile substances based on their vaporization and dehydration enthalpies [2]. DSC measures the thermal behavior of the herbal drugs, including changes in melting point, heat capacity, and glass transition temperature, which can be correlated with the moisture content or presence of impurities. The moisture content can be estimated by analyzing the difference in the melting point between the hydrated and anhydrous forms of the herbal drugs. The impurities can be identified by analyzing changes in heat capacity or melting behavior [2].

8.3 DIFFERENTIAL THERMAL ANALYSIS

Differential thermal analysis (DTA) measures the difference in temperature between samples and controls. In DTA, the temperature difference between a sample and an inert reference is monitored while both are subjected to an identical thermal program [9–11]. It has been shown to be a useful technique for identifying the thermal properties of active compounds, such as melting points and decomposition temperatures. The use of DTA in herbal drugs can provide valuable insights into the thermal stability and degradation of bioactive compounds, which is essential for ensuring the safety and efficacy of herbal products. Additionally, DTA can help optimize the formulation and processing of herbal medicines by providing insight into the optimal temperature and time for extraction and storage. The use of DTA in herbal medicine has become an indispensable tool for analyzing the thermal behaviors of active components and herbal products [2].

8.3.1 PRINCIPLE

DTA is based on the principle of measuring the temperature difference (ΔT) between a sample and a reference substance under identical thermal conditions. During an increase in temperature, the sample undergoes thermal events (e.g., melting, degradation) and absorbs or releases heat. This leads to a deviation in temperature from the inert reference. This temperature difference is recorded as a function of time or temperature. For endothermic events (e.g., melting), the sample absorbs heat and ΔT increases negatively, whereas for exothermic events (e.g., crystallization), the sample releases heat and ΔT increases positively [10, 11].

8.3.2 ADVANTAGES OF DTA

- DTA is a simple and cost-effective thermal technique.
- It enables the simultaneous analysis of the sample and reference.
- Identifies both exothermic and endothermic transitions.
- Useful for analysis of thermal stability.
- Applicable to a wide range of materials, including inorganics, organics, active molecules, isolated phytoconstituents, and complex herbal drugs.

8.3.3 Instrumentation

A schematic representation of the DTA instrument is shown in Figure 8.2. A typical DTA setup includes the following components [10, 11]:

- **Sample and reference holders**: Crucibles to hold the test sample and inert reference (e.g., alumina, titania, carborundum). Crucibles are made up of aluminum, silver, gold, platinum, graphite, and stainless steel. The reference material should be inert over the temperature range studied, unreactive to the crucible or thermocouple, and, preferably, similar in thermal properties to the sample.
- **Furnace**: To provide controlled heating or cooling.
- **Sensors:** Thermocouples and platinum resistance thermometers are used as sensors to measure the temperature of both the sample and reference.
- **Temperature programmer**: Controls the heating/cooling rate. The temperature regulation for isothermal work must be very good, while a wide range of heating rates should be available and well controlled.
- **Atmosphere control**: The atmosphere surrounding the sample affects the thermogram of the sample. Most DTA instruments have facilities for establishing an atmosphere of a chosen gas around the sensor assembly. Inert gases (nitrogen, argon, or helium) and reactive gases (air and oxygen) are used in most of the DTA instruments.
- **Pressure system**: If DTA studies at high pressure are required, high-pressure cells may be employed, or the entire system may be modified to withstand higher pressures up to 7 MPa. The increased pressure is useful to raise the vaporization temperature of a material, or to increase the rate of reaction.
- **Data recorder**: It plots temperature difference (ΔT) versus temperature.

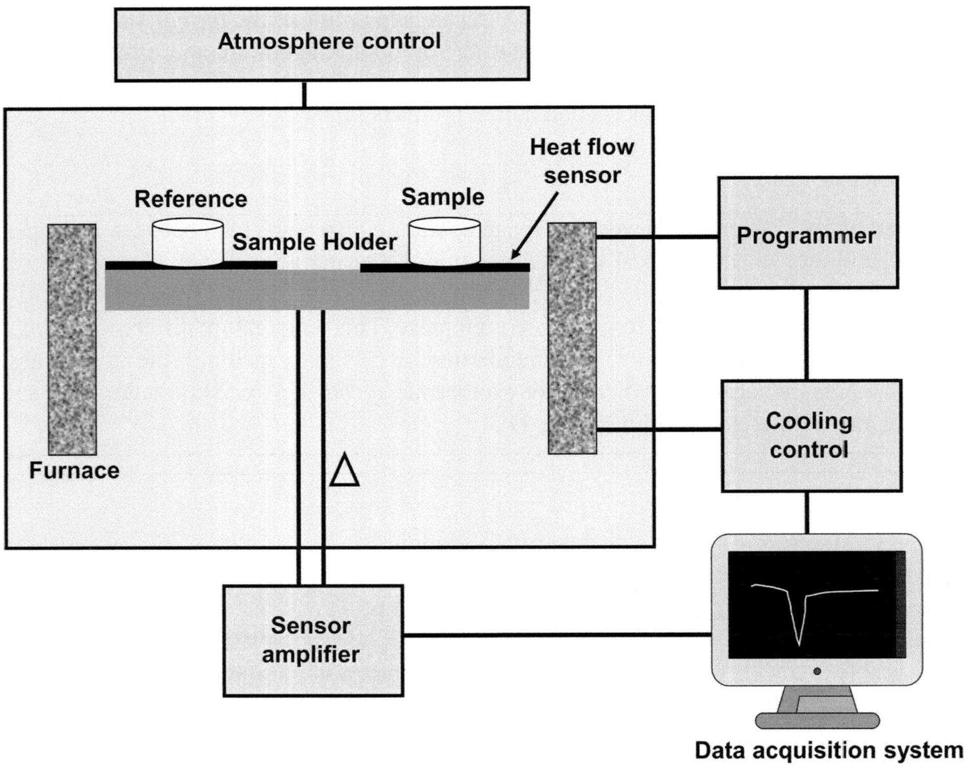

FIGURE 8.2 Schematic representation of DTA

8.3.4 Procedure

The DTA involves the following steps [10].

- **Sample preparation:** The sample is finely powdered and dried, and 2–30 mg of the sample is taken in a clean crucible. The powdered sample is directly loaded into the crucible. Voluminous powders and fibrous materials are difficult to retain within the sample pan before sealing. They may be kept in the center of a small piece of thin foil of the same metal as the crucible (e.g., aluminum or platinum). Then, this is folded over them to contain the sample into a smaller bulk, and the wrapped sample is then loaded into the pan and flattened before sealing. Another method is to use a washer or inner lid to retain the sample before lidding. Thin film samples are cut in such a way as to fit inside the crucible. Liquid samples may be directly introduced into the sample pan using a micropipette, syringe, or dropper. Thick semisolid samples (paste, waxes) are placed into the cell with a spatula and then flattened against the bottom of the pan using a rod with a flat end. Volatile samples should be run in sealed pans or sealed glass capillaries, and, for good contact, the surrounding space should be filled with conducting material.
- **Reference material:** Alumina or other reference materials are placed in an identical crucible.
- **Loading of crucibles:** Both sample and reference crucibles are loaded into the DTA instrument.
- **Setting of heat program:** The heating rate and range with the atmospheric gas are defined in this step.
- **Sample analysis**: The analysis starts and the system records temperature differences between the sample and reference during heating. Then the system is allowed to cool and cleaned for the next sample.
- **Data analysis and interpretation:** The DTA thermogram is analyzed for endothermic or exothermic events (e.g., decomposition, melting).

8.3.5 Challenges

- **Phytochemical complexity:** Herbal drugs are composed of numerous phytochemicals (flavonoids, alkaloids, terpenoids, etc.) that may degrade or interact at different temperatures. This can lead to overlapping thermal events and poorly resolved DTA peaks, making interpretation difficult.
- **Sample heterogeneity:** Crude herbal powders and their extracts often have inconsistent particle size, composition, and density, which can affect heat transfer and baseline stability during DTA analysis.
- **Limited quantitative accuracy:** DTA primarily provides qualitative or semi-quantitative data rather than quantitative data.
- **Thermal lag:** At very high heating rates, thermal lag can distort peak positions.
- **Decomposition before melting:** Some of the thermolabile phytoconstituents in herbal drugs decompose before melting. This complicates the identification of phase transitions like melting or crystallization.
- **Oxidation and volatilization:** Some components may oxidize or volatilize under heating, especially if an inert atmosphere is not used, leading to misleading thermal signals.
- **Lack of specificity:** Sometimes, the overlapping or broad thermal peaks in herbal extracts may complicate data interpretation.
- **Batch-to-batch variability:** The chemical composition of herbal drugs and extracts differs from batch to batch due to variations in geographical origin, processing, harvesting time, species, and processing conditions. This leads to inconsistent thermal behavior.
- **Moisture content:** Herbal drugs are mostly hygroscopic and contain bound or free water. The presence of moisture causes the appearance of early endothermic peaks (due to evaporation or dehydration) that can mask other important transitions.

8.3.6 Application to Herbal Drugs

DTA is particularly useful in the characterization and quality control of herbal drugs. It evaluates decomposition temperatures and thermal resistance of herbal powders and extracts, thus investigating their thermal stability. DTA is also useful for identifying polymorphic transitions in phytoconstituents. It identifies dehydration events in hygroscopic plant materials, which are essential for moisture content analysis. It is also used for compatibility analysis among herbal drugs and excipients during formulation development. The thermal profile of isolated phytoconstituents and complex herbal drugs can be recorded for the authentication of herbal drugs and adulterant identification.

Thermoanalytical characterization of plant drugs of different particle sizes and hydroalcoholic extract of *Urtica dioica* L. was carried out using DTA analysis. The DTA curve showed three exothermic peaks for all powdered herbal plant drugs and two peaks for extracts, indicating that some of the phytoconstituents of the powdered plant drug are not extracted. The exothermic peaks provide differences in terms of released heat among the samples. In all samples, the second exothermic peak appeared near 450°C, and represented higher values of enthalpy than the first peak. The thermal properties were found to be dependent on the particle size of powdered plant drugs. The enthalpy values are found to be higher in samples with larger particle sizes, and the samples with smaller particle sizes showed lower heat decomposition rates than the others. The spray-dried extract showed lower enthalpy in the first event and greater enthalpy in the second event compared to the oven-dried extract. Thus, the extracts and herbal powdered materials are distinguished by DTA thermal data [12].

The thermal properties of magnesium oxide (MgO) nanoparticles bio-synthesized from *Hevea brasiliensis*, *Chromolaena odorata*, and *Elaeis guineensis* leaf extracts were reported by DTA analysis. The DTA thermogram reflected variations in energy release. The *C. odorata*-based MgO nanoparticles showed the most thermally stable exothermic peaks. These outcomes reflected the influence of plant extracts on the thermal properties of nanoparticles. The differences in the thermal properties are attributed to the presence of unique phytochemical composition of each plant [13].

In a recent study, DTA was used to study the thermal analysis of curcumin-loaded nanostructured lipid carriers (NLCs). The TGA and DTA studies are carried out to highlight the thermal stability of CUR-loaded NLCs. CUR showed a sharp endothermic peak at 180°C. The CUR-loaded NLCs showed a small endothermic peak at 130°C corresponding to the first decomposition stage, a sharp endothermic peak at 270°C corresponding to the second decomposition stage, and one sharp exothermic peak at 290°C corresponding to the last decomposition stage. The presence of surfactants and lipid mixtures with curcumin molecules in the crystal lattice arrangement has strong hydrophobic bonding, resulting in higher and sharper peaks. The entrapment of CUR in the NLCs improved its thermal stability [14].

In one of the recent studies, thermal analysis and compatibility study of oral mucoadhesive hydrogel formulations using *Schinopsis brasiliensis* Engler was carried out. The extract and excipients were analyzed using DTA and TGA. The TGA and DTA thermograms for the extract-excipient binary mixtures with carbopol or pluronic F127 hydrogel maintained both the thermal properties of the extract and the excipient, representing compatibility between them. However, the suppression of the endothermic peak was noticed with other binary mixtures of extract with other gelling agents (carboxymethyl cellulose, chitosan, hydroxyethyl cellulose, hydroxypropyl methyl cellulose), representing incompatibility [15].

Curcumin-loaded chitosan nanoparticles were prepared using the ionic gelation method, and their anticancer activity was investigated against the Vero cell line. The nanoparticles were evaluated using thermal analysis, such as DTA and TGA. The DTA results of the nanoparticle showed a strong endothermic peak at 107°C, representing their thermal stability at higher temperatures. The TGA thermogram also demonstrated an initial slight weight loss followed by substantial weight loss at 107°C, representing thermal stability up to that temperature [16].

In a recent study, rutin-based solid dispersion (SD) was prepared using Poloxamer 407 and Poloxamer 188 as surfactant-based carrier matrices to overcome the poor aqueous solubility and permeability during oral administration. The developed SD and its excipients were characterized by DTA for their thermal property. The DTA data revealed that rutin existed as partially amorphous in the prepared solid dispersion. Pure rutin exhibited a sharp endothermic peak at 189°C due to its melting point, representing its crystallinity. The Poloxamer 407 and Poloxamer 188 displayed an endothermic peak at 69°C and an exothermic peak at 167°C, representing recrystallization of material upon heating. The DTA results of all SD were found to be partially amorphous, as the thermogram of rutin no longer appeared in the formulation. However, the thermogram of the physical mixture of drug and excipient showed three notable peaks referring to the individual components, though the height of the peaks was significantly decreased. During the heat of physical mixture, all components melt together, resulting in a thermogram similar to SD due to in situ development of SD by fusion [17].

In another study, curcumin-loaded liposomal solid gels were developed for hepatitis C Virus. The thermal properties and miscibility were characterized by DTA analysis. The curcumin showed a sharp endothermic peak at 174°C. The excipients D-α-tocopheryl polyethylene glycol 1000 succinate, cholesterol, and sucrose showed respective endothermic peaks. All the sharp peaks confirmed their melting point and crystalline property. The DTA thermogram of hydroxypropyl methylcellulose demonstrated broader endothermic peaks, whereas the soy phosphatidylcholine and egg phosphatidylcholine showed wide endothermic peaks. The DTA thermograms of curcumin-loaded liposomal solid gels represented a single melting endothermic peak around 200°C, representing a single-phase transition of the mixtures. The liposomal solid gels were found to be in a more ordered form, with behavior like a crystalline form. Further, none of the DTA thermograms of liposomal gels indicated a single endothermic peak of curcumin, suggesting amorphous modifications and curcumin did not exist as phase-separated in the mixtures. The DTA data revealed effective molecular dispersion of curcumin in the developed liposomal gel. In contrast, the thermograms of all physical mixtures exhibited peaks of almost every single formulation component, suggesting that a simple mixing is not sufficient to produce homogenous and molecular dispersion of curcumin in the solid gels [18].

8.4 THERMOGRAVIMETRIC ANALYSIS

Thermogravimetric analysis (TGA) investigates the temperature-based weight loss of drugs. It provides information on thermal stability and material composition. It is an essential tool for studying the thermal behavior of isolated phytoconstituents and herbal drugs, especially crystallinity, thermal degradation, moisture content, and volatile oil analysis. It helps in developing rational herbal products with excellent drug-excipient compatibility and stability [2, 19]. The principles, advantages, instrumentation, procedure, challenges during analysis, and their application for crystallinity and polymorphism analysis of herbal drugs are described in Chapter 6. In this chapter, their applications for investigating the thermal properties, stability, and drug-excipient compatibility are described with updated examples.

In addition to crystallinity and polymorphism, TGA is useful for analyzing the moisture content, volatile content, thermal properties, purity, stability, and drug-excipient compatibility of herbal drugs and their formulations. Most herbal drugs contain water in their crystalline structure. The level of water can be quantified from the percentage weight loss due to dehydration [2, 15]. It is also used for the investigation of volatile components, the investigation of major components, and the moisture content of herbal drugs [2]. It provides information on the thermal stability of herbal drugs, which is required for their storage and processing [2].

The modified Ozawa method is utilized as an assisting technique to distinguish diverse granulometry powders by explaining the kinetic data from the TGA curve. Thermal analysis using the Ozawa technique establishes an analytical tool for assessing the physical stability of herbal drugs. The TGA method allows the discernment of various powder particle sizes of herbal drugs [2]. By investigating the weight loss under different heating rates, valuable insights into the kinetics of decomposition can be obtained, which is required for quality control, process optimization, and formulation development in the herbal industry. The particle sizes of *Tabebuia caraiba* (Mart.) Bur. powder was reported using the Ozawa technique. The TGA analysis was carried out at various heating rates. The kinetic data results showed different activation energy values and the same reaction order, consistent with the particle size of the powdered sample in a specific model. The herbal drugs having a larger particle size require additional energy to advance the enthalpic proceedings associated with their thermal decomposition. Thus, the TGA analysis allows the discrimination of particle sizes of herbal drugs [20].

Tucumã fat from *Astrocaryum vulgare* was explored as a pharmaceutical excipient for the development of a nanostructured lipid carrier. Its thermal property was evaluated using DSC and TGA. The TGA study was carried out under a nitrogen and synthetic air flow atmosphere. The Tucumã fat showed a mass loss of 98.83% with an onset temperature of 384°C due to the decomposition of triglycerides within 380–480°C in a nitrogen environment. The residual mass of 1.17% demonstrated the maximum lipidic content of fat. Under a synthetic air flow atmosphere, the mass loss was found to be 97.69% at an onset temperature of 355°C. The reduction in onset temperature in TGA in synthetic air compared to nitrogen is attributed to the degradation of unsaturated fatty acids that occurred earlier in the presence of an oxidative atmosphere. A small event was reported at 445°C in the curve derivative, due to the combustion of Tucumã fat [6].

In one of the recent studies, the compatibility of *Schinopsis brasiliensis* Engler extract and excipients [carboxymethyl cellulose (CMC), Carbopol (CAR), hydroxypropyl methyl cellulose (HPMC), pluronic F127 (PLU), and hydroxyethyl cellulose (HIDRO)] in oral mucoadhesive hydrogel formulations has been reported using TGA. The TGA curves for the API- binary mixtures with carbopol (CAR) or pluronic hydrogel (PLU) maintained the thermal events of both excipient and API, suggesting drug-excipient compatibility, hence used for hydrogel formulation [15].

TGA analysis of herbal drugs allows the investigation of the major components of herbal drugs. By comparing the thermal properties of multiconstituent herbal extracts with those of pure phytoconstituents, it is possible to identify major constituents and estimate their amount [2]. In a study, TGA analysis of Eucalyptus essential oil from *Eucalyptus camaldulensis*, *Eucalyptus pellita*, and commercial essential oil samples was made and results compared. The TGA thermograms of *Eucalyptus camaldulensis* essential oil, commercial oil, and its major phytoconstituent cineol were found to be similar within 55–145°C. The TGA curve similarity reflected cineole as its major active constituent of *Eucalyptus camaldulensis* essential oil and commercial essential oil. However, the TGA thermogram of the *Eucalyptus pellita* essential oil showed a different pattern due to altered phytochemical compositions [21].

TGA allows the investigation of the thermal stability of herbal drugs and allows their quality control evaluations. Fatty acids are important components of herbal drugs, and their content is related to their thermal stability. TGA can estimate the amount of fatty acids in the herbal extract. By investigating the thermal properties of fatty acids in the extract, it is possible to understand the relationship between fatty acid content and the thermal stability of herbal drugs [2].

The thermal stability of oils is inversely proportional to the degree of unsaturation, as the unsaturated fatty acids possess a lower boiling point than their saturated counterparts. TGA analysis of oils from Brazilian Cerrado plants was made due to their high oleic acid content. The Pequi and baru oils showed relatively higher stability than amburana oil due to their lower content of volatile short-chain fatty acids [22].

In a recent study, the thermal properties of *Malania oleifera* seed oil were studied using TGA analysis. The initial weight loss within 27–107°C is attributed to the loss due to the presence of low-boiling-point solvent residues. The oil showed thermal stability up to 187°C with negligible weight loss. The thermal event was found to have suddenly changed around 200°C, and thermal degradation started within 230–476°C due to the oxidation and decomposition of organic matter [7].

In another study, the thermal stability of *Piper longum*-derived theranostic carbon nano dots was studied by TGA. The TGA thermogram showed a five-step degradation event up to 800°C. The carbon dots showed thermal stability up to 196.75°C with an initial negligible weight loss. The initial weight loss of 2.68% was reported for the evaporation of water molecules and weakly bonded chemical groups. The continuous weight loss within 200–800°Cwas reported for the denaturation of strongly attached functional groups that are linked via covalent bonds [23].

Thermoanalytical evaluation of powdered plant drug of different particle sizes and hydroalcoholic extract of *Urtica dioica* L. was made by TGA analysis. The TGA thermogram for powdered plant material showed a five-step decomposition pattern, whereas the extract showed a six-step degradation pattern. The first weight loss for powdered plant drugs occurred within 35.0–107.0°C due to loss of volatiles and water. The main degradation step occurred after 300°C, corresponding to macro-and micro compound degradations. The powdered plant drugs with very low particle size showed greater loss of weight and a higher amount of residue than the others. The spray-dried extract showed a greater amount of residue than the oven-dried extract, suggesting greater physical stability due to its higher level of dehydration than the oven-dried extract [12].

In a recent study, curcumin-loaded NLCs were prepared to improve bioavailability and thermal stability. The TGA analysis of pure curcumin and nanoformulation was carried out to study their thermal stability. The TGA thermogram of curcumin-loaded NLC indicated a loss of weight at 133°C, while weight loss in pure curcumin was found to occur at 100°C. The decomposition temperature of curcumin-loaded NLCs was quite different from that of pure curcumin. The encapsulation of curcumin into NLCs increased the thermal stability [14].

8.5 HOT STAGE MICROSCOPY

Hot stage microscopy integrates optical microscopy with a temperature-controlled system to regulate the temperature of the sample under investigation. It offers real-time analysis of the morphology, physical properties, phase transitions, melting, and crystallinity of herbal drugs. It also provides important information on the polymorphism and stability of phytoconstituents and herbal drugs [24–26]. The principles, advantages, instrumentation, procedure, and challenges during analysis of hot stage microscopy and their application for crystallinity and polymorphism analysis of herbal drugs are described in Chapter 6. In this chapter, their applications for investigating the thermal properties, stability, and drug-excipient compatibility are described.

In addition to crystallinity analysis and polymorphism by hot stage microscopy, it is used for particle size distribution and characterization of bioactives in the final formulation [24]. Further, it can be used in addition to DSC to investigate possible drug-excipient compatibility. The visual capabilities offer an advantage as the interactions can be visually observed, which was limited in DSC studies [24].

The thermal property and melting behavior of gallic acid-loaded cellulose acetate electrospun nanofibers was studied by hot stage microscopy. It was observed that the melting temperature of the neat electrospun cellulose acetate mat started at 212°C. However, the onset of the melting point of gallic acid-loaded electrospun cellulose acetate fiber mats was found to be decreased with increasing concentration of gallic acid from 2.5% to 10% [27].

In a recent study, the berberis extracts based on solid lipid nanoparticles (SLN) were developed and evaluated for wound healing. The thermal properties of vacuum oven-dried, freeze-dried extracts and SLN were studied by hot-stage microscopy, DSC, and TGA. The DSC and TGA thermograms of vacuum-dried extract showed broad and sharp peaks, indicating the melting process of crystalline constituents. However, only the broad peaks were observed in the freeze-dried sample, indicating a lack of crystallinity. The results are found to be in accordance with the hot-stage microscopy outcomes of two extracts [28].

8.6 DYNAMIC MECHANICAL ANALYSIS OR DYNAMIC MECHANICAL THERMAL ANALYSIS

DMA or DMTA is a non-destructive mechanical technique that analyzes the viscoelastic and deformation properties of samples under varying temperature and frequency conditions [29–31]. Through cyclic application of a non-destructive stress or strain, a comprehensive understanding of the viscoelastic properties of materials can be obtained. Parameters such as crosslinking density, dynamic/complex viscosity, dynamic fragility, storage/loss compliance, creep compliance/stress–relaxation modulus, and the non-Arrhenius variation of relation times with temperature are studied using DMA [32]. When the temperature increases, molecular motion in polymers increases, changing their mechanical behavior. It characterizes the linear viscoelastic properties of a substance through application of a small oscillating stress or strain, selected from within the linear viscoelastic region. An increase in stress results in a proportional increase in strain, and thus the relationship between the applied stress and resultant strain is constant, despite the variance of strain with time. After the application of constant stress, viscoelastic substances show a time-dependent, recoverable deformation property in both elastic and viscous materials. Viscoelastic substances will continuously deform as a function of time after application of a constant stress. Upon removal of the applied stress, some of the applied energy will have been lost due to viscous flow; however, energy will also have been stored and used to recover from the stress-induced deformation [29].

8.6.1 Principle

DMA works by applying a small, oscillatory force to a material sample and measuring the resulting strain. The response of the material is analyzed to determine its elastic (storage modulus) and viscous (loss modulus) properties. The storage modulus (E' or G') indicates the elastic (energy-storing) component, and the loss modulus (E' or G') indicates the viscous (energy-dissipating) component. The ratio of loss modulus to storage modulus (E''/E') or tan δ (delta) represents damping behavior [31].

8.6.2 Advantages of DMA

- **Sensitive identification of transitions**: DMA accurately identifies glass transition temperature (Tg) and other thermal properties in crude herbal drugs, extracts, isolated phytoconstituents, excipients, and herbal formulations, even in small quantities.
- **Characterization of viscoelastic properties:** It offers a detailed insight into the thermomechanical properties of herbal drugs and their formulations over a range of temperatures and frequencies.
- **Compatibility and stability studies**: It is also helpful for identifying the drug-excipient compatibility by analyzing the changes in mechanical behavior, which helps in optimizing formulation stability.
- **Phase property evaluation**: DMA differentiates between amorphous and crystalline phases, helping in detecting amorphization or polymorphic transitions of herbal constituents or extracts.
- **Non-destructive and requires a small sample of low quantity:** DMA requires very small amounts of sample and often allows non-destructive testing.
- **Real-time curing and gelation analysis**: It monitors curing behavior in semi-solid or gel-based herbal formulations, which is helpful in investigating gel formation, biopolymer interactions, and drying kinetics.

- **Quantification of damping behavior**: DMA measures tan δ, which is useful for evaluating material resilience, stiffness, or flexibility in topical herbal formulations or films.

8.6.3 Instrumentation

The DMA instrument is schematically shown in Figure 8.3. The major components of the DMA instrument are as follows:

- **Sample holder**: It holds the sample in place, and various modes of deformation (tensile, compression, shear) are applied to the sample.
- **Oscillatory force generator/drive motor**: It applies an oscillatory (sinusoidal) force or displacement to the sample and precisely controls the amplitude and frequency of the applied deformation.
- **Displacement sensor/strain transducer**: It measures the strain (deformation) of the sample in response to the applied force. It is highly sensitive and enables the identification of subtle transitions and behavior.
- **Force sensor or load cell**: It measures the stress applied to the specimen and ensures accurate measurement of loss modulus (E''), storage modulus (E'), and damping factor (tan δ).
- **Furnace**: It controls the temperature of the sample chamber and provides a uniform environment. It enables dynamic thermal scanning from cryogenic temperatures to high temperatures and provides linear heating/cooling ramps for thermal studies.
- **Frequency generator/controller**: It adjusts the oscillation frequency over a broad range and used to study the time–temperature–frequency dependence of viscoelastic properties of samples.
- **Data acquisition and control system**: It records and analyzes the mechanical response (stress–strain phase shift).

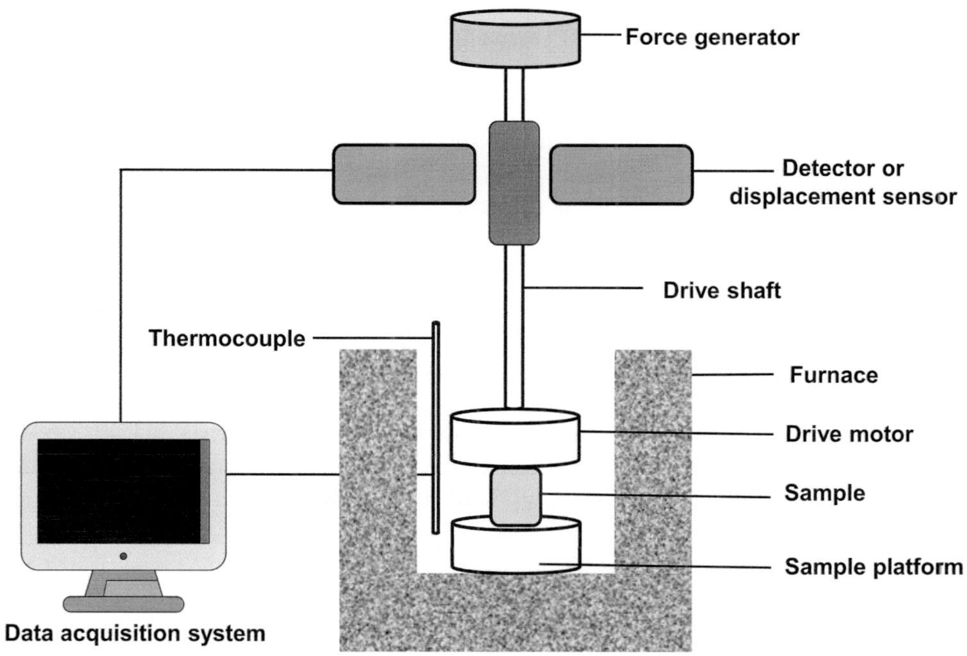

FIGURE 8.3 Schematic representation of DMA

8.6.4 Procedure

- **Sample preparation:** The herbal sample is prepared in a suitable form (film, tablet, pellet, or unprocessed) to ensure its thickness and dimensions.
- **Selection of appropriate fixture and mounting:** An appropriate fixture is selected based on the type of sample (tensile for films, compression for tablets, and three-point bending for stiff samples).
- **Setting of test parameters:** Then, the DMA parameters (temperature range, heating rate, frequency, strain amplitude) are set in the software.
- **Baseline correction:** Sometimes, the baseline correction is carried out without a sample to correct the thermal behavior of the instrument.
- **Running of experiment:** The oscillatory force is applied on the sample, and strain response, moduli (E', E''), and tan δ over the temperature range are recorded. The glass transition temperatures are monitored.
- **Data acquisition and analysis:** The data is analyzed by software to study changes in moduli, or other transitions to investigate viscoelastic properties, thermal stability, and mechanical integrity.

8.6.5 Challenges

- **Chemical complexity:** Herbal drugs are composed of numerous phytochemicals that may lead to overlapping transitions, making interpretation difficult.
- **Difficulty in sample preparation due to heterogeneity:** Crude herbal powders and their extracts often have inconsistent particle size, composition, and density, which makes it difficult to prepare uniform, reproducible samples (tablets, films, or composites) from herbal raw materials.
- **Moisture content:** Herbal drugs are mostly hygroscopic and contain bound or free water. The presence of moisture causes significant alteration of their viscoelastic properties. Herbal drugs often require a controlled humidity environment and pre-conditioning, which increases the experimental complexity.
- **Thermal degradation:** Some of the thermolabile phytoconstituents in herbal drugs decompose before meaningful mechanical transitions occur. This limits the usable temperature range for DMA.
- **Low mechanical strength:** Herbal drugs and formulations sometimes possess poor mechanical strength, leading to sample failure or slippage during testing. This limits the use of certain deformation modes (e.g., bending or tension).
- **Incompatibility with conventional fixtures:** The irregular shapes, powders, or sticky pastes may not be compatible with standard DMA fixtures. Such materials require custom sample holders or support matrices (e.g., embedding in polymer films) for accurate mechanical analysis.
- **Batch-to-batch variability:** The chemical composition of herbal drugs and extracts differs from batch to batch due to variations in geographical origin, processing, harvesting time, species, and processing conditions. This leads to inconsistent thermo-mechanical behavior.

8.6.6 Application of Herbal Drugs

In herbal drugs, DMA is useful for studying the mechanical properties of powdered drugs, extract, and their formulations. Such analysis enables the investigation of stability, structural integrity, and performance under various conditions. Such information is important for evaluating the safety and efficacy of herbal drugs. Additionally, DMA is used to optimize processing and development of herbal formulations by providing information on the optimal temperature, pressure, and time required to develop products with the desired mechanical properties. Overall, DMA has become an essential tool for investigating the mechanical properties of herbal drugs [2].

DMA is one of the most powerful techniques for investigating the behavior of polymer composites and natural fiber composites. DMA investigates the viscoelastic and deformation properties of natural fiber composites. DMA is used in characterizing composite structures and damping as a function of temperature, frequency, stress, time, atmosphere, or a combination of these parameters. DMA depends on morphology, the physical or structural arrangement of phases, and the nature of constituents [32].

Recently, a comparative study on the mechanical properties of unidirectional and flax fiber reinforced thermoplastic polypropylene, thermoset, and polylactic acid composites, with those of carbon and glass fiber reinforced epoxy composites, was evaluated by DMA. The DMA analysis of composites reinforced with flax fiber showed improved damping compared to composites reinforced with synthetic carbon and glass fiber [32].

In another study, the dynamic mechanical properties and nanomechanical properties of rubber–wood–plastic composites were evaluated by DMA. DMA results showed that the composites treated with multiple coupling agents (combination of maleic anhydride polyethylene (MAPE) and bis(triethoxysilylpropyl) tetrasulfide, and combination of MAPE and vinyltrimethoxysilane) showed greater storage modulus than both untreated and single-coupling-agent-treated composites due to their superior interfacial bonding quality. The treated composites showed the shifting of the relaxation peak and Tg toward higher temperatures, representing the enhancement of interfacial interaction and adhesion [33]. The thermomechanical properties of nanocellulose elastomer composites were studied by dynamic mechanical analysis. Cellulose nanofiber and cellulose nanocrystals can be easily derived from cellulose sources, are ubiquitous and renewable, and are a potential filler for elastomeric nanocomposites instead of widely used conventional fillers such as carbon black and silica. The influence of cellulose nanofibers and cellulose nanocrystals on the dynamic performance of various elastomers (natural rubber, styrene-butadiene rubber, nitrile rubber, and polyurethane) was studied [34].

8.7 THERMOMECHANICAL ANALYSIS

TMA analyzes the changes in dimensions of a sample as a function of temperature or time under a constant load [31, 35]. It is especially useful for understanding mechanical integrity, thermal stability, and process-related transformations in drugs, polymers, films, and drug formulations. In the case of herbal drugs, TMA is used to study the expansion and contraction of herbal products under different temperature and pressure conditions. Such information is needed to investigate the stability and performance of isolated active phytoconstituents and herbal products [2]. TMA is used to optimize the formulation and processing of herbal drugs by providing information on the optimal temperature, pressure, and time required to develop formulations with the desired physicochemical properties, which is crucial for ensuring their safety and therapeutic efficacy [2].

8.7.1 Principle

TMA works on the principle that materials undergo contraction, expansion, or deformation when subjected to changes in temperature and/or mechanical load. These changes are recorded as a function of time and temperature. By recording these dimensional changes precisely, TMA provides valuable information on the physical and thermal properties (coefficient of thermal expansion, stress relaxation, softening point, glass transition temperature, creep behavior, sintering or curing behavior) of samples [31, 35].

8.7.2 Advantages

- **Requirement of small sample:** TMA requires a minimal quantity of sample, making it ideal for precious or rare herbal drugs.
- **Non-destructive:** TMA allows non-destructive analysis in certain modes, preserving valuable herbal specimens.
- **Early detection of phase transitions and degradation:** It can detect dimensional instability before visible signs of degradation, aiding in preformulation studies.
- **Detection of glass transition temperature and softening points:** TMA can identify Tg or softening behavior of herbal drugs and formulations, which is required for the investigation of storage stability.
- **Measurement of thermal expansion**: It evaluates the dimensional changes of herbal drugs to assess their thermal expansion or contraction, helping in packaging and shelf life prediction.
- **Study of moisture-dependent properties**: Many of the herbal drugs are hygroscopic in nature. TMA investigates the influence of moisture on mechanical stability.
- **Characterization of herbal-excipient interactions**: Identifies incompatibilities or interactions between herbal actives and excipients by observing unexpected shrinkage or deformation during heating.
- **Evaluation of herbal gels**: TMA helps to investigate the deformation, flexibility, and dimensional stability of topical herbal gel at storage temperatures.
- **Investigating curing and sintering properties**: Useful in studying thermal curing of herbal gums, resins, or polymer–herb matrices, particularly for dosage forms involving natural binders or adhesives.

8.7.3 Instrumentation

The TMA instrument is schematically shown in Figure 8.4. The major components of the TMA instrument are as follows [31, 35]:

- **Sample holder/sample stage**: It provides a platform to hold the sample in place. Various fixture types (expansion mode, penetration mode, tension mode, compression mode) are applied to the sample. The sample holder is designed to hold film, fiber, solid, or powdered samples.
- **Displacement sensor/probe**: It comprises a precision quartz or metal probe that contacts the sample. It detects vertical and horizontal displacement of the sample. It is connected to a linear variable differential transformer (LVDT) or capacitive sensor for accurate displacement measurements.
- **Furnace/temperature control system**: It controls the temperature of the sample chamber and provides a uniform environment. It provides linear heating/cooling ramps and isothermal holds for thermal studies.
- **Load application system:** It provides a static or dynamic force to the sample in the mN to N range. It is often controlled by motor driven or electromagnetic force mechanisms.
- **Gas purging system (optional):** The gas purging system allows inert gas (e.g., nitrogen, argon) or air to be purged through the sample chamber and prevents oxidative degradation during heating. It offers a controlled atmosphere for sensitive herbal drugs.
- **Data acquisition and control system**: It records and analyzes the mechanical response and provides temperature versus displacement plotting, thermal expansion coefficients, Tg, softening point, and detection of sintering behavior.

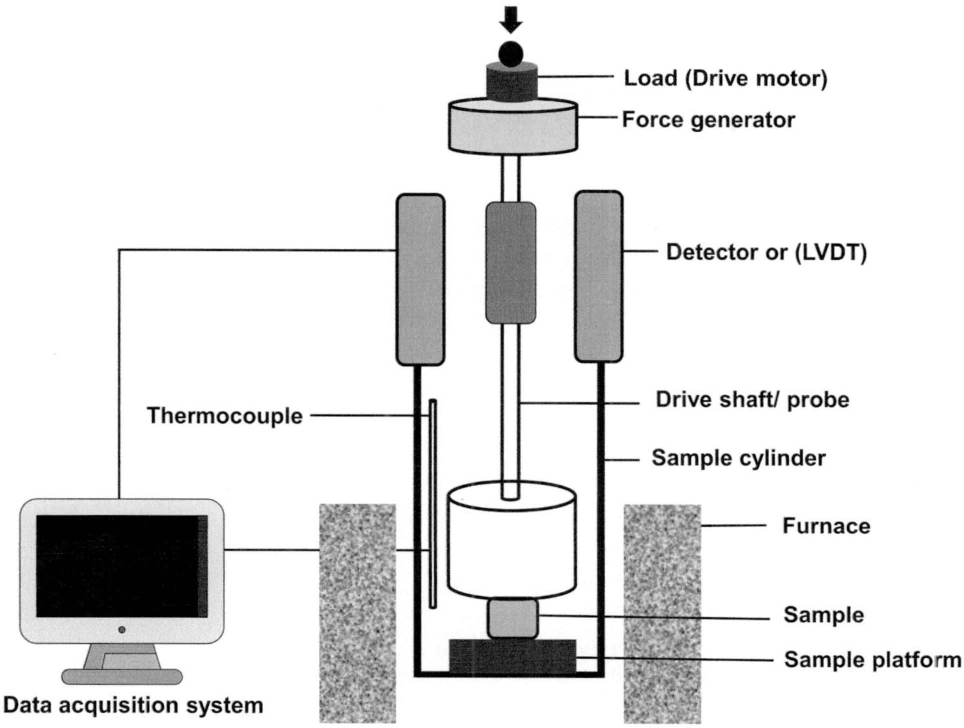

FIGURE 8.4 Schematic representation of TMA

8.7.4 Procedure

TMA involves following steps for analysis of herbal samples [31, 35].

- **Sample preparation:** The herbal sample is prepared in a suitable form (film, tablet, pellet, or unprocessed) to ensure its thickness and dimensions. The sample is ensured for smooth surface and uniform thickness.
- **Instrument setup:** TMA is allowed for stabilization, and the appropriate test mode is applied (expansion mode, penetration mode, compression mode, tension mode).
- **Sample loading:** The sample is placed on the sample stage, and the probe is lowered to make light contact with the sample surface. Then, a small preload force is applied to ensure sample contact without deformation.
- **Setting of test parameters:** Then, TMA parameters (temperature range, heating rate, load, and atmosphere) are set in the software.
- **Running of experiment:** The thermal scan is started, and the dimensional changes (expansion, shrinkage, penetration) with an increase in temperature are recorded. The thermal expansion curves, softening point, Tg, and dimensional stability are recorded in real time.
- **Data acquisition and analysis:** The data is analyzed by software to study the coefficient of thermal expansion, glass transition temperature, and softening or melting point.

8.7.5 Challenges

- **Chemical complexity:** Herbal drugs are composed of numerous phytochemicals that may lead to overlapping transitions, making it difficult to identify clear transitions such as Tg or softening point in TMA curves.
- **Difficulty in sample preparation:** Crude herbal powders and their extracts often are heterogeneous in nature (irregular in shape and texture), which makes it difficult to prepare uniform, flat, and reproducible samples for TMA analysis.
- **Moisture sensitivity:** Herbal drugs are mostly hygroscopic and contain bound or free water. The presence of moisture significantly alters their thermal expansion and softening behavior. Hence, it often requires controlled humidity conditions and pre-conditioning, which increases complexity.
- **Thermal degradation:** Thermolabile phytoconstituents degrade at moderate temperatures, preventing the study of thermal profile in the full temperature range. Thermal degradation leads to the development of artifacts, which leads to irregular probe response.
- **Low mechanical strength:** Herbal drugs and formulations sometimes possess poor mechanical strength, leading to breaking or deformation under even small loads. It limits the applicability of tension or penetration modes.
- **Fixture compatibility:** Some herbal preparations are pastes, sticky extracts, and resinous in nature, which are not compatible with standard sample holders. Such materials require custom sample holders for accurate mechanical analysis.
- **Batch-to-batch variability:** The chemical composition of herbal drugs and extracts differs from batch to batch due to variations in geographical origin, processing, harvesting time, species, and processing conditions. This leads to inconsistent thermo-mechanical properties.
- **Difficulty in data interpretation:** Overlapping thermomechanical profiles from multiple constituents makes it difficult to interpret TMA results.

8.7.6 Application to Herbal Drugs

TMA is used for evaluating the thermal behavior and mechanical properties of herbal drugs under controlled temperature and force. It is used to study shrinkage, expansion, glass transition, compaction behavior, and softening point of herbal powders or extracts, as well as herbal formulations. It aids in evaluating the physical stability of plant-based drugs and formulations and understanding how they respond to thermal stress, which is important during processing, storage, and formulation development. In herbal granules or tablets, TMA identifies compressibility, disintegration, and deformation behavior under heat, contributing to quality control and formulation optimization. It is also used to study the Tg in amorphous herbal drugs. Overall, TMA offers important insights into the thermo-mechanical properties of herbal drugs.

In a study, residues high in lignin and ash generated from the simultaneous saccharification and fermentation of corn stover were thermally pretreated in an inert (N_2) environment to investigate the effect of temperature and time on their softening points. The untreated and treated residues were characterized by TMA. It was observed that thermal pre-treatment raises the softening point from 187°C to 250°C. The increase in softening point was identified due to the formation of covalent bonds in the lignin structures and the removal of low molecular weight volatile intermediates [36].

8.8 EVOLVED GAS ANALYSIS

EGA involves the measurement of gases that are evolved during the thermal degradation of a sample [37–40]. In case of herbal drugs, EGA is used to identify volatile constituents of herbal extracts and to examine thermal stability and degradation pathways of plant bioactives. Such information ensures the safety and efficacy of herbal drugs and formulations. EGA is used to optimize the formulation and processing conditions of herbal drugs by providing information on the optimal temperature, extraction time, and storage [2].

8.8.1 Principle

EGA is based on the principle that when a sample is subjected to heat, it may undergo desorption, decomposition, oxidation, or other thermal events, leading to the release of gaseous products. Then the evolved gases are analyzed in real time using appropriate gas-detecting analytical instruments (gas chromatography, mass spectrometry, Fourier transform infrared spectroscopy, and photoionization detectors). Such analysis is often coupled with TGA, DSC, and DTA.

8.8.2 Advantages

- **Non-destructive and coupling with other analytical tools:** EGA is often combined with other analytical tools (TGA-MS, TGA-FTIR, TGA-GC) for simultaneous weight loss and gas evolution analysis [40–42]. It provides a comprehensive understanding of structural and thermal information from a single experiment.
- **Minimal sample requirement:** It requires small amounts of herbal drugs, making it ideal for rare or costly phytoconstituents, standardized herbal extract, and formulations.
- **Identification of volatile phytoconstituents:** It enables the identification of volatile constituents, essential oil, terpenes, and so forth, in herbal drugs upon heating and aids in fingerprinting and quality control of herbal drugs.
- **Detection of residual moisture and solvents:** EGA is very sensitive to moisture, ethanol, methanol, and other solvents of extraction. Hence, it is useful for residual solvent analysis in extract and herbal formulations. It is useful for ensuring regulatory compliance with the limits of residual solvent.
- **Understanding thermal decomposition pathways:** Since EGA causes temperature-dependent release of gases, it helps in studying the thermal stability and decomposition mechanism of phytoconstituents.
- **Identification of purity and contamination:** EGA enables the identification of unwanted impurities, or adulterants, based on evolved gases not associated with the authentic herbal profile.

8.8.3 Instrumentation

The EGA instrument is schematically shown in Figure 8.5. The major components of the TMA instrument are as follows:

- **Thermal analyzer (TGA/DSC/DTA unit):** This is the core of the heating system that produces gases due to thermal heating of samples. The unit heats the sample under a programmed temperature ramp, and the evolved gas is directed to the detector.
- **Gas transfer interface/heated transfer line:** It is a thermally insulated heated tube that connects the thermal analyzers to the gas detection instrument. It transports the generated gas to the detector. Inert gas (e.g., He, N_2) is often used as a carrier gas to transport evolved gases to the detector.
- **Gas detection system:** Various gas detection instruments, such as MS, FTIR, and GC, including the thermal conductivity detector, photoionization detector, or flame ionization detector, are used for the analysis of evolved gas.
- **Data acquisition and control system:** It regulates the heating program and gas flow and enables data collection. It synchronizes data between the thermal analyzer and the gas detector, produces a mass-loss-versus-temperature curve and gas evolution profiles, and helps in the identification of compounds in real time.

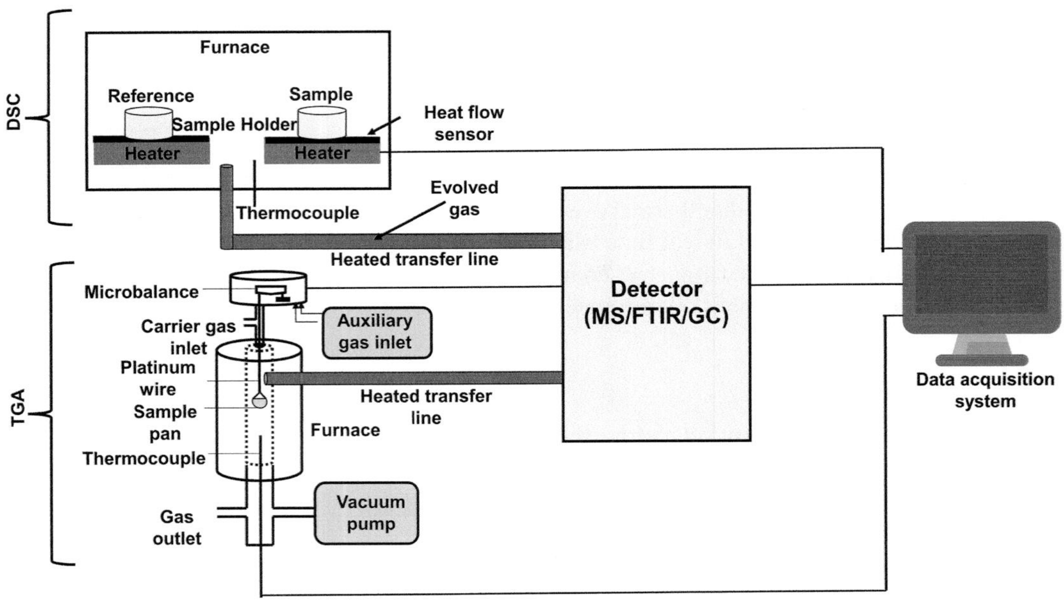

FIGURE 8.5 Schematic representation of EGA

8.8.4 Procedure

- **Sample preparation and loading:** The herbal sample is dried and powdered to produce a uniform sample and loaded into the TGA/DSC crucible (platinum, alumina, or ceramic).
- **Instrument setup:** The thermal analyzer is turned on, and the heated gas transfer line is activated to prevent gas condensation. Then the system is purged with inert carrier gas (e.g., nitrogen or helium) to maintain a stable atmosphere.
- **Setting of experimental conditions:** Then, various parameters (temperature range, heating rate, load, gas flow rate, atmosphere, and detection mode) are selected in the software.
- **Calibration of instrument:** It is an optional step; however, it is required for producing consistent data. The instrument is calibrated for the thermal analyzer and detectors using standard materials or known gases (e.g., CO_2, H_2O, methanol) for accurate identification and quantification of gases.
- **Running of the experiment:** The thermal scan is started.
- **Data acquisition and analysis:** The data is analyzed by software to study decomposition temperatures of phytoconstituents, percentage of residual solvent, and volatile matters of herbal drugs for their authentication.

8.8.5 Challenges

- **Chemical complexity:** Herbal drugs and extracts are composed of numerous phytochemicals. Their simultaneous decomposition may release overlapping gases, leading to difficulty in the interpretation of the gas profile.
- **Low concentration of evolved gases:** Sometimes, the volatile components of herbal drugs are present in trace amounts, leading to weak gas signals due to the concentrations below the lower limit of detection.
- **Thermal degradation:** Thermolabile phytoconstituents degrade at moderate temperatures, forming complex secondary products rather than simple volatiles. This makes it hard to correlate original constituents to evolved gases.

- **Difficulty in sample preparation:** Crude herbal powders and their extracts often are heterogeneous in nature (varied moisture content, particle size), which influences gas evolution.
- **Condensation and adsorption into the transfer lines:** Sometimes, high-boiling-point volatile substances or sticky materials may condense or adsorb inside transfer lines, causing loss of signal or sample contamination in subsequent runs.
- **Instrument compatibility and calibration:** Coupling TGA with MS, FTIR, or GC requires precise alignment, calibration, and temperature control.
- **Batch-to-batch variability:** The chemical composition of herbal drugs and extracts differs from batch to batch due to variations in geographical origin, processing, harvesting time, species, and processing conditions. This leads to inconsistent outcomes among batches.
- **Interference from matrix or excipients:** In herbal formulations, excipients and adulterants can interfere with gas detection and mask true herbal signals.

8.8.6 Application to Herbal Drugs

EGA is used for characterization of volatile phytoconstituents (essential oils, terpenes, flavonoids), moisture content, residual solvent analysis, fingerprinting for quality control and authentication, thermal degradation and stability studies, identification of decomposition products, process optimization, stability, and investigating the interaction of components in herbal formulation.

In a study, the EGA of de-oiled microalgal biomass of *Chlorella variabilis* was studied. The analysis of evolved gases during pyrolysis was investigated using a TGA-MS analysis. EGA results showed the abundance of most gases between 200–300°C and 400–500°C. It has been noticed that most of the nitrogenous compounds were produced within the range of 200–550°C, which is mainly occupied by the volatilization zone. Higher ammonia production was observed between 200°C and 300°C. Nitrile group containing compound hydrogen cyanide was observed between 200°C and 500°C. Nitrogen-containing compounds released from the pyrolysis of de-oiled microalgal biomass are mainly associated with the decomposition of proteins present in it [43].

In another study, EGA of coffee was reported by single photon ionization-mass spectrometry. Coffee is a chemically well-known food product of scientific and commercial interest. The roasting process of coffee beans (Robusta, and Arabica) was simulated in a thermogravimetry (TG) coupled to single photon ionization time-of-flight mass spectrometry (TG-SPI-TOFMS). The chemical composition of the evolved roasting gases was investigated online. Roasted and ground coffee powders of different types and brands, as well as instant coffee, were successfully investigated. Diterpenes kahweol and cafestol were detected among multiple roasted products. These compounds can be of particular interest for the quality control of coffee. It is shown that kahweol can be used as a tracer compound to discriminate Arabica coffee species from Robusta species [44].

8.9 CONCLUSION

Thermal analytical techniques are widely used for the characterization of physical, chemical, and mechanical properties of herbal drugs. The thermal properties of multicomponent-based herbal extracts are influenced by their chemical composition, the extraction method used, and the presence of residual solvents. The investigation of thermal properties of herbal drugs using techniques like DSC, TGA, DTA, and hot-stage microscopy offers comprehensive insights into their physicochemical properties, thermal stability, quality, and drug-excipient compatibility in the finished herbal formulation. Nowadays, these thermal techniques have been hyphenated with various analytical methods (TGA with gas chromatography: TGA-GC, TGA with Fourier transform infrared spectroscopy-mass spectrometry: TGA-FTIR-MS, TGA with gas chromatography-mass spectroscopy: TGA-GC-MS) for providing a comprehensive understanding of the physico-chemical properties of herbal drugs and their products [2]. Collectively, thermal analytical methods offer valuable information on thermal properties, phase transitions, degradation temperatures, moisture content, volatile material content, and crystallinity, which are important for understanding the quality, purity, integrity, and stability of herbal drugs and their formulations. These techniques also allow the standardization, quality control, and shelf life prediction of herbal drugs and finished formulations for their safe and therapeutic applications. Considering the growing demand for herbal drugs and their formulations, thermal analysis stands out as a robust and indispensable approach in ensuring product consistency, efficacy, and regulatory compliance. However, the thermal property evaluations of herbal drugs are relatively difficult compared to those of isolated phytoconstituents and synthetic drugs due to their chemical complexity.

REFERENCES

1. Singadi, R., et al., Application of differential scanning calorimetry to study the interpretation on herbal medicinal drugs: A review. *TMR Pharmacology Research*, 2022. 2(2): p. 6.
2. Pirsaheb, M., et al., Thermal analysis techniques in herbal medicine: A comprehensive review on unveiling integrity and quality for future perspectives. *Journal of Analytical and Applied Pyrolysis*, 2023. 175: p. 106192.
3. Aswathy, K., et al., Formulation and in-vitro characterization of fast-disintegrating herbal extract sublingual immunotherapy tablet for peanut-induced allergic asthma. Saudi *Journal of Biological Sciences*, 2022. 29(3): p. 1283–1297.
4. Manoharan, S., P. Balakrishnan, and L.K. Sellappan, Fabrication of highly flexible biopolymeric chitosan/agarose based bioscaffold with Matricaria recutita herbal extract for antimicrobial wound dressing applications. *International Journal of Biological Macromolecules*, 2024. 281: p. 136195.
5. Malik, Z., et al., Solid lipid nanoparticles for delivery of 5-fluorouracil and andrographolide: A combination approach for the treatment of cancer. *BioNanoScience*, 2024. 14(3): p. 3059–3077.
6. Gomes, A.T.A., et al., Tucumã (Astrocaryum vulgare) fat: An Amazonian material as a pharmaceutical input for lipid nanoparticle production. *Journal of Thermal Analysis and Calorimetry*, 2022. 147(1): p. 355–365.
7. Su, J.-L., et al., Fatty acid composition and thermal characteristics of Malania oleifera seed oil. *Advanced Composites and Hybrid Materials*, 2022. 5(2): p. 1268–1279.
8. Chaudhari, V.S., U.S. Murty, and S. Banerjee, Nanostructured lipid carriers as a strategy for encapsulation of active plant constituents: Formulation and in vitro physicochemical characterizations. *Chemistry and Physics of Lipids*, 2021. 235: p. 105037.
9. Brown, M.E., Differential thermal analysis (DTA) and differential scanning calorimetry (DSC), in *Introduction to Thermal Analysis: Techniques and Applications*. 1988, Springer. p. 23–49.
10. Haines, P., M. Reading, and F. Wilburn, Differential thermal analysis and differential scanning calorimetry, in *Handbook of Thermal Analysis and Calorimetry*. 1998, Elsevier. p. 279–361.
11. Devi, N., et al., A review on differential thermal analysis. *Chemistry Research Journal*, 2021. 6(4): p. 71–80.
12. Cuinica, L.G. and R.O. Macêdo, Thermoanalytical characterization of plant drug and extract of Urtica dioica L. and kinetic parameters analysis. *Journal of Thermal Analysis and Calorimetry*, 2018. 133(1): p. 591–602.
13. Otabor, G.O., et al., Comparative thermogravimetric and differential thermal analysis of bio-synthesized MgO-nanoparticles using Chromolaena odorata, Hevea brasiliensis, and Elaeis guineensis leaf extracts. *Tropical Journal of Chemistry*, 2025. 1(1): p. 1–7.
14. Pishnamazi, M., S.M. Alshahrani, and A. Alkhammash, Exploring SLT80 nanostructured lipid carriers as a novel delivery system for curcumin: Thermal analysis of experimental and theoretical approach. *Case Studies in Thermal Engineering*, 2025. 70: p. 106066.
15. Sampaio, T.P.D., et al., Biological evaluation and compatibility study of oral mucoadhesive hydrogel formulations using Schinopsis brasiliensis Engler. extract and excipients. *Journal of Thermal Analysis and Calorimetry*, 2023. 148(1): p. 141–158.
16. Prasad, M., A. Salar, and R.K. Salar, In vitro anticancer activity of curcumin loaded chitosan nanoparticles (CLCNPs) against Vero cells. *Pharmacological Research-Modern Chinese Medicine*, 2022. 3: p. 100116.
17. Yusuf, H., et al., A lyophilized surfactant-based rutin formulation with improved physical characteristics and dissolution for oral delivery. *Saudi Pharmaceutical Journal*, 2023. 31(6): p. 1077–1083.
18. Yusuf, H., et al., Formulation design and cell cytotoxicity of curcumin-loaded liposomal solid gels for anti-hepatitis C virus. *Advances in Pharmacological and Pharmaceutical Sciences*, 2022. 2022(1): p. 3336837.

19. Alexander, K.S., A.T. Riga, and P.J. Haines, Thermoanalytical instrumentation and applications, in *Ewing's Analytical Instrumentation Handbook*, Fourth Edition. 2019, CRC Press. p. 433–490.
20. Correia, L.P., et al., Herbal medicine physical quality evaluation by thermal analysis using adapted Ozawa method. *Journal of Thermal Analysis and Calorimetry*, 2015. 122(1): p. 207–214.
21. Siqueira, G., et al., Thermoanalytical evaluation of essential oils of the leaves from Eucalyptus spp susceptible and resistant to Glycaspis brimblecombei. *Brazilian Journal of Thermal Analysis*, 2016. 5.
22. Garcia, C., et al., Thermal stability studies of some cerrado plant oils. *Journal of Thermal Analysis and Calorimetry*, 2007. 87(3): p. 645–648.
23. Mohapatra, D., et al., Bioengineered dual fluorescent carbon nano dots from Indian long pepper leaves for multifaceted environmental and health utilities. *Environmental Science and Pollution Research*, 2023. 30(18): p. 52182–52208.
24. Kumar, A., P. Singh, and A. Nanda, Hot stage microscopy and its applications in pharmaceutical characterization. *Applied Microscopy*, 2020. 50(1): p. 12.
25. Panna, W., P. Wyszomirski, and P. Kohut, Application of hot-stage microscopy to evaluating sample morphology changes on heating. *Journal of Thermal Analysis and Calorimetry*, 2016. 125(3): p. 1053–1059.
26. Vitez, I.M., et al., The evolution of hot-stage microscopy to aid solid-state characterizations of pharmaceutical solids. *Thermochimica Acta*, 1998. 324(1–2): p. 187–196.
27. Phiriyawirut, M. and T. Phaechamud, Gallic acid-loaded cellulose acetate electrospun nanofibers: Thermal properties, mechanical properties, and drug release behavior. *Open Journal of Polymer Chemistry*, 2012: p. 21–29.
28. Sharma, N., et al., Blending ethnomedicine with modern technology – From conventional to tailored products: Modulating biopharmaceutical properties of berberis extract by solid lipid nanoparticles for wound healing. *Journal of Functional Biomaterials*, 2023. 14(8): p. 418.
29. Jones, D.S., et al., Pharmaceutical applications of dynamic mechanical thermal analysis. *Advanced Drug Delivery Reviews*, 2012. 64(5): p. 440–448.
30. Menard, K.P. and N. Menard, *Dynamic Mechanical Analysis*. 2020, CRC press.
31. Bair, H.E., et al., Thermomechanical analysis (TMA) and thermodilatometry (TD), in *Thermal Analysis of Polymers: Fundamentals and Applications*. 2009, John Wiley & Sons, Inc. p. 319–385.
32. Saba, N., et al., A review on dynamic mechanical properties of natural fibre reinforced polymer composites. *Construction and Building Materials*, 2016. 106: p. 149–159.
33. Zhou, Y., et al., Nanomechanical and dynamic mechanical properties of rubber – wood – plastic composites. *Nanotechnology Reviews*, 2021. 11(1): p. 167–175.
34. Gopalakrishnan, J., Characterization of nanocellulose elastomer composites using dynamic mechanical analysis, in *Elastomeric Nanocellulose Composites*. 2024, Elsevier. p. 375–396.
35. James, J., Thermomechanical analysis and its applications, in *Thermal and Rheological Measurement Techniques for Nanomaterials Characterization*. 2017, Elsevier. p. 159–171.
36. Howe, D., et al., Thermal pretreatment of a high lignin SSF digester residue to increase its softening point. *Journal of Analytical and Applied Pyrolysis*, 2019. 142: p. 103691.
37. Brown, M.E., Evolved gas analysis (EGA), in *Introduction to Thermal Analysis: Techniques and Applications*. 1988, Springer. p. 86–98.
38. Mullens, J., EGA-evolved gas analysis, in *Handbook of Thermal Analysis and Calorimetry*. 1998, Elsevier. p. 509–546.
39. Lephardt, J.O., Fourier transform infrared evolved gas analysis (FT-IR-EGA). *Applied Spectroscopy Reviews*, 1982. 18(2): p. 265–303.
40. Materazzi, S. and S. Vecchio, Recent applications of evolved gas analysis by infrared spectroscopy (IR-EGA). *Applied Spectroscopy Reviews*, 2013. 48(8): p. 654–689.
41. Materazzi, S. and R. Risoluti, Evolved gas analysis by mass spectrometry. *Applied Spectroscopy Reviews*, 2014. 49(8): p. 635–665.
42. Risoluti, R., et al., FTIR-evolved gas analysis in recent thermoanalytical investigations. *Applied Spectroscopy Reviews*, 2017. 52(1): p. 39–72.
43. Maurya, R., et al., Non-isothermal pyrolysis of de-oiled microalgal biomass: Kinetics and evolved gas analysis. *Bioresource Technology*, 2016. 221: p. 251–261.
44. Fischer, M., et al., Evolved gas analysis by single photon ionization-mass spectrometry: A tool to distinguish different types of coffee. *Journal of Thermal Analysis and Calorimetry*, 2014. 116(3): p. 1461–1469.

9 Instrumental Methodology for Estimation of Volatile Components and Residual Solvent in Herbal Drugs and Formulation

ABBREVIATIONS

Abbreviation	Description
GC	Gas Chromatography
GC-MS	Gas Chromatography-Mass Spectrometry
GC-HS	Headspace Gas Chromatography
ICH	International Council for Harmonisation
USP	United States Pharmacopoeia
EP	European Pharmacopoeia
NIST	National Institute of Standards and Technology

9.1 INTRODUCTION

Volatile compounds are important constituents present in herbs and herbal drugs, which contribute to their aroma, flavor, and often their therapeutic properties. On the other hand, residual solvents, which may remain from the use of organic solvents during extraction or processing, can pose health risks if not properly removed. Therefore, accurate estimation of both is essential for ensuring the safety, efficacy, and regulatory compliance of herbal extracts and phytopharmaceuticals. Analytical techniques (Figure 9.1) such as gas chromatography-mass spectrometry (GC-MS) and headspace GC (HS-GC) are widely used for analyzing volatile and semi-volatile compounds due to their high sensitivity and reliability. These methods are often combined with sample preparation tools like solid-phase microextraction (SPME) to enhance detection without introducing solvent interference. Monitoring these components not only helps in detecting harmful residues but also plays a critical role in quality control, batch standardization, and confirming the authenticity of herbal formulations [1, 2] (https://downloads.regulations.gov/FDA-2020-D-1301-0007/attachment_1.pdf).

9.2 GAS CHROMATOGRAPHY

Gas chromatography (GC) is a widely applied analytical technique used to separate and quantify volatile and semi-volatile compounds in complex herbal mixtures. In herbal drug analysis, GC plays a vital role in profiling essential oils, detecting residual solvents, and ensuring the quality and consistency of herbal formulations for quality control purposes.

To improve selectivity and reliability, GC is often coupled with either a headspace sampler (GC-HS) or a mass spectrometer (GC-MS). GC-HS is especially advantageous for herbal matrices,

DOI: 10.1201/9781003540823-9

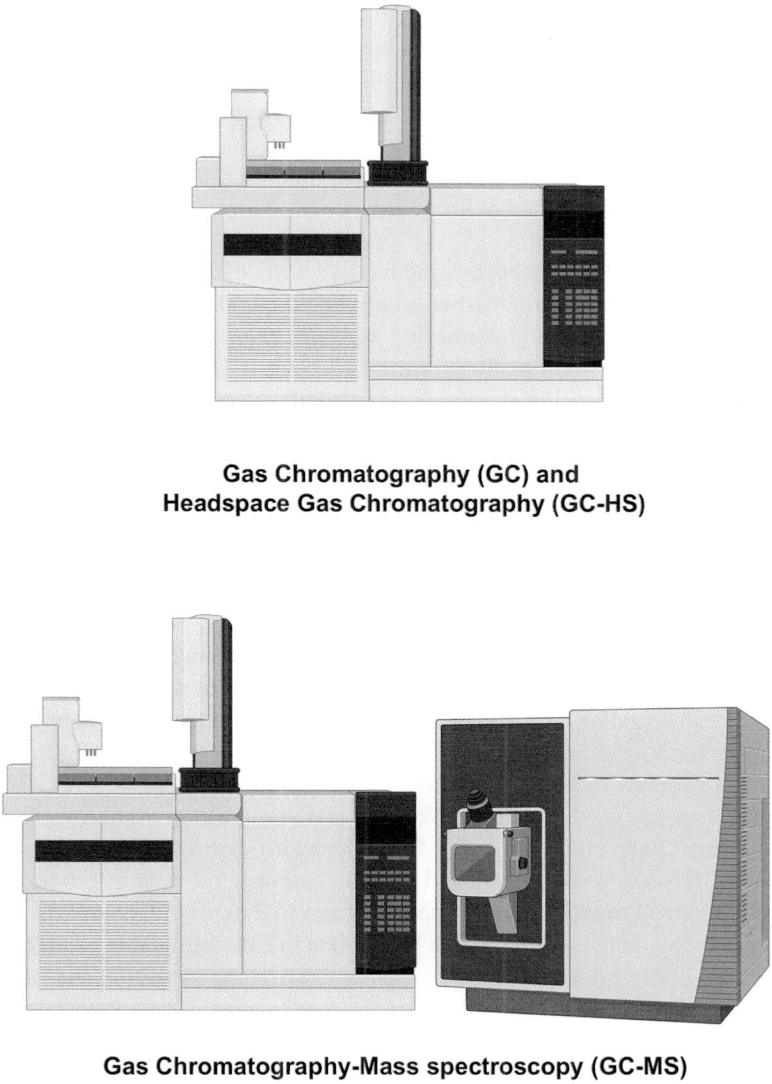

FIGURE 9.1 Instruments for structural elucidation and phytochemical integrity analysis of herbal drugs and formulations

as it enables the extraction and analysis of volatile compounds directly from the vapor phase above the sample. This reduces matrix interferences and minimizes the need for extensive sample preparation, making it particularly suitable for analyzing delicate plant materials and formulations without degrading sensitive constituents.

When combined with mass spectrometry, GC provides not only quantitative data but also structural identification of compounds through their mass spectra. This is highly beneficial for complex herbal mixtures where multiple volatiles may co-elute or occur at trace levels. Collectively, these techniques establish a powerful analytical platform for ensuring the safety, consistency, and regulatory compliance of herbal medicines in line with international quality standards.

While Chapter 5 describes the principles, advantages, instrumentation, procedures, challenges, and applications of GC-MS for structural elucidation and phytochemical integrity analysis, this chapter focuses on its applications in studying the amount of volatile components present and the contents of residual solvents in extracts.

GC-MS plays a crucial role in the analysis of herbal drugs, particularly for estimating volatile components, residual solvents, and pesticide residues. GC-MS effectively separates, identifies, and quantifies volatile compounds such as essential oils and aroma chemicals, which are important for the quality and efficacy of herbal products. It is also suitable for detecting minute quantities of residual solvents left after extraction or during processing so that the product meets regulatory guidelines. Moreover, GC-MS is a very sensitive technique used for screening and quantification of pesticide residues to ensure that herbal drugs are free from any contaminants, thus safeguarding public health.

In a study, a validated GC-MS/MS screening method was utilized to analyze pesticide residues present in the root and rhizome of the herbal drugs. The method was able to detect 296 pesticide residues simultaneously for regulatory monitoring purposes to ascertain their safety as well as compliance of herbal drugs in the market. To minimize the matrix interference effect, the developed method included optimized sample extraction and a dual-stage cleanup process, as complied with stringently by several international bodies or guidelines for validation of the method. When applied to 47 commercial herbal samples, 10 different pesticides were detected, mainly in *Rehmannia glutinosa* and *Cnidium officinale* from Korea, with most residues failing to comply with the regulatory limits, except for one sample, which marginally exceeded the standard for pendimethalin [3].

Recent research also established a reliable GC-MS/MS technique for the quantification of patchoulic alcohol present in the fresh aerial parts of *Pogostemon cablin* (Herba Pogostemonis), for the quality control of Herba Pogostemonis formulation. Since the herb contains a complex combination of volatile oils, conventional gas chromatography often produces unreliable results due to overlapping peaks and interference. To overcome this issue, freeze-drying followed by the supercritical CO_2-based extraction method was employed, which helped to retain the integrity of volatile compounds. Unlike traditional methods, tandem mass spectrometry was used for the precise detection of patchoulic alcohol by targeting a specific ion, that is, m/z 138 [4].

Another study focused on investigating the chemical composition and biological activities of *Crithmum maritimum* L. essential oil by GC-MS and its hydro-distillation residual water by HPLC-DAD-MS/MS. The GC-MS analysis revealed that the essential oil is rich in γ-terpinene (33.6%) and sabinene (32.0%), followed by thymol methyl ether (15.7%). The essential oil exhibited anti-inflammatory activity by reducing nitric oxide production in LPS-stimulated macrophages, without causing cytotoxicity. It also inhibited *Candida albicans* germ tube formation and reduced viability and biofilm biomass [5].

9.3 HEADSPACE GAS CHROMATOGRAPHY

Headspace gas chromatography (GC-HS) is an analytical technique used for the qualitative and quantitative analysis of volatile and semi-volatile organic compounds (VOCs) present in complex solid or liquid matrices. Headspace sampling is a sample introduction technique used together with GC for analyzing volatile compounds. Instead of injecting the sample (either solid or liquid) directly into the chromatographic system, it analyzes only the vapor phase (headspace) present above the sample. This selective method allows the extraction of only volatile compounds while avoiding non-volatile compounds from the complex plant matrices. Therefore, provides an efficient and better chromatographic profile by minimizing matrix interferences.

9.3.1 Principle

The main principle of GC-HS is based on the partitioning of volatile compounds between two phases: the sample matrix phase (solid or liquid) and the vapor phase (headspace) inside a sealed vial. Once the sample vial is sealed and gently heated to a specific temperature, the volatile compounds present in the sample begin to vaporize. As the temperature reaches the equilibrium state, depending on their volatility and partitioning behavior, the volatile compound gets distributed between the sample phase and the headspace. Then, a minimal measured volume of the headspace vapor containing volatile compounds is drawn and introduced into the GC column for separation, followed by analysis.

9.3.2 Advantages

- **Selective analysis:** Only the vapor phase above the sample is introduced into the GC, permitting selective analysis of the volatile and semi-volatile compounds while filtering out the non-volatile matrix components.
- **Minimized matrix interference:** The possible interferences are removed as the non-volatile portion of the sample never enters the GC system, thus enhancing the reliability and clarity of chromatographic data.
- **Cleaner chromatograms:** Lower baseline noise and minimized interference can lead to improved signal-to-noise ratios and improved peak resolution.
- **Minimal sample preparation:** There is no or minimal requirement for sample preparation by extraction, filtration, or chemical derivatization. It generally consists of weighing the sample, sealing the vial, and carrying out the analysis.
- **Preserves instrument integrity:** Non-volatile residues do not enter the column and detector, which helps prevent contamination, prolongs the lifespan of the instrument, and reduces the need for maintenance.
- **Highly reproducible results:** Consistent incubation, automated sample handling, and uniform injection volumes ensure high reproducibility throughout multiple runs.
- **Enhanced safety:** Using sealed vials minimizes the analyst's contact with volatile or hazardous solvents and also protects the sample from degradation or evaporation.
- **Time and cost efficiency:** Automation and reduced sample preparation increase throughput, resulting in decreased labor needs and lower operational expenses.
- **Regulatory compliance:** It is generally accepted for testing of residual solvent in compliance with ICH Q3C, USP <467>, and EP 2.4.24 guidelines.
- **Applicable to complex matrices:** It is suitable for solid, liquid, and semi-solid samples from pharmaceuticals, food, cosmetics, polymers, and environmental matrices.
- **Adaptability to various detectors:** It can be coupled with a flame ionization detector (FID) for quantification or with mass spectrometry (MS) for both qualitative and quantitative analysis of unknown or trace-level volatile compounds.
- **Non-destructive sampling:** The original sample remains largely intact after analysis, allowing for repeat testing or complementary analyses if needed.

9.3.3 Instrumentation

A schematic representation of a GC-HS instrument is shown in Figure 9.2. It consists of several major integrated units, which include the following [6–8]:

a) **Gas flow control system:** An inert gas, such as helium or nitrogen, is used to carry vaporized analytes through the system. Flow rate and pressure are carefully regulated to ensure consistent flow and efficient separation. Capillary columns typically operate at lower flow rates than packed columns [6].
b) **Headspace sampling system:** The sample is kept in a sealed glass vial and heated in a headspace sampler oven so that the volatile analytes can enter the headspace. Then, a preset amount (in milliliters) of the sample vapor is withdrawn by a manual device or an autosampler for analysis.
c) **Headspace interface and transfer line:** The headspace transfer line allows the headspace vapor to enter the GC column, either automatically or manually, for steady sample delivery.

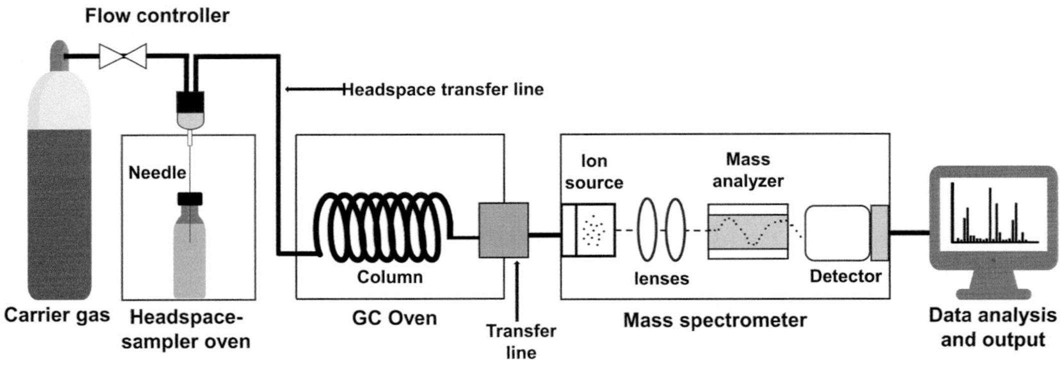

FIGURE 9.2 Schematic representation of GC-HS

d) **GC column and oven:** The analytical GC column is placed in a temperature-programmed GC oven, where the volatile compounds are separated based on their volatility as well as the interaction of compounds with the stationary phase. Gradual heating provided by the GC oven precisely controls the temperature and helps to optimize compound elution successively through the column, thus improving the resolution.
e) **Interface and transfer line:** A heated interface ensures smooth transfer of analytes from the GC to the MS without condensation or loss. This connection maintains the sample in its vapor phase until ionization.
f) **Ionization and mass analysis:** In the mass spectrometer, analytes are typically ionized by electron ionization (EI), generating charged fragments. These ions are separated based on their mass-to-charge ratio using analyzers like quadrupoles or time-of-flight (TOF) systems. Each analyte produces a characteristic fragmentation pattern [7].
g) **Detector and data processing system:** After volatile compounds are separated in the column, a detector (commonly a mass spectrometer) ionizes the compounds and measures their mass-to-charge ratios for identification and quantification. The resulting data are then processed using specialized software on the computer, which displays, analyzes, and interprets the data to provide qualitative and quantitative information about the resultant compounds [8].

9.3.4 Procedure

- Herbal drugs or extracts are finely powdered, accurately weighed, and placed into headspace vials for analysis [9].
- After adding a suitable diluent, the vials are sealed and incubated at a controlled temperature with agitation to allow the volatile compounds to accumulate in the headspace.
- The headspace is then injected into a GC system equipped with a capillary column and coupled to a mass spectrometer for the profiling and quantification of volatile components.

9.3.5 Challenges

- **Band broadening and separation efficiency:** If the vapor band of the sample injected into the column is already broad, then the separation of the analyte will be impaired due to the initial broadening of the analyte mixture.
- **Sample requirement for analysis:** All components of a sample must be volatile; otherwise, non-volatile solid residues may deposit in the inlet system. Due to high temperature

conditions in this zone, these residues may eventually decompose, generating volatile byproducts which enter the column and appear in the chromatogram, giving the false impression that these compounds were originally present in the original sample. Moreover, these residues may affect subsequent analysis due to catalytic or adsorption reactions.

- **Pre-treatment of non-volatile samples:** To address this problem, an indirect method is used while analyzing complex solid samples or samples having non-volatile components by first extracting the target compound, followed by injecting a fraction of the resultant solution into the gas chromatograph.
- **Solvent selection and simultaneous extraction issues:** Solvent extraction of low molecular weight compounds from a complex solid matrix is beneficial only if the high molecular weight or non-volatile components are insoluble in the solvent; otherwise, these undesired constituents will also be co-extracted and injected into the gas chromatograph.
- **Desorption artifacts from polymer adsorbents:** Tenax, a porous polymer adsorbent, may release artifacts at elevated desorption temperatures, resulting in false peaks in the chromatogram.

9.3.6 Application to Herbal Drugs

GC-HS is a selective and rapid technique that allows precise quantification of volatile essential oils, aromatic compounds, and residual solvents carried over by extraction or processing by only analyzing the volatile fraction present in the gas phase above a sample. This approach preserves the integrity of the complex herbal matrices. It also minimizes sample preparation, making it feasible for routine quality control analysis and thus ensuring batch-to-batch consistency in herbal drug manufacturing. It provides detailed chemical profiles of the herb or herbal drugs, allowing detection of even trace-level contaminants. It plays an important role in herbal drug standardization and ensuring their regulatory compliance with various regulatory bodies or guidelines. It ultimately supports the safe amalgamation of traditional herbal remedies into modern healthcare systems [10, 11].

Recently, residual solvent contamination in herbal medicinal products (HMPs) marketed in Ghana was investigated using validated orthogonal headspace gas chromatography methods as per USP <467> guideline. The developed method enabled the detection and quantification of 28 residual solvents belonging to classes 1 and 2. Among the 7 tested HMPs, one product (S/003) was found to be contaminated with chloroform (class 2B) at a 498.33 ppm level, leading to a daily exposure of 41.82 mg/day, which is above the permissible limit of 0.6 mg/day [12].

In another study, residual solvent analysis in the standardized ethanolic extract of *Piper longum* was studied by the GC-HS method. The level of ethanol was found to be very low and below the permissible level. The GC-HS chromatogram of the extract showed no peak of ethanol, reflecting that the ethanol content in the extract is below the detection and quantification limit (LOD = 3.253 ppm, LOQ = 9.859 ppm) [13].

9.3.7 Conclusion

GC-MS and GC-HS standard analytical tools are commonly used for the determination of volatile constituents and residual solvents present in herbal drugs. These techniques offer excellent selectivity, sensitivity, and resolution for complex phytochemical matrices. These methods enable precise characterization of essential oils, volatile compounds, and trace residual solvent residues, thus ensuring the quality of herbal drugs, their quality and compliance with regulatory standards. Recent developments suggest the incorporation of green or solvent-free extraction techniques, namely SPME and HS-SPME, to decrease the generation of sample artifacts. Also, for quick quality assessment of herbal drugs, the combination of real-time and on-site monitoring devices is essential. Furthermore, the use of chemometric tools along with machine learning is projected to assist in the recognition of patterns, the detection of adulteration, and tracing of the source. Moreover, the development of wide-ranging reference libraries for volatile components coupled with advanced hyphenated techniques will improve authentication, standardization, and the relationship between phytochemical composition and therapeutic activities.

REFERENCES

1. Wang, S., et al., Headspace solid-phase micro-extraction gas chromatography/mass spectrometry (HS-SPME-GC/MS)-based untargeted metabolomics analysis for comparing the volatile components from 12 panax herbal medicines. *Phyton*, 2022. 91(7).
2. Tholl, D., et al., Practical approaches to plant volatile analysis. *The Plant Journal*, 2006. 45(4): p. 540–560.
3. Yang, S.-H., Y. Shin, and H. Choi, Simultaneous analytical method for 296 pesticide multiresidues in root and rhizome based herbal medicines with GC-MS/MS. *PLoS One*, 2023. 18(7): p. e0288198.
4. Zhao, Z., et al., Determination of patchoulic alcohol in herba pogostemonis by GC-MS-MS. *Chemical and Pharmaceutical Bulletin*, 2005. 53(7): p. 856–860.
5. Alves-Silva, J.M., et al., Chemical composition of Crithmum maritimum L. essential oil and hydrodistillation residual water by GC-MS and HPLC-DAD-MS/MS, and their biological activities. *Industrial Crops and Products*, 2020. 149: p. 112329.
6. Poole, C.F. and S.A. Schuette, *Contemporary Practice of Chromatography*, Vol. 5. 2012, Elsevier.
7. Gross, J.H., *Mass Spectrometry: A Textbook*. 2006, Springer Science & Business Media.
8. Kind, T., et al., FiehnLib: Mass spectral and retention index libraries for metabolomics based on quadrupole and time-of-flight gas chromatography/mass spectrometry. *Analytical Chemistry*, 2009. 81(24): p. 10038–10048.
9. Gnana Raja, M., G. Geetha, and A. Sankaranarayanan, A concise study of organic volatile impurities in ten different marketed formulations by [GC/HS-FID/MS] gas chromatography technique. *Journal of Analytical & Bioanalytical Techniques*, 2014. 5(4).
10. Zhang, J., et al., Fast analysis of principal volatile compounds in crude and processed Atractylodes macrocephala by an automated static headspace gas chromatography-mass spectrometry. *Pharmacognosy Magazine*, 2014. 10(39): p. 249.
11. Mazumder, J., D. Pathak, and R. Kumria, Analytical method validation of an herbal formulation by headspace gas chromatography using QbD. *International Journal of Pharmaceutical Sciences and Drug Research*, 2014. 6(4): p. 291–302.
12. Opuni, K.F., et al., Monitoring of residual solvent contamination in herbal medicinal products in Ghana: A pilot study. *Scientific African*, 2021. 13: p. e00825.
13. Mohapatra, D., et al., Quality by design – based development and optimization of fourth-generation ternary solid dispersion of standardized Piper longum extract for melanoma therapy. *Drug Delivery and Translational Research*, 2023. 13(12): p. 3094–3131.

10 Instrumental Methodology for Morphological Analysis of Crude Herbal Drugs and Their Formulations

ABBREVIATIONS

Abbreviation	Description
AFM	Atomic Force Microscopy
AI	Artificial Intelligence
AM-AFM	Amplitude Modulation Atomic Force Microscopy
AuNPs	Gold Nanoparticles
DLS	Dynamic Light Scattering
EPS	Extracellular Polymeric Substances
G	Gravitational Force (Unit in Centrifugation)
GT	Gum Tragacanth
HPMP	High-Protein Milk Powder
HR-SEM	High-Resolution Scanning Electron Microscopy
HR-TEM	High-Resolution Transmission Electron Microscopy
KPFM	Kelvin Probe Force Microscopy
LS	Longitudinal Section
MFM	Magnetic Force Microscopy
MPC	Milk Protein Concentrate
NSOM	Near-Field Scanning Optical Microscopy
Omics	Integrated Biological Profiling Approaches (e.g., Genomics, Proteomics, Metabolomics)
PSPD	Position-Sensitive Photodetector
Ra	Average Surface Roughness
RH	Relative Humidity
Rq	Root Mean Square Surface Roughness
RLS	Radial Longitudinal Section
SEM	Scanning Electron Microscopy
SMP	Skimmed Milk Powder
SNEDDS	Self-Nanoemulsifying Drug Delivery System
SPM	Scanning Probe Microscopy
STM	Scanning Tunneling Microscopy
TEM	Transmission Electron Microscopy
TLS	Tangential Longitudinal Section
TS	Transverse Section

DOI: 10.1201/9781003540823-10

UV	Ultraviolet
WPI	Whey Protein Isolate
X, Y, Z	Cartesian axes (lateral and vertical dimensions)
Z-piezo	Piezoelectric actuator controlling vertical movement

10.1 INTRODUCTION

The growing global interest in herbal medicine necessitates rigorous quality control measures to ensure efficacy, safety, and authenticity. Morphological analysis stands as a foundational pillar in this effort, offering direct insights into the structural characteristics of crude herbal drugs and their derived formulations [1]. While macroscopic examination provides initial clues regarding identity, it often falls short in distinguishing closely related species, detecting subtle adulterations, or characterizing the intricate details of processed materials [2]. The inherent complexity and variability of natural products, coupled with potential for misidentification, substitution, and contamination, underscore the indispensable role of advanced instrumental methodologies in modern herbal quality assurance. Microscopic analysis moves beyond the naked eye, allowing for the visualization of cellular features, tissue organization, and specific inclusions that are characteristic of particular plant species. This level of detail is crucial for authenticating botanical materials, identifying foreign matter, and assessing the quality of powdered drugs and their formulations. However, even conventional optical microscopy, while powerful, has its inherent resolution limits [3]. To truly delve into the ultrastructure of cells, probe surface topography at the nanoscale, or understand the physical interactions at the atomic level, more sophisticated instruments are required [4].

This chapter explores a range of instrumental methodologies that transcend the limitations of traditional macroscopic and basic microscopic techniques. We explore the principles, instrumentation, advantages, uses, sample preparation, challenges, and specificities of the optical microscope, high-resolution transmission electron microscope (HR-TEM), high-resolution scanning electron microscope (HR-SEM), and atomic force microscope (AFM)/scanning probe microscope (SPM), as shown in Figure 10.1. These instruments, each offering unique features and perspectives, collectively provide a powerful toolkit for comprehensive morphological analysis, ensuring the integrity and quality of herbal materials from raw drug to finished product.

10.2 OPTICAL MICROSCOPY

Optical microscopy, often considered the cornerstone of botanical quality control, utilizes visible light to magnify and resolve features that are invisible to the unaided eye. Its accessibility, relative ease of use, and ability to observe a wide array of characteristic cellular structures make it an indispensable first-line tool for the morphological analysis of crude herbal drugs and their formulations [5].

10.2.1 Principle

The fundamental principle of optical microscopy relies on the interaction of light waves with the specimen. A light source illuminates the sample, and light passing through or reflected from the specimen is then magnified by a series of lenses. The objective lens, positioned close to the specimen, gathers light and produces a magnified intermediate image. This intermediate image is then further magnified by the eyepiece (ocular lens), which presents a virtual, magnified image to the observer's eye [6].

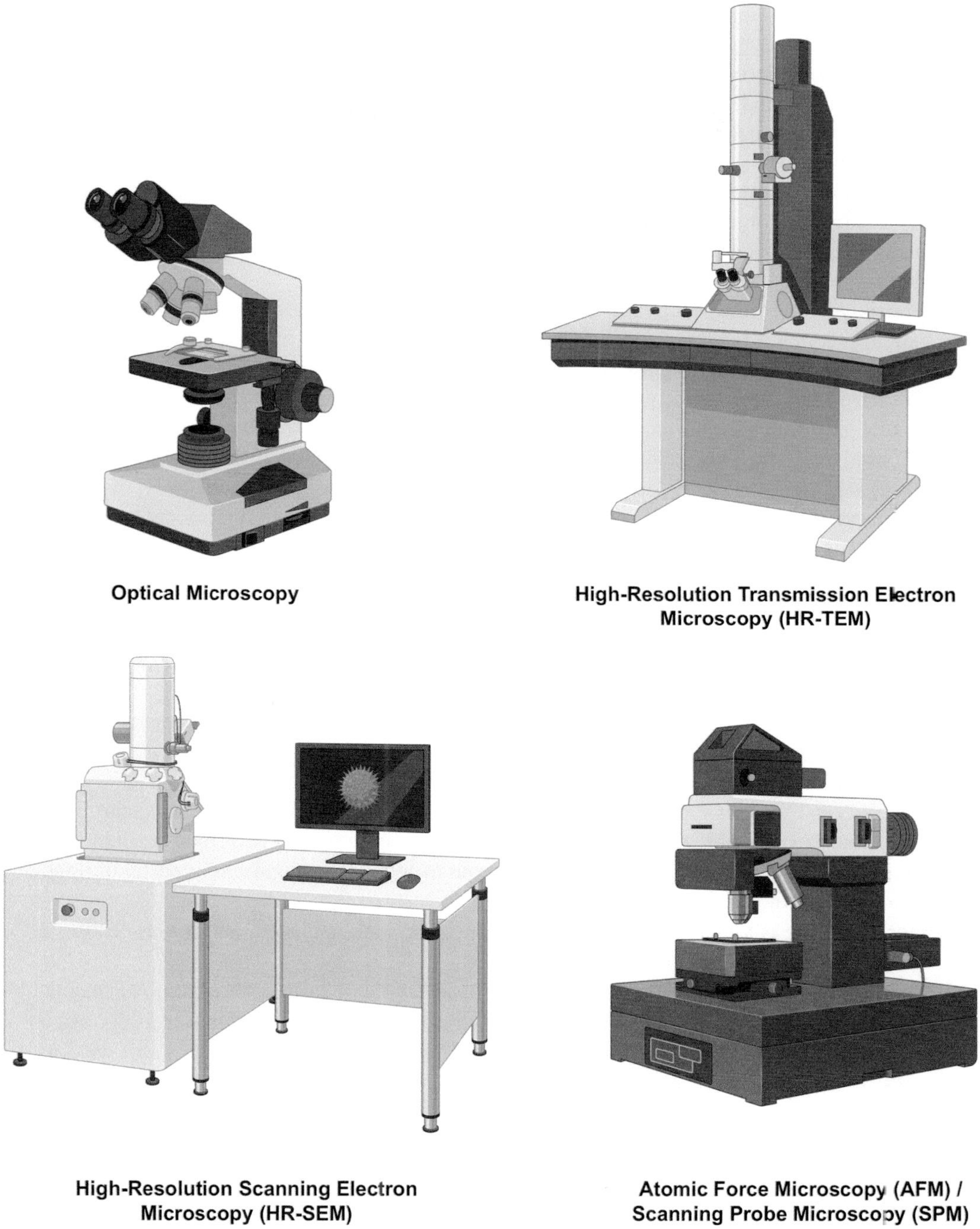

FIGURE 10.1 Instruments for morphological analysis of crude herbal drugs and their formulations

Various techniques enhance the contrast and visualization of different specimen features:

- **Bright-field microscopy:** The most common type, where light passes directly through the specimen, and the image is formed by the differential absorption and scattering of light by various parts of the sample. Staining is often required for transparent biological specimens.
- **Dark-field microscopy:** Illuminates the specimen with oblique light, so only scattered light enters the objective lens. This results in a bright image against a dark background, ideal for visualizing unstained, living specimens or fine structures that would otherwise be difficult to see.
- **Phase contrast microscopy:** Converts subtle phase shifts in light passing through transparent specimens into brightness differences, making unstained biological samples visible without the need for dyes [7].
- **Polarized light microscopy:** Utilizes polarized light to reveal structures that are birefringent (e.g., starch grains, calcium oxalate crystals, some fibers), which can rotate the plane of polarized light. This technique is invaluable for identifying specific inclusions in herbal drugs [8].
- **Fluorescence microscopy:** Employs ultraviolet (UV) light to excite fluorescent molecules within the specimen, causing them to emit light at a longer, visible wavelength. This is useful for detecting autofluorescent components (e.g., chlorophyll, some alkaloids) or for targeted staining with fluorescent probes [8].

10.2.2 Instrumentation

A standard compound optical microscope consists of several essential components that work together to magnify and visualize specimens (Figure 10.2).

Light source: The light source, usually a halogen or LED lamp, provides the necessary illumination. Positioned beneath the stage is the condenser, a lens system focuses light onto the specimen; it often includes an iris diaphragm to control the angle and intensity of light, thereby affecting contrast and resolution.

Stage: The stage is a flat platform where the specimen slide is placed, and it may feature a mechanical stage for precise slide movement.

Lens: Objective lenses, mounted on a rotating nosepiece, typically offer a range of magnifications such as 4X, 10X, 40X, and 100X, with higher magnifications like 100X often requiring immersion oil to enhance resolution. Eyepieces, or ocular lenses, are used to view the magnified image, commonly providing 10X or 15X magnification; the total magnification is calculated by multiplying the objective lens magnification by that of the eyepiece.

Focusing knobs: Focusing knobs, including both coarse and fine adjustments, are used to bring the specimen into clear focus.

Body tube: The body tube connects the eyepiece to the objective lenses, ensuring proper optical alignment. Finally, the base and arm provide the microscope with structural stability and support [9].

10.2.3 Advantages

The several advantages of microscopy for herbal analysis are as follows:

- Compared to electron microscopy or AFM, optical microscopes are relatively inexpensive to purchase and maintain, making them accessible to a wider range of laboratories.
- Operation is simple, requiring no training.

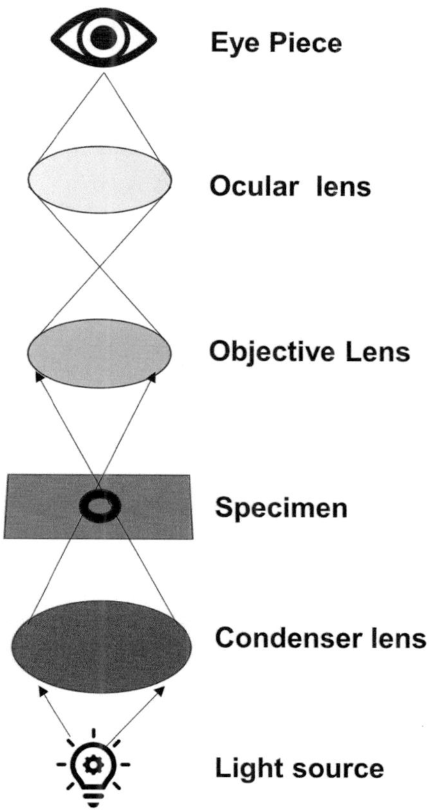

FIGURE 10.2 Working of the optical microscopy instrument

- The sample need not be destroyed and can often be preserved. Sectioning, however, is a destructive process.
- Capable of observing a wide range of cellular features and tissue organizations, including parenchyma, collenchyma, sclerenchyma, xylem, phloem, epidermal cells, stomata, trichomes, starch grains, calcium oxalate crystals (raphides, druses, prisms), aleurone grains, and oil droplets.
- Allows for direct visual inspection and immediate interpretation of morphological features.
- Serves as an excellent first step in the authentication process, quickly identifying gross morphological discrepancies or obvious adulterants.

10.2.4 Procedure

Sample preparation for optical microscopy varies depending on the nature of the herbal material and the specific features to be examined.

- **Crude Drugs (Whole or Cut):**
  - **Sectioning:** For detailed examination of tissue organization, transverse (TS), longitudinal (LS), or tangential (TLS/RLS) sections (10–100 µm thick) are prepared using a hand microtome, freezing microtome, or sliding microtome. Free-hand sectioning with a sharp razor blade is common for rapid [10].
  - **Maceration:** For separating individual cells or tissues (e.g., fibers, vessels), the material is treated with chemicals (e.g., nitric acid and potassium chlorate) to dissolve the middle lamella, allowing for dissociation [11].

- **Mounting:** Sections or macerated material are typically mounted in water, glycerine, or chloral hydrate solution on a glass slide and covered with a coverslip. Glycerine helps prevent drying and provides a clearer view.
- **Powdered drugs:** A small amount of powder is placed on a glass slide, a drop of mounting medium (e.g., water, glycerine, chloral hydrate) is added, and it is gently covered with a coverslip, avoiding air bubbles.
- **Staining techniques:** Various staining techniques are employed in microscopy to enhance the visibility of specific cellular components in plant tissues. These techniques are summarized in Tables 10.1 and 10.2.
- **Nano Formulations:**
    - **Dilution/dispersion:** Nano formulations (e.g., suspensions, nanoemulsions, transferosomes) are diluted in distilled water or a suitable buffer to reduce aggregation and ensure uniform distribution.
    - **Drop casting**: A small aliquot is placed on a slide and gently spread to form a thin film.

Fixation or Staining (if required)

10.2.5 Challenges

Despite its widespread utility, optical microscopy still has its limitations, which are highlighted as follows:

TABLE 10.1
Staining Reagent for Plant Materials

Stain	Target Component	Color Change	Purpose/Use
Lugol's Iodine (iodine solution)	Starch grains	Blue-black	Identifies and assesses the presence of starch
Phloroglucinol + Conc. HCl	Lignified tissues (xylem, sclereids, fibers)	Red	Highlights lignified structures in plant tissues
Sudan III or IV	Fatty substances (oils, resins)	Orange-red	Detects lipids and resinous substances
Fehling's solution	Reducing sugars	Brick-red precipitate	Indicates the presence of reducing sugars
Toluidine blue O	Various plant tissue components	Differential staining	General stain for tissue analysis (e.g., lignin, pectin)

TABLE 10.2
Additional Staining Techniques in Plant Microscopy

Technique	Target	Applications
Toluidine blue O	General staining of plant cells	Useful for distinguishing lignin, pectin, and other cell wall components
Sudan III/IV	Lipophilic components in lipid-based nanocarriers	Stains oils, fats, and resins orange-red
Fluorescent dyes	Targeted labeling or observation of autofluorescent phytoconstituents	Enables detailed visualization under fluorescence microscopy
Mounting	Sample preparation for microscopic observation	Involves adding a drop of mounting medium, placing a coverslip, and allowing it to stabilize before viewing

- The main limitation of optical microscopy is its inability to visualize structures smaller than approximately 0.2 μm, making it impossible to resolve cell organelles (e.g., ribosomes, endoplasmic reticulum), viruses, or fine details of cell walls at the ultrastructural level [12].
- At higher magnifications, only a thin plane of the specimen is in focus, requiring constant fine adjustment.
- While different focal planes can give an impression of depth, the images are fundamentally 2D, making it difficult to fully appreciate complex 3D structures or surface topography [13].
- Staining, dehydration, and sectioning processes can introduce artifacts (e.g., shrinkage, swelling, precipitation) that do not reflect the true morphology of the living tissue [14].
- Interpretation of morphological features can sometimes be subjective, requiring experienced microscopists for accurate identification.
- Highly opaque or thick samples are challenging to analyze without extensive preparation.

10.2.6 Application to Herbal Drugs

Optical microscopy remains paramount for routine quality control and authentication of herbal drugs:

- **Identification and authentication of leaves:** Identifying diagnostic microscopic features such as stomatal type (e.g., diacytic in Mentha, anomocytic in Digitalis), trichome morphology (e.g., glandular in Hyoscyamus, stellate in Verbascum), palisade ratio, vein islet number, and epidermal cell characteristics are essential for crude drug authentication. For example, Senna leaves are identified by unicellular, non-glandular trichomes and paracytic stomata. Comparison with authenticated reference materials – such as trichome types in Cannabis or Verbascum, palisade ratio in Senna leaves, stomatal index, presence or absence of lignified cells, and starch grain morphology in roots and rhizomes – further supports accurate identification [15, 16].
- **Characterization of barks:** Identifying cork cells, stone cells (sclereids), phloem fibers, medullary rays, and characteristic inclusions like calcium oxalate prisms in Cinchona bark or crystal sheaths in Cascara bark.
- **Analysis of roots and rhizomes:** Observing starch grains (their shape, size, hilum, and stratification are species specific, for example, concentric in potato, eccentric in rice), vessel types, parenchyma, and characteristic inclusions.
- **Detection of adulterants:** For instance, the presence of specific trichomes not found in the authentic drug, or starch grains of different shapes/sizes than expected, points toward adulteration. Microscopic analysis can easily differentiate between authentic saffron stigmas and adulterants like dyed silk fibers or colored starch granules [17].
- **Detection of adulteration:** Identifying foreign plant parts (e.g., stems in leaf drugs, leaves in bark drugs), or non-botanical materials (e.g., sand, synthetic fibers, inorganic fillers). Starch from alternative sources (e.g., maize starch as an adulterant in rice starch) can be differentiated by granule morphology.
- **Quality control of powdered drugs:** Microscopic examination of powdered herbal drugs reveals the presence of specific diagnostic characters (e.g., fragments of vessels, fibers, sclereids, parenchyma cells, crystals, starch grains, epidermal cells, trichomes) that allow for identification and assessment of purity. Ensures the presence of expected diagnostic fragments and the absence of foreign materials. For example, ginger powder should show abundant simple oval or pear-shaped starch grains and oleo-resin cells [18].
- **Evaluation of processing effects:** Observing changes in cellular structure due to drying, grinding, or extraction processes, which can impact drug quality

- **Evaluation of formulations:** Assessing the uniformity of particle size in powders, observing the distribution of active ingredients or excipients in tablets, capsules, or creams, and identifying any foreign contaminants in finished products. Microscopic examination of herbal tablets can reveal the uniformity of powder mixing, the presence of excipients, and any signs of degradation or foreign particles. In herbal ointments or creams, it can assess particle size distribution and homogeneity. Optical microscopy can also be used for morphology analysis of advanced vesicular formulations, such as transferosomes. The phospholipid bilayer of transferosome is clearly observed using optical microscopy (Figure 10.3) [19].

10.3 HIGH-RESOLUTION TRANSMISSION ELECTRON MICROSCOPY

High-resolution transmission electron microscopy (HR-TEM) revolutionizes morphological analysis by pushing the boundaries of resolution far beyond the optical microscope. Instead of light, it employs a beam of electrons to illuminate the specimen, allowing for the visualization of cellular and subcellular structures at the nanometer to even the atomic scale. This capability is indispensable for understanding the intricate architecture of plant cells, the localization of secondary metabolites, and the characteristics of nanoparticles in herbal systems.

10.3.1 Principle

HR-TEM operates on the principle that electrons, being charged particles, can be focused by electromagnetic lenses much like light is focused by glass lenses. A high-energy electron beam, generated

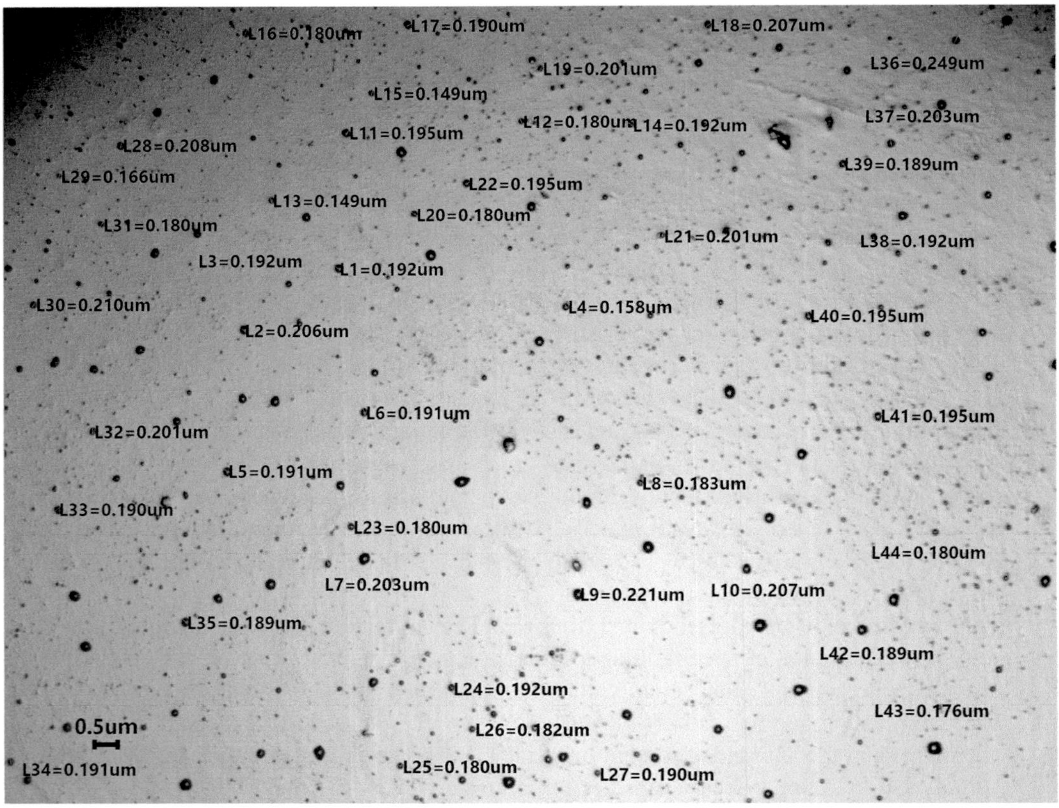

FIGURE 10.3 Optical microscopy of transferosomes

by an electron gun (e.g., thermionic emitter or field emission gun), passes through a very thin section of the specimen. As electrons traverse the sample, some are scattered by the atoms within the specimen, while others pass through unscattered. The scattered electrons are removed by an aperture, leading to differential contrast in the image. The unscattered electrons and those scattered at small angles are focused by a series of electromagnetic lenses to form a highly magnified image on a fluorescent screen or a digital detector [20].

The high resolution of TEM (typically around 0.1–0.2 nm for HR-TEM) is attributed to the very short de Broglie wavelength of electrons accelerated to high voltages (e.g., 80–300 kV). According to the Rayleigh criterion, resolution is inversely proportional to wavelength. The shorter wavelength of electrons compared to visible light allows for significantly higher resolving power [20].

Contrast in TEM images primarily arises from the following:

- **Amplitude contrast:** Differences in the number of electrons scattered and absorbed by different regions of the specimen. Denser regions scatter more electrons, appearing darker in the image. This is often enhanced by heavy metal stains.
- **Phase contrast (in HR-TEM):** At very high resolutions, contrast can also arise from differences in the phase of electron waves passing through the sample. This allows for the direct imaging of atomic lattices.

10.3.2 Instrumentation

A HR-TEM consists of several components within a high vacuum system to prevent electron scattering by air molecules. The electron gun (e.g., tungsten filament, LaB_6 crystal, or field emission gun) produces a stable, high-intensity electron beam, which is focused by condenser lenses into a fine, parallel beam to illuminate the specimen. The specimen stage holds the ultrathin sample on a small grid and allows precise positioning and tilting. The objective lens, the most critical component, forms the initial magnified image and defines the resolution. This image is further enlarged by intermediate lenses and projected by projector lenses onto a fluorescent screen or digital detector, such as a CCD or CMOS camera, for visualization and image capture. Apertures help control the beam, eliminate scattered electrons, and enhance contrast. The vacuum system, including rough and high-vacuum pumps like turbomolecular and ion pumps, ensures optimal electron beam propagation [21]. The schematic representation of HRTEM is shown in Figure 10.4.

10.3.3 Advantages

- Provides direct visualization of structures at the nanometer scale, including cell organelles and fine details of cell walls, membranes, and even macromolecules. Atomic resolution is achievable with HR-TEM.
- Allows for in-depth study of the internal architecture of cells, understanding the spatial relationships between organelles, and identifying pathological changes at a subcellular level.
- Elemental analysis (with EDX/EELS): Many TEMs are equipped with energy dispersive X-ray spectroscopy (EDX) or electron energy loss spectroscopy (EELS) detectors, enabling elemental mapping and analysis of specific regions within the sample, identifying the distribution of minerals or heavy metals.
- It is Ideal for characterizing the size, shape, crystallinity, and dispersion of nanoparticles within plant tissues or herbal formulations, which are increasingly common in nanomedicine [22].
- Electron diffraction patterns can be obtained from crystalline regions, providing information on crystal structure and orientation.

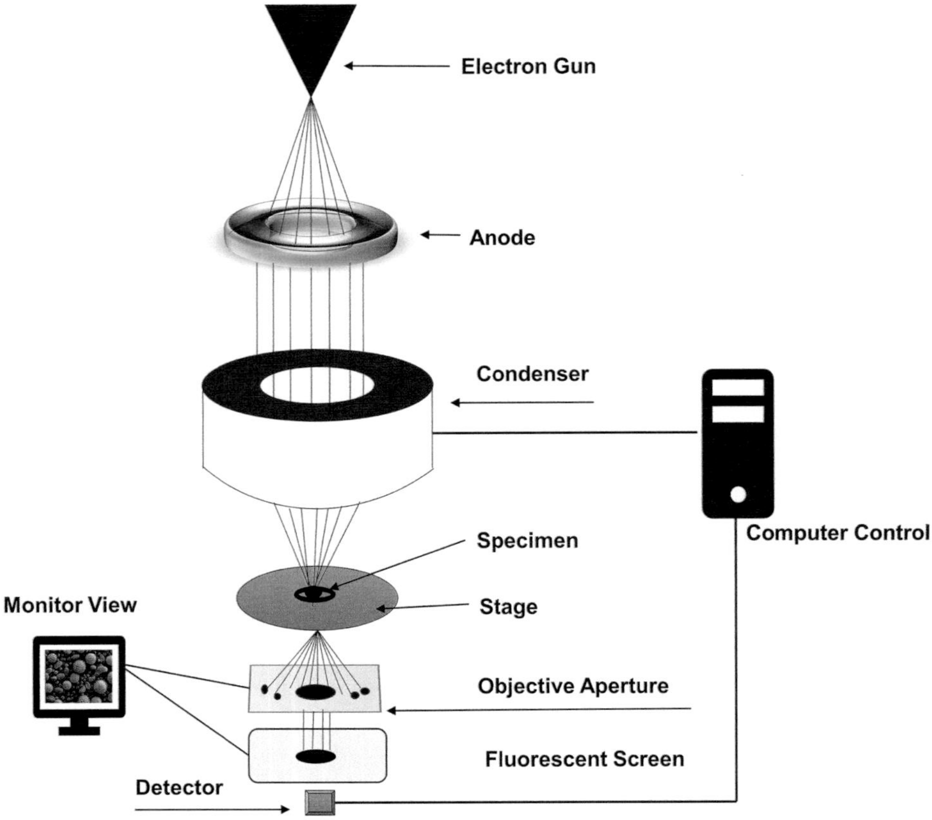

FIGURE 10.4 Instrumentation of HR-TEM

10.3.4 Procedure

- Sample preparation for HR-TEM is critical, complex, and time consuming, as the samples must be extremely thin (50–100 nm) and able to withstand the high vacuum and electron beam.
- TEM sample preparation involves rapid and thorough chemical fixation using agents such as glutaraldehyde, which cross-links proteins, followed by osmium tetroxide to stabilize lipids and membranes.
- After fixation, the sample is washed multiple times with a buffer solution to remove excess fixatives.
- Dehydration is then carried out through a graded series of ethanol or acetone concentrations to gradually replace water, preventing structural collapse.
- The dehydrated tissue is subsequently infiltrated with a liquid resin – such as Spurr's, Araldite, or Epon – which replaces the dehydrating agent and is polymerized by heat to form a hardened block.
- This block is then sectioned into ultrathin slices (typically 50–100 nm) using an ultramicrotome equipped with a diamond knife. The sections are floated onto a water surface and mounted onto electron-transparent grids made of copper, nickel, or carbon-coated materials.
- Post-staining with heavy metal salts such as uranyl acetate and lead citrate enhances contrast by scattering electrons effectively. After air drying or desiccation, the prepared grid

is inserted into the TEM, where the electron beam is aligned and focused to examine and capture high-resolution images of the sample's internal structures.

- For herbal formulations with nanoparticles: Preparation can be simpler, sometimes involving direct deposition of a diluted suspension onto a carbon-coated grid, followed by drying and potentially negative staining (e.g., with phosphotungstic acid) for contrast.

10.3.5 Challenges

Despite its high magnification power, HR-TEM has significant drawbacks, which include the following:

- Extremely expensive to purchase, operate, and maintain, limiting its accessibility to highly specialized research institutions.
- The preparation process completely destroys the original sample. The high vacuum can lead to sample desiccation and potential structural alterations, though fixation aims to minimize this.
- Fixation, dehydration, embedding, and staining can all introduce artifacts that may be misinterpreted as genuine biological structures. Carefully controlled experiments and interpretation are necessary.
- Only very small pieces of tissue can be processed at a time, making it challenging for a broad survey of heterogeneity.
- The high-energy electron beam can cause radiation damage to the specimen, especially sensitive biological materials, leading to structural changes or degradation during observation.

10.3.6 Application to Herbal Drugs

HR-TEM provides unprecedented insights into the cellular and subcellular aspects of herbal materials:

- **Ultrastructural anatomy:** Detailed mapping of cell wall layers, plasmodesmata, pit fields, and the organization of parenchyma, collenchyma, and sclerenchyma cells at a microscopic level that optical microscopy cannot achieve, for instance, studying the detailed morphology of secondary cell walls in wood fibers used in traditional medicine [23].
- **Localization of metabolites:** Visualizing the accumulation of secondary metabolites within specific organelles (e.g., electron-dense inclusions in vacuoles indicating phenolic compounds or tannins, lipid droplets in the cytoplasm storing essential oils). This helps us understand the biosynthesis and storage mechanisms.
- **Assessment of cellular integrity:** Evaluating the impact of various processing methods (e.g., drying, grinding, extraction solvents) on the integrity of plant cells and organelles, ensuring optimal preservation of active compounds. For example, comparing cells from fresh versus spray-dried herbal extracts. In a recent study, TEM was employed to investigate the ultrastructural features of *Lycium barbarum* (goji berry) fruit tissues. Ultrathin sections were prepared using standard fixation, embedding, and staining protocols, and examined under a Tesla BS 500 TEM at 120 kV. TEM revealed key cellular components within hypodermal and mesocarp tissues, including large vacuoles, nuclei, and chromoplasts at various developmental stages. Early-stage chromoplasts retained thylakoid structures, while mature vesicular and reticulotubular chromoplasts exhibited electron-transparent vesicles and fibrillar carotenoid accumulations, respectively. Phytoferritin deposits appeared as osmiophilic granules, and remnants of degraded protoplasts were also observed. The endocarp cells surrounding the seed were notable for their lignified

secondary walls, forming a distinct reticulate-puzzle architecture. TEM provided critical insight into plastid differentiation, cell wall architecture, and subcellular organization, underscoring its value in plant anatomical and developmental studies [24].

- **Nanoparticle characterization in nanomedicine:** It is crucial for verifying the size, shape, crystallinity, and dispersion of metallic nanoparticles (e.g., silver nanoparticles synthesized using plant extracts) or encapsulated herbal compounds (e.g., liposomes loaded with curcumin) within formulations or after uptake by cells. HR-TEM can reveal if nanoparticles are internalized into plant cells or accumulate on surfaces. In a recent study, a novel and environmentally friendly synthesis method was developed using *Stachys lavandulifolia* extract as both silver nanoparticles (Ag NPs) onto magnetic Fe_3O_4 nanoparticles. The formation of Ag NPs was initially confirmed through TEM and EDX analysis, verifying their successful synthesis without the use of toxic reductants or additional stabilizers. HR-TEM further revealed that the Fe_3O_4 *S. lavandulifolia*/Ag nanocomposite consisted of uniformly dispersed, spherical nanoparticles, encapsulated by a thin biomolecular layer derived from the plant extract. This layer was found to contribute to both the stabilization and the reduction processes. The successful surface modification and nanoparticle deposition were further supported by complementary techniques such as XRD, XPS, and FT-IR. Overall, TEM and HRTEM played a critical role in confirming the morphology, dispersion, and structural integrity of the biosynthesized magnetic nanocatalyst [25]. In a recent study, HR-TEM reported the morphology of carbon dots from *Tinospora cordifolia* leaves–derived carbon dots. The photomicrograph depicts nanometric size (<10 nm) and near-spherical nature of the synthesized carbon dots and the diffused SAED pattern reflected amorphous nature (Figure 10.5) [26].

Interactions with pathogens: HR-TEM is used to study the ultrastructural changes in plant cells infected by fungi or bacteria, providing insights into plant-pathogen interactions relevant to herbal quality and safety. In a recent study, TEM was used in characterizing the morphology and antibacterial mechanism of self-assembled berberine-based nanostructures. TEM revealed that BA-BBR nanoparticles were spherical with diameters around 100 nm, while WOG-BBR formed nanofiber bundles measuring 50–100 nm in width and several micrometers in length. TEM also confirmed

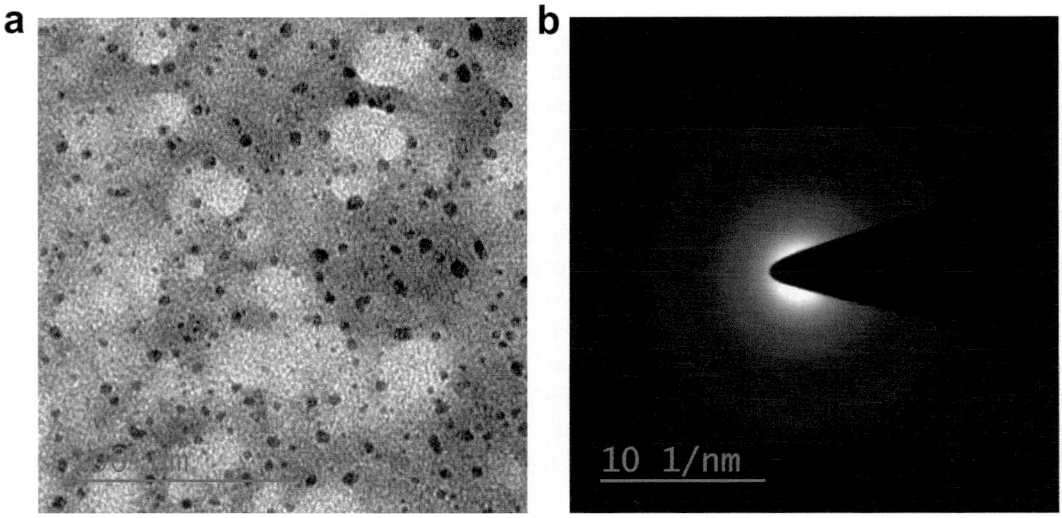

FIGURE 10.5 HR-TEM photomicrograph and SAED pattern of *Tinospora cordifolia* leaves–derived carbon dots (a) HR-TEM photomicrograph at 100 nm scale (b) SAED pattern

bacterial membrane disruption upon treatment with BA-BBR NPs, showing severe structural damage, cytoplasmic leakage, and cell lysis, whereas WOG-BBR NFs caused minimal membrane alteration. These results highlighted TEM's key role in visualizing both nanostructure formation and their interaction with bacterial cells at the ultrastructural level [27].

Also, research was conducted to investigate the antibacterial mechanism of action of *Annona squamosa* Linn. extract, a traditional medicinal plant known for its antimicrobial properties. Using *Bacillus cereus* and *Campylobacter jejuni* as representative Gram-positive and Gram-negative bacteria, respectively, TEM provided crucial insights into the ultrastructural alterations induced by the bioactive reticuline-containing fraction of the extract. At inhibitory concentrations, TEM images revealed significant morphological changes, including peptidoglycan layer disruption and cell wall separation, particularly in *B. cereus*. Bactericidal concentrations further demonstrated severe cell envelope damage, cytoplasmic depletion, and intracellular content leakage in both bacterial strains. These findings suggest that the extract exerts its antibacterial effect by compromising cell wall integrity, ultimately leading to osmotic lysis. The study highlights the indispensable role of TEM in visualizing nanoscale cellular damage and elucidating the bactericidal mechanisms of plant-derived antimicrobial agents [28].

Quality control of extracts: It is used to examine the morphology of precipitations or aggregates in concentrated herbal extracts, understanding their composition at a nanoscale level. In a recent study, platinum nanoparticles were synthesized via a green method using *Bidens tripartitus* extract. TEM played a crucial role in confirming the nanoscale structure and morphology of the synthesized particles. TEM analysis revealed that the platinum nanoparticles were predominantly below 10 nm in size, with some showing irregular rod-like shapes and slight agglomeration, likely due to van der Waals interactions. These observations provided critical insight into the particle size distribution and structural features, underscoring the effectiveness of TEM in characterizing biosynthesized nanomaterials [29].

10.4 HIGH-RESOLUTION SCANNING ELECTRON MICROSCOPY

High-resolution scanning electron microscopy (HR-SEM) offers a powerful approach to visualize the surface topography and morphology of solid specimens with a remarkable depth of field and high resolution, far exceeding that of optical microscopy. Unlike TEM, which images transmitted electrons to reveal internal structure, SEM images electrons emitted from the surface, providing a detailed three-dimensional-like view. This makes it exceptionally valuable for characterizing the external features of crude drugs, powders, and formulated products. The principles, advantages, instrumentation, procedure, challenges during analysis, and application of HR-SEM in crystallinity analysis have been described in Chapter 6. However, this chapter highlights its application in the morphological analysis of crude herbal drugs and their formulations.

HR-SEM is a powerful tool for the authentication and quality evaluation of herbal drugs and formulations. By using an electron beam and electromagnetic lenses in a vacuum, HR-SEM generates high-resolution, three-dimensional images with greater depth of field and magnification than light microscopy. This enables detailed visualization of surface features such as stomata, trichomes, cuticular striations, lenticels, pollen grains, seeds, and bark patterns, which are critical for botanical identification. In powdered drugs, HR-SEM provides insights into particle size, shape, roughness, and aggregation, directly influencing flow, compressibility, and dissolution. It also aids in examining fracture surfaces of brittle materials like resins, as well as visualizing crystallization patterns of secondary metabolites and inorganic inclusions. Furthermore, HR-SEM is invaluable for evaluating dosage forms, including tablet coating integrity, capsule surface texture, and excipient distribution, while also detecting contaminants such as dust, foreign particles, or microorganisms. Collectively, it strengthens the authentication, standardization, and safety of herbal medicines [30].

SEM can distinguish between different plant species or varieties by their unique surface features, for example, comparing the pattern of epidermal cells, the density and morphology of trichomes (e.g., glandular vs. non-glandular, branching patterns), and the types of stomata on leaves. *Cannabis sativa* trichome morphology is a classic example. In a recent study, researchers highlighted the lack of anatomical and morphological data for several medicinal plant species, which poses a significant challenge to quality control and standardization efforts. The study aimed to identify key histological features for the botanical identification of herbal drugs across different morphological groups, such as root, bark, leaf, flower, fruit, and seed. Important diagnostic traits included stomatal types, epidermal trichomes, papillae, calcium oxalate crystals, and cell organization. For example, *Jateorhiza palmata* (calumba root), a well-known digestive bitter, was examined using light and scanning electron microscopy. The analysis revealed key features such as sclereids in the cortical parenchyma, bordered pitted xylem vessels, calcium oxalate crystals, and abundant starch grains – characteristics that are critical for accurate identification and quality assessment of the plant material [30].

Detailed analysis of the particle size and shape of powdered crude drugs (e.g., rhizomes, roots, barks) and herbal extracts is critical for pharmaceutical applications as particle morphology affects flow properties, solubility, and bioavailability, for example, the irregular, angular particles of a ground root drug versus the more spherical particles of a spray-dried extract. While optical microscopy shows overall shape, SEM can reveal intricate surface details of starch granules (e.g., lamellae, fissures, pits) that are specific to botanical origin (e.g., potato starch vs. corn starch vs. rice starch granules).

It also assess the surface characteristics of tablets (e.g., roughness, porosity, uniformity of coatings), the appearance of granules in encapsulated products, and the dispersion of active ingredients in complex matrices like herbal creams or transdermal patches. In a research study, SEM was employed to investigate the surface morphology and distribution of silver nanoparticles (AgNPs) synthesized using *Solanum trilobatum* leaf extract. The SEM images revealed well-dispersed, predominantly spherical nanoparticles with some degree of aggregation, measuring approximately 50–70 nm in size. This analysis provided critical information on particle uniformity, surface texture, and aggregation behavior, which are essential parameters for evaluating nanoparticle stability and suitability for biomedical applications. The high-resolution imaging capability of SEM made it possible to visually confirm the successful formation of nanostructures. A similar application of SEM was reported in a study using *Coccinia grandis* leaf extract, where monodispersed AgNPs were also observed, further validating the reliability of SEM in characterizing nanoparticles produced via green synthesis methods [31].

HR-SEM is also used in identifying foreign materials based on their distinct surface morphology, which might be otherwise difficult to detect with optical microscopy, for example, distinguishing plant fibers from synthetic fibers or identifying non-botanical fillers [26]. A recent study demonstrates that SEM is a valuable tool for authenticating medicinal plants by revealing distinct micro-morphological features such as pollen structure, spine type, and exine sculpturing. These characteristics, observed across 16 medicinal species, aid in differentiating genuine plant material from adulterants. For instance, the presence of pollinia in *Wattakaka volubilis* and unique exine patterns in *Berberis* species highlight SEM's effectiveness in species-level identification. Thus, SEM supports accurate authentication, ensuring the safety and efficacy of herbal medicines [32].

Examining the surface of raw materials for microbial contamination or damage and assessing the physical stability and integrity of final products after storage or stress testing is another application area of HR-SEM.

10.5 ATOMIC FORCE MICROSCOPY/SCANNING PROBE MICROSCOPY

Atomic force microscopy (AFM) is a powerful type of scanning probe microscopy (SPM) that allows for imaging and characterization of surfaces at the nanoscale, down to atomic resolution. Unlike electron microscopes, which use electron beams, AFM uses a physical probe to "feel" the surface, making it suitable for a wide range of samples, including non-conductive biological materials, without the need for extensive preparation or vacuum environments. It provides not only topographical information but also insights into various surface properties like roughness, elasticity, adhesion, and friction [33].

10.5.1 Principle

The core principle of AFM involves a sharp probe (tip), typically made of silicon or silicon nitride, located at the free end of a flexible cantilever. This cantilever is mounted on a piezoelectric scanner that precisely controls its movement in three dimensions (X, Y, Z). As the tip is brought into close proximity to the sample surface, interatomic forces (van der Waals, electrostatic, capillary, etc.) between the tip and the sample cause the cantilever to deflect [34, 35].

A laser beam is focused onto the back of the cantilever, and its reflection is detected by a position-sensitive photodetector (PSPD). As the cantilever deflects, the position of the reflected laser spot on the PSPD changes. This change is converted into an electrical signal, which is fed into a feedback loop. The feedback loop then adjusts the height of the piezoelectric scanner (Z-axis) to maintain a constant force (or constant amplitude in tapping mode) between the tip and the sample as the tip scans across the surface in a raster pattern (X-Y plane). The precise adjustments made to the Z-piezo translate directly into a topographical map of the sample surface [34, 35].

AFM can operate in several modes:

- **Contact mode:** The tip is in constant physical contact with the sample surface. Cantilever deflection is used to create the topographic image. It is good for rigid samples but can cause sample damage or tip wear on soft materials [35].
- **Non-contact mode:** The tip oscillates just above the sample surface (typically 1–10 nm away). Changes in the resonant frequency or amplitude of oscillation, caused by van der Waals forces, are used to map topography. This mode minimizes sample damage but is sensitive to environmental factors [35].
- **Tapping mode (intermittent contact mode):** The cantilever oscillates at its resonant frequency, and the tip intermittently "taps" the surface. Changes in the oscillation amplitude (or phase) upon tapping are used in the feedback loop to maintain a constant tapping force and generate a topographic image. This mode is widely used for biological and soft samples as it reduces shear forces and minimizes sample damage compared to contact mode. It also provides phase imaging, which can reveal differences in material properties [35].

SPM is a broader category that includes AFM, along with other techniques like scanning tunneling microscopy (STM), which requires conductive samples, and near-field scanning optical microscopy (NSOM). AFM is the most versatile of the SPM techniques, especially for biological samples.

10.5.2 Instrumentation

An AFM system comprises several key components, as shown in Figure 10.6, which work in coordination to achieve high-resolution surface imaging. At the core is the cantilever and tip assembly – a micro-fabricated beam with a sharp pyramidal or conical tip used to scan the sample surface. Movement of the tip in the X, Y, and Z directions is precisely controlled by a piezoelectric scanner, which consists of a ceramic material that responds to applied voltage with minute dimensional changes. A diode laser directs a beam onto the back of the cantilever, and the reflected beam is captured by a PSPD, typically a four-quadrant photodiode. This detector measures both vertical and lateral deflections of the cantilever. The signal from the PSPD is processed by a feedback loop system, which adjusts the voltage to the Z-piezo in real time to maintain constant force or amplitude during scanning. To ensure accurate and noise-free data acquisition, the AFM is housed within a vibration isolation system and an acoustic isolation enclosure, which minimizes environmental interference.

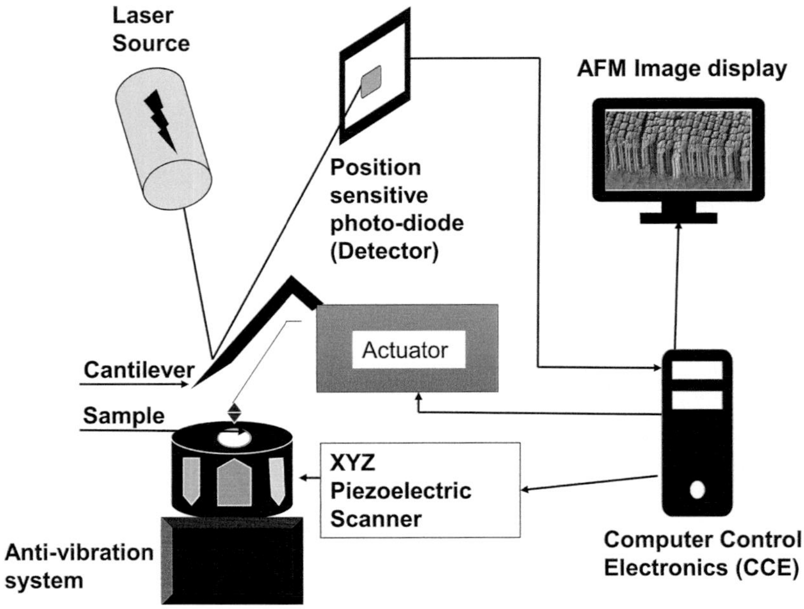

FIGURE 10.6 Instrumentation of AFM

Finally, a computer equipped with specialized software controls the instrument, collects data, and processes the resulting images for analysis [35, 36].

10.5.3 Advantages

AFM offers unique advantages for surface characterization:

- Can achieve atomic resolution for crystalline materials and sub-nanometer resolution for biological samples, providing incredibly detailed topographical maps.
- Unlike SEM, AFM does not require samples to be electrically conductive, eliminating the need for sputter coating and its associated artifacts. This is a huge advantage for biological and polymeric materials.
- Can operate in air, liquid (water, buffer solutions, organic solvents), and vacuum, making it ideal for studying biological processes in their native or near-native environments.
- Beyond topography, AFM can quantitatively measure various surface properties, including the following:
  - **Surface roughness:** Root-mean-square roughness (Rq), average roughness (Ra).
  - **Adhesion forces:** Between the tip and sample.
  - **Friction forces:** Lateral forces as the tip scans.
  - **Elasticity/ stiffness:** By analyzing force-distance curves (indentation).
  - **Magnetic and electrical properties:** With specialized tips and modes (MFM, KPFM).
- Especially in tapping or non-contact modes, the forces applied to the sample are extremely low, minimizing damage to delicate biological structures.
- Provides quantitative height information, allowing for precise measurements of feature dimensions (height, width, depth) and roughness parameters.

10.5.4 Procedure

Sample preparation for AFM is generally simpler than for electron microscopy, but samples must be relatively flat and securely mounted.

The steps for sample preparation are as follows:

1. The sample surface should be as flat as possible within the scan area to ensure the tip can effectively track the topography without crashing or losing contact.
2. The sample is mounted (e.g., with double-sided tape, epoxy, or vacuum grease) onto a clean glass slide or a metal puck that fits onto the AFM stage.
3. The sample surface should be clean and free of dust or loosely adhering particles, which can contaminate the tip or interfere with imaging.

For biological samples

1. Cells or subcellular components can be directly deposited from a suspension onto a freshly cleaved mica surface (atomically flat) or treated glass slide, then allowed to dry or imaged in liquid.
2. For imaging individual molecules (e.g., proteins from herbal extracts), they can be chemically immobilized onto a functionalized surface.
3. For imaging in air, samples may need gentle drying. However, AFM's ability to operate in liquid environments means hydration can be maintained for native-state imaging.

For powders

1. A small amount of powder can be dispersed in a solvent and a drop placed on a flat substrate and allowed to dry slowly to form a thin, relatively uniform layer. Individual particles can then be imaged. Alternatively, powder can be pressed onto an adhesive surface.

10.5.5 Procedure

1. The prepared sample is placed on the AFM stage.
2. An appropriate cantilever and tip are selected based on the desired resolution and mode of operation (e.g., soft cantilever for biological samples in tapping mode).
3. Carefully, the tip is engaged with the sample surface, typically automatically controlled by the AFM software.
4. Imaging parameters are scan size, scan speed, set point (desired force or amplitude), and feedback loop gains, which are set.
5. The scan is initiated, and the software displays the topographic image in real time.
6. The parameters are adjusted as needed to optimize image quality.
7. The images, force curves, or other data are acquired, and post-processing analysis (e.g., roughness calculation, feature measurement) is performed for detailed morphological investigation.

10.5.6 Challenges

- AFM typically scans relatively small areas, usually ranging from a few nanometers up to about 100–150 micrometers. This makes it difficult to survey large areas or analyze highly heterogeneous samples quickly.
- AFM is significantly slower than SEM, with image acquisition often taking minutes to hours, depending on scan size and resolution. This can be problematic for dynamic processes or routine high-throughput analysis.
- The shape of the AFM tip can influence the image, making features appear wider than they are. This "tip convolution" effect can be significant, especially when imaging features similar in size to the tip radius.
- AFM is extremely sensitive to external vibrations, acoustic noise, and temperature fluctuations, requiring a highly controlled environment and vibration isolation.

- The tips are very delicate and can be easily damaged or contaminated, requiring careful handling and frequent replacement.
- Operating an AFM effectively and interpreting the data accurately requires considerable expertise and training.
- Extremely rough surfaces can be challenging to image because the tip may not be able to track the topography effectively or may crash into steep features.
- AFM is a surface imaging technique and does not provide information about the internal structure of the sample.

10.5.7 Application to Herbal Drugs

AFM's unique capabilities make it a valuable tool for advanced morphological and biophysical characterization of herbal materials at the nanoscale.

Nanoscale cell wall architecture: It provides detailed imaging of the surface topography of plant cell walls, revealing the organization of cellulose microfibrils, pectin networks, and other polymeric components at high resolution. This can provide insights into cell wall biosynthesis, degradation, and interaction with various compounds, for example, studying the effects of extraction on the integrity of cell wall components. In a recent study, AFM was employed to investigate the morphology of gum tragacanth (GT)–stabilized green gold nanoparticles (AuNPs) developed as carriers for the hydrophobic therapeutic compound Naringin. AFM imaging revealed that the synthesized AuNPs were spherical and uniformly distributed in the nanoscale range, with stabilization provided by the multifunctional groups of GT. Upon loading Naringin, the AuNPs demonstrated enhanced bactericidal activity, confirmed through both tetrazolium microplate assays and AFM-based morphological analysis. AFM further revealed increased membrane disruption in treated bacterial cells, supporting the improved antimicrobial efficacy of the naringin-loaded nanocarrier system. These findings underscore AFM's critical role in characterizing nanocarrier morphology and assessing nanoscale interactions in drug delivery systems [37].

Starch granule characterization: AFM provides unprecedented visualization of the surface texture, lamellae, and growth rings of individual starch granules from different herbal sources. AFM can precisely measure the dimensions, height variations, and surface roughness of starch granules, aiding in their differentiation and quality assessment. In recent studies, AFM has been extensively applied to investigate the surface and internal structures of starch granules with nanometer-scale resolution. Unlike conventional microscopy, AFM requires minimal sample preparation and enables imaging under near-native conditions without coating or dehydration. AFM revealed distinct surface topographies in native starch, including globular and beaded-chain structures, and detected structural disruptions following plasticizer treatment. It also identified raised nodules on wheat and potato starch, correlating with blocklet structures. Using force modulation mode, AFM visualized internal granule features such as growth rings and viscoelastic contrasts, highlighting regional differences in granule composition. These findings demonstrate AFM's critical role in revealing both surface and internal architectures of starch from various botanical sources, offering insights into their morphology and structural integrity [38].

Surface roughness of powders and formulations: It is used to quantify the surface roughness of finely milled herbal powders, which influences their flowability, packing density, and dissolution rates in formulations, and assess the uniformity and smoothness of pharmaceutical coatings on herbal tablets or films containing herbal extracts.

In a recent study, microscopic morphological developments such as lactose crystallization, particle swelling, and changes in surface roughness were investigated as a function of moisture sorption in skimmed milk powder (SMP), milk protein concentrate (MPC), and whey protein isolate (WPI) powders. AFM was applied to identify high-resolution microstructural changes in high-protein milk powder (HPMP) particles. A novel sample preparation technique was developed to achieve a uniform, single-particle layer on a mica surface, enabling effective AFM analysis. Using amplitude modulation (AM)-AFM, it was observed that increasing relative humidity (RH) during equilibration led to distinct topographical changes and increased surface roughness. In SMP, surface transformations were primarily due to lactose crystallization and the subsequent stacking of crystal layers. MPC displayed characteristic surface dimples and folds, possibly resulting from shrinkage and compaction. As RH increased, MPC particles exhibited more pronounced folds and jagged textures, particularly at 85% RH. WPI powders, in contrast, had smoother surfaces with broken fragments, indicating higher friability, and showed minimal structural change with moisture sorption. Overall, AM-AFM provided detailed three-dimensional imaging at nano- and micrometer scales, highlighting its value in evaluating moisture-induced surface transformations in HPMPs [39].

- **Interaction of herbal compounds with biological surfaces:** It is used for studying the adhesion forces between active compounds from herbal extracts and target biological membranes or proteins at a single-molecule level, which is critical for understanding their mechanism of action.
- **Characterization of herbal nanomedicines:** It is used for precisely measuring the size, shape, aggregation state, and surface properties of nanoparticles (e.g., liposomes, polymeric nanoparticles, inorganic nanoparticles synthesized via green routes) used to encapsulate or deliver herbal active ingredients. AFM can also assess the stability and integrity of such nanocarriers. AFM imaging was employed to investigate the surface morphology of lyophilized Baicailin-loaded SNEDDS nanodroplets. Non-contact mode revealed uniformly dispersed, smooth, circular particles on mica surfaces, indicating no aggregation. The mean particle height was 15.4 ± 9.0 nm – smaller than the DLS-measured diameter – suggesting droplet flattening or AFM tip convolution. The study confirms AFM as a powerful tool for nanoscale morphological analysis of emulsion-based drug delivery systems [40].

In a recent study, the morphology of transferosomal vesicles was studied by AFM. The transferosomes are found to be spherical and uniform in size. The 2D and 3D AFM morphology showed all the dimensions below 200 nm and the suitability of vesicles for transdermal permeation (Figure 10.7).

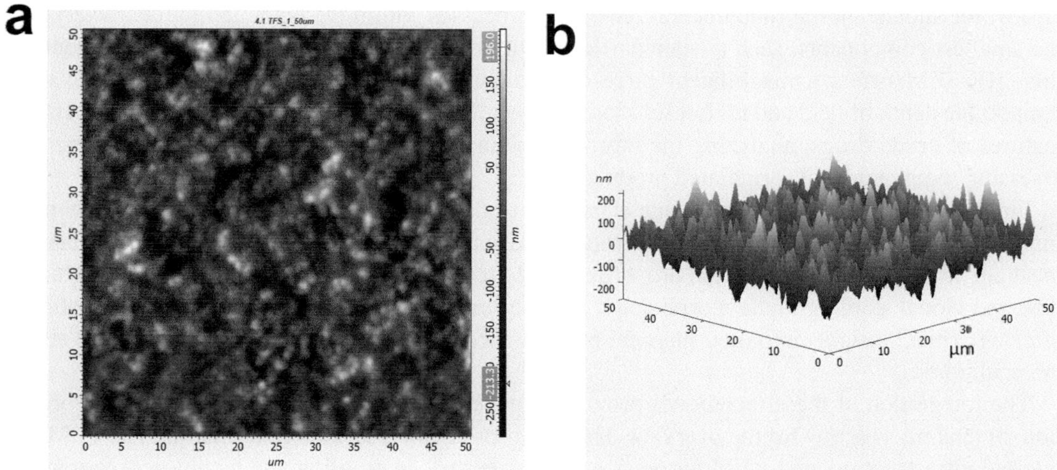

FIGURE 10.7 Morphology analysis of transferosome by AFM (a) 2D SPM photomicrograph of transferosomes and (b) 3D SPM photomicrograph of transferosomes

The vesicles are then incorporated into a hydrogel to produce a transgelosome. The transgelosome demonstrated excellent therapeutic activity for effective melanoma therapy [19].

Elasticity and mechanical properties of herbal tissues: It is used for measuring the local stiffness (Young's modulus) of plant cells or tissues, which can be correlated with their physiological state, maturity, or the effects of environmental stressors. This can be relevant for assessing the quality and mechanical handling properties of crude drugs [41].

Biofilm formation on herbal surfaces: It is used for investigating the morphology of microbial biofilms on herbal raw materials at the nanoscale, which is important for understanding contamination and storage challenges [29]. Recent studies have emphasized the pivotal role of advanced microscopic techniques – particularly AFM and SPM – in elucidating the nanoscale adhesive forces that govern bacterial biofilm formation, structure, and disruption. These methods provide high-resolution insights into the mechanical properties of biofilms, such as surface stiffness, adhesion force mapping, and topographical imaging, which are critical for understanding bacterial adhesion and resistance mechanisms. AFM and SPM have enabled visualization of receptor–ligand interactions at the biofilm interface and helped in assessing the effectiveness of antibiofilm compounds in both clinical and industrial settings [42].

10.6 CONCLUSION

The morphological analysis of crude herbal drugs and their formulations is a complex yet indispensable aspect of quality control in phytomedicine. As this chapter has demonstrated, no single instrumental technique can provide a complete picture of the multifaceted structural characteristics of these intricate natural products. Instead, a synergistic approach, leveraging the unique strengths of various microscopic methodologies, offers the most comprehensive and insightful understanding.

Optical microscopy, with its accessibility and broad utility, remains the foundational tool for initial identification, macroscopic feature correlation, and routine quality assessment. It excels at revealing overall tissue organization, cellular inclusions, and distinguishing features at the microscale, serving as the first line of defense against misidentification and gross adulteration.

Stepping beyond the limitations of light, electron microscopy (HR-TEM and HR-SEM) opens new frontiers. HR-TEM provides unparalleled insights into the ultrastructure of plant cells, resolving organelles, cell wall layers, and the precise localization of metabolites at the nanometer scale. It is critical for understanding fundamental biological processes within the plant and for characterizing the smallest components, such as nanoparticles, within complex herbal matrices. Complementing this, HR-SEM offers a magnificent three-dimensional perspective of surface topography with remarkable depth of field and high resolution. It is invaluable for characterizing the intricate surface features of crude drugs, analyzing the morphology and particle size distribution of powders, and assessing the integrity of formulated products.

Finally, AFM/ SPM brings a unique capability to the forefront: atomic to nanoscale resolution imaging of non-conductive samples in various environments, coupled with the ability to quantify mechanical and surface interaction properties. AFM reveals ultra-fine details of cell wall architecture, individual starch granule morphology, and the surface characteristics of herbal formulations, offering critical insights into their physical performance and potential biological interactions at the molecular level.

The integration of these techniques provides a powerful continuum of morphological information: optical microscopy for the overview, HR-SEM for detailed surface examination, HR-TEM for internal ultrastructure, and AFM for precise nanoscale surface profiling and biophysical characterization. For instance, an unknown powdered herbal drug might first be examined by optical microscopy to identify characteristic cell fragments and starch grains. If further detail on particle surface morphology is needed for processing or formulation, HR-SEM would be employed. To understand how specific constituents are encapsulated within the cell or if nanoparticles are present, HR-TEM would be the method of choice. And for an extremely high-resolution look at the surface topography of a cell wall or a formulated herbal film, AFM would provide the most precise data.

As herbal medicine continues to evolve, embracing advanced drug delivery systems and nanotechnologies, the role of these instrumental methodologies will become even more critical.

REFERENCES

1. Klein-Junior, L.C., et al., Quality control of herbal medicines: From traditional techniques to state-of-the-art approaches. *Planta Medica*, 2021. 87(12/13): p. 964–988.
2. Ahmad, I., M.S. Ahmad Khan, and S.S. Cameotra, Quality assessment of herbal drugs and medicinal plant products. *Encyclopedia of Analytical Chemistry*, 2014. 46: p. 1–17.
3. Lindenmaier, M.P., M.W. Bernart, and J.A. Brinckmann, Advanced methodologies for the quality control of herbal supplements and regulatory considerations. *Phytochemical Analysis*, 2025. 36: p. 1–17.
4. Joshi, V.C. and I.A. Khan. Microscopy techniques for the identification and authentication of botanicals, in *IV International Conference on Quality and Safety Issues Related to Botanicals*. 2005. p. 720.
5. Diaspro, A. and C. Usai, Optical microscopy. *Wiley Encyclopedia of Biomedical Engineering*, 2006. 6: p. 1–8.
6. Johnson, H. and D. Parsons, Enhanced contrast in electron microscopy of unstained biological material: III. In-focus phase contrast of large objects. *Journal of Microscopy*, 1973. 98(1): p. 1–17.
7. Cui, Y., et al., Multiscale microscopy to decipher plant cell structure and dynamics. *New Phytologist*, 2023. 237(6): p. 1980–1997.
8. Poole, J.J. and L.B. Mostaço-Guidolin, Optical microscopy and the extracellular matrix structure: A review. *Cells*, 2021. 10(7): p. 1760.
9. McDonald, J., Optical microscopy, in *Microelectronics Failure Analysis: Desk Reference*. 2019, ASM International. p. 42–61.
10. Bondada, B.R. and M. Keller, Not all shrivels are created equal – morpho-anatomical and compositional characteristics differ among different shrivel types that develop during ripening of grape (Vitis vinifera L.) berries. *American Journal of Plant Sciences*, 2012. 3(7): p. 879–898.
11. Pohl, R.W., Controlled maceration of grass leaves in 40–80% nitric acid for preparation of epidermis for slides. *Stain Technology*, 1967. 42(4): p. 195–197.
12. Smolyaninov, I.I., Optical microscopy beyond the diffraction limit. *HFSP Journal*. 2008. 2.
13. Fischer, R.S., et al., Microscopy in 3D: a biologist's toolbox. *Trends in Cell Biology*, 2011. 21(12): p. 682–691.
14. Jimson, S., et al., Artifact in histological section. *Biomedical and Pharmacology Journal*, 2016. 9(2): p. 843–845.
15. Moon, H.-K., et al., Phylogenetic significance of leaf micromorphology and anatomy in the tribe Mentheae (Nepetoideae: Lamiaceae). *Botanical Journal of the Linnean Society*, 2009. 160(2): p. 211–231.
16. Turner, J.C., J.K. Hemphill, and P.G. Mahlberg, Cannabinoid composition and gland distribution in clones of Cannabis sativa L. (Cannabaceae). *Bulletin on Narcotics*, 1978. 30: p. 55–65.
17. Stoyanov, A., A joint macro and microscopic authentication key for pharmacognostic analysis of croci sativi stigma and its common adulterants on the Bulgarian market. *Обучение по природни науки и върхови технологии*, 2024. 33(1–2): p. 7–16.
18. Govindaraghavan, S. and N.J. Sucher, Quality assessment of medicinal herbs and their extracts: Criteria and prerequisites for consistent safety and efficacy of herbal medicines. *Epilepsy & Behavior*, 2015. 52: p. 363–371.
19. Mohapatra, D., et al., Quality-by-design-based development of ultradeformable nanovesicular transgelosome of standardized Piper longum extract for melanoma. *Nanomedicine*, 2023. 18(14): p. 963–985.
20. Agrawal, S., Spectroscopy techniques for rare-earth-activated phosphors, in *Rare-Earth-Activated Phosphors*. 2022, Elsevier. p. 147–175.
21. Javed, Y., et al., TEM for atomic-scale study: Fundamental, instrumentation, and applications in nanotechnology, in *Handbook of Materials Characterization*. 2018, Springer. p. 147–216.
22. Porter, A.E., et al., Direct imaging of single-walled carbon nanotubes in cells. *Nature Nanotechnology*, 2007. 2(11): p. 713–717.
23. Tippo, O., *An Introduction to Plant Anatomy*. 1950, JSTOR.

24. Konarska, A., Microstructural and histochemical characteristics of Lycium barbarum L. fruits used in folk herbal medicine and as functional food. *Protoplasma*, 2018. 255(6): p. 1839–1854.
25. Shahriary, M., et al., In situ green synthesis of Ag nanoparticles on herbal tea extract (Stachys lavandulifolia)-modified magnetic iron oxide nanoparticles as antibacterial agent and their 4-nitrophenol catalytic reduction activity. *Materials Science and Engineering: C*, 2018. 90: p. 57–66.
26. Mohapatra, D., et al., Tinospora cordifolia leaves derived carbon dots for cancer cell bioimaging, free radical scavenging, and Fe3+ sensing applications. *Journal of Fluorescence*, 2022. 32(1): p. 275–292.
27. Li, T., et al., Natural berberine-based Chinese herb medicine assembled nanostructures with modified antibacterial application. *ACS Nano*, 2019. 13(6): p. 6770–6781.
28. Dholvitayakhun, A., et al., Using scanning and transmission electron microscopy to investigate the antibacterial mechanism of action of the medicinal plant Annona squamosa Linn. *Journal of Herbal Medicine*, 2017. 7: p. 31–36.
29. Dobrucka, R., Synthesis and structural characteristic of platinum nanoparticles using herbal bidens tripartitus extract. *Journal of Inorganic and Organometallic Polymers and Materials*, 2016. 26(1): p. 219–225.
30. Serrano, R., G. Da Silva, and O. Silva, Application of light and scanning electron microscopy in the identification of herbal medicines. *Microscopy: Science, Technology, Applications and Education*, 2010. 3: p. 182–190.
31. Vanaja, M., et al., Herbal plant synthesis of antibacterial silver nanoparticles by Solanum trilobatum and its characterization. *International Journal of Metals*, 2014. 2014(1): p. 692461.
32. Ahmed, S.N., et al., Comparative light and scanning electron microscopy in authentication of adulterated traded medicinal plants. *Microscopy Research and Technique*, 2019. 82(7): p. 1174–1183.
33. Bian, K., et al., Scanning probe microscopy. *Nature Reviews Methods Primers*, 2021. 1(1): p. 36.
34. Loos, J., The art of SPM: Scanning probe microscopy in materials science. *Advanced Materials*, 2005. 17(15): p. 1821–1833.
35. Bhushan, B. and O. Marti, Scanning probe microscopy – principle of operation, instrumentation, and probes, in *Springer Handbook of Nanotechnology*. 2004, Springer. p. 325–369.
36. Bhushan, B. and O. Marti, Scanning probe microscopy – principle of operation, instrumentation, and probes, in *Springer Handbook of Nanotechnology*. 2010, Springer. p. 573–617.
37. Rao, K., et al., Gum tragacanth stabilized green gold nanoparticles as cargos for Naringin loading: A morphological investigation through AFM. *Carbohydrate Polymers*, 2017. 174: p. 243–252.
38. Chakraborty, I., et al., Advanced microscopy techniques for revealing molecular structure of starch granules. *Biophysical Reviews*, 2020. 12(1): p. 105–122.
39. Mishra, V.S., et al., Topographical changes in high-protein, milk powders as a function of moisture sorption using amplitude-modulation atomic force microscopy. *Food Hydrocolloids*, 2022. 127: p. 107504.
40. Jakab, G., et al., Optimization of quality attributes and atomic force microscopy imaging of reconstituted nanodroplets in baicalin loaded self-nanoemulsifying formulations. *Pharmaceutics*, 2018. 10(4): p. 275.
41. Peaucelle, A., AFM-based mapping of the elastic properties of cell walls: At tissue, cellular, and subcellular resolutions. *Journal of Visualized Experiments: JoVE*, 2014(89): p. 51317.
42. Dash, S., et al., Probing the surface-attached in vitro microbial biofilms with atomic force (AFM) and scanning probe microscopy (SPM), in *Analytical Methodologies for Biofilm Research*. 2021, Springer. p. 223–241.

11 Instruments for Elemental Analysis of Herbal Drugs and Formulations

ABBREVIATIONS

Abbreviation	Description
AAS	Atomic Absorption Spectroscopy
ICP-MS	Inductively Coupled Plasma-Mass Spectrometry
FDA	Food and Drug Administration
ICP-AES	Inductively Coupled Plasma Atomic Emission Spectroscopy
MS	Mass Spectrometer
RF	Radiofrequency
HPLC-ICP-MS	High-Performance Liquid Chromatography-Inductively Coupled Plasma-Mass Spectrometry
LC-ICP-MS	Liquid Chromatography-Inductively Coupled Plasma-Mass Spectrometry
API	Ayurvedic Pharmacopoeia of India
US-EPA	United States Environmental Protection Agency

11.1 INTRODUCTION

Elemental analysis is important for the standardization, quality control, and safety assessment of medicinal plants, their extracts, and herbal drugs. Medicinal plants during cultivation may accumulate both necessary and toxic elements from various sources, such as soil, water, and the environment, and become contaminated. Hence, there must be well-defined limits of elemental composition decided for medicinal plants.

The elemental content related to the active phytocompound of the medicinal plant increases both the beneficial and the hazardous effects of the herbal product. Depending upon factors such as soil type, pH of soil, temperature, and rainfall, the elemental content varies across the diverse geographical locations. So, the environment of the geographical location in which the medicinal herbs grow plays a vital role and is to be taken into consideration. Furthermore, monitoring of elemental composition during cultivation, processing, and formulation using analytical tools such as AAS and ICP-MS (Figure 11.1) safeguards against possible contamination and preserves therapeutic integrity, making them an indispensable tool for ensuring the safety and efficacy of herbal medicines.

11.2 ATOMIC ABSORPTION SPECTROSCOPY

AAS is a well-established analytical tool well known for its capability for the detection and quantification of metallic elements that are present at trace or ultra-trace levels in various sample types. It is utilized for its high sensitivity, selectivity, and reproducibility, thus making it an important standard method for research and regulatory aspects. The technique is comparatively fast and cost effective compared to other advanced spectrometric techniques (such as ICP-MS) and also capable

DOI: 10.1201/9781003540823-11

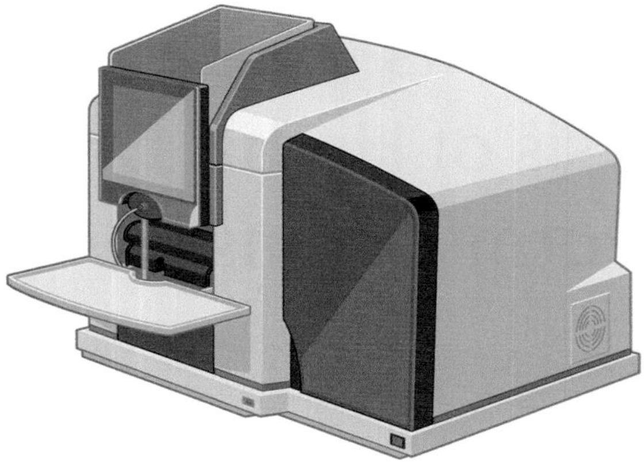

Atomic Absorption Spectroscopy (AAS)

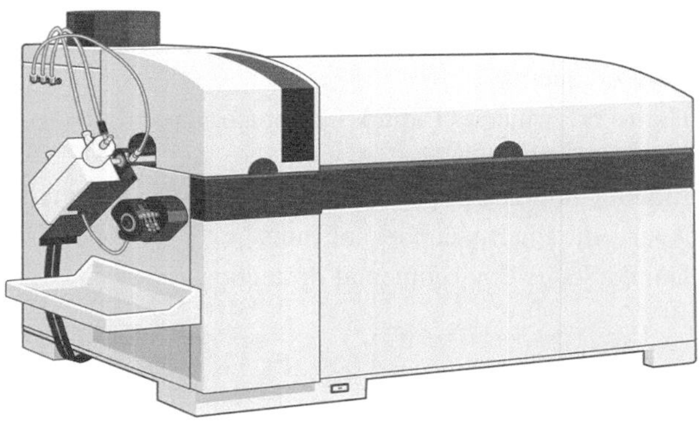

Inductively Coupled Plasma Mass Spectrometry (ICP-MS)

FIGURE 11.1 Instruments for structural elucidation and phytochemical integrity analysis of herbal drugs and formulations

of analyzing complex matrices requiring nominal sample preparation steps. AAS is available in various setups, namely flame AAS (used for routine sample analysis), graphite furnace AAS (used for detection at ultra-trace levels), and hydride generation AAS (mainly used in the case of volatile elements), thus permitting flexibility based on different analytical needs. Its robust capability has led to its wide utilization in areas such as environmental testing of pharmaceutical drugs, including assessment of the safety and the quality control of plant extracts, herbal drugs, and herbal formulations.

11.2.1 Principle

AAS is based on the principle that in the ground state, free atoms absorb light of a specific wavelength characteristic of each element. When a sample solution comprising the component of interest is introduced into a graphite or flame furnace, the thermal energy changes the analyte into free, ground-state atoms by a process called atomization.

A hollow cathode lamp (or any other suitable radiation source) emits light at a specific wavelength corresponding to the target element. In the atomizer, as this light passes through the cloud of atoms in the ground state, a portion of it is absorbed. This follows Beer-Lambert law, which states that the amount of light absorbed is directly proportional to the concentration of the element in the sample.

The equation is as follows:

$$A = k \times C \tag{11.1}$$

where,
A = absorbance
k = proportionality constant (which depends on path length and molar absorptivity)
C = concentration of the sample or analyte

The instrument records the absorbance of the sample and, using a standard calibration curve of various known concentrations, detects the metal content/concentration in the sample.

11.2.2 Advantages

- **High sensitivity:** AAS is a well-known and accepted technique for determining elements up to ultra-trace (sub-µg/mL) and trace (µg/mL) levels with great accuracy and acceptable precision, making it suitable for trace analysis in herbal drugs.
- **High selectivity:** The element-specific hollow cathode lamp confirms negligible spectral interference, improving accuracy for the target analyte analysis.
- **Versatility:** It is employed for the determination of over 70 to 80 metals and metalloids, both qualitatively and quantitatively, present in various matrices such as plant extracts, soils, and biological fluids.
- **Cost effectiveness:** Flame AAS is cost effective compared to other advanced techniques, such as ICP-MS, used for single-element or few-element analysis.
- **Simple calibration:** The calibration process is easy, utilizing aqueous solutions of standards prepared in the same matrix as the sample reduces complexity.
- **Wide applicability:** It is a widely used, robust technique with global acceptance in quality control, regulatory testing, and research.

11.2.3 Instrumentation

A schematic representation of the AAS instrument is shown in Figure 11.2. It consists of several key components that work together to perform the selective and sensitive analysis of elements. It includes the following:

- **Radiation source:** Provides light at the characteristic wavelength of the element being analyzed. They are of the following types:
  a. Hollow cathode lamp: It is the most commonly used radiation source, and it consists of a cathode, which is made of the element to be analyzed.
  b. Electrodeless discharge lamp: It is used for elements which generally have low emission efficiency, such as and Se.
- **Chopper:** It is a rotating wheel or modulator placed between the lamp and atomizer. It periodically interrupts the light beam, converting it to an alternating signal. This helps us differentiate the analytical signal from background emission.
- **Nebulizer and atomizer:** The sample, generally in solution form, is introduced through a nebulizer, which generates a fine mist. The mist is further carried to an atomizer, either

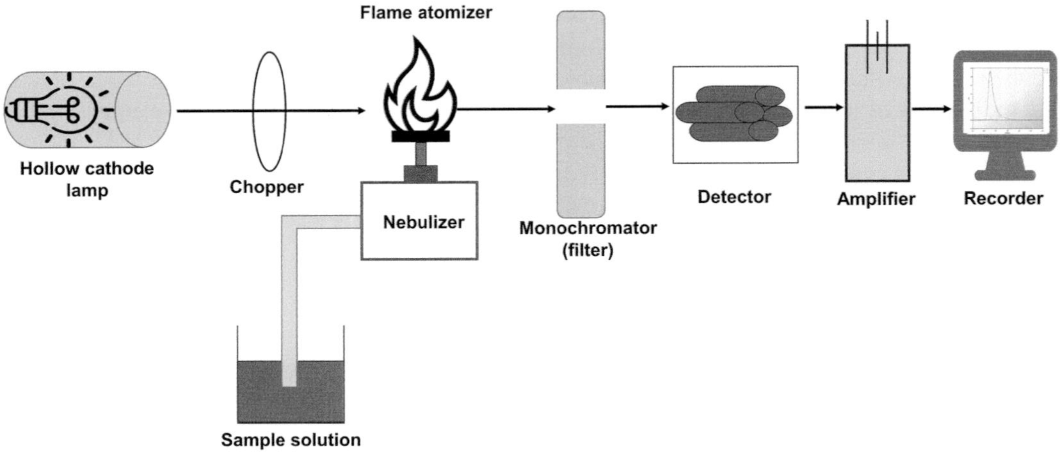

FIGURE 11.2 Instrumentation of AAS

a flame or a graphite furnace. Here, the sample gets vaporized, thus dissociating into free ground-state atoms for analysis.

There are different types of atomizers:

a. Flame atomizer: It requires fuel (e.g., acetylene) and an oxidant (air/nitrous oxide) for achieving ppm level sensitivity.
b. Graphite furnace atomizer: It electrically heats a graphite tube suitable for sensitivity at ppb levels.
c. Hydride generation system: It converts specific elements such as, Se, and Sb into volatile hydrides.
d. Cold vapor technique: It is employed for the determination of mercury content.

- **Monochromator:** This optical device isolates the specific analytical wavelength absorbed by the sample from the radiation source, minimizing scattered light or spectral interferences. Diffraction grating monochromators are the most commonly used.
- **Detector:** The detector measures and converts transmitted light into an electrical signal, which is proportional to the amount of light being absorbed. A photomultiplier tube is the most commonly used detector because of its high sensitivity.
- **Signal amplifier and readout system:** The electrical signal is amplified by a signal amplifier for ease of measurement and processing. The recorder or readout system records the absorbance data for quantification of the element.

11.2.4 Procedure

The AAS involves the following steps [1–3]:

- The herbal drug or plant material is dried and finely ground into powder. The powder is accurately weighed and transferred into reaction vessels for digestion.
- A specific amount of 65% nitric acid (HNO_3), a primary oxidizing agent, is added to the sample. It is heated to about 120°C to 150°C to allow initial digestion of the sample.
- Then, 30% perchloric acid ($HClO_4$) or hydrogen peroxide (H_2O_2) is added dropwise to allow complete digestion of the sample.
- The heating is continued until a clear solution of the sample is obtained.
- The solution is filtered using Whatman filter paper number 42 (having particle retention of 2.5 µm).

- It is allowed to cool and further diluted using deionized water to adjust the volume up to 50 mL.
- As an alternative to this wet digestion method (most commonly used method), nowadays microwave-assisted digestion is the preferred technique used for digestion.
- Before analysis of the sample, a blank is run containing all the reagents except the sample for background correction.
- Similar to sample preparation, elemental standards are prepared for calibration of the instrument.
- The standard stock solutions of elements (Cd, Hg, As, and Pb) at a concentration of 1000 ppm are prepared, and standard working solutions are prepared by further diluting the respective stock standard solutions with 0.2% nitric acid (HNO_3).
- For analysis, in case of cadmium (Cd) hollow cathode lamps, for arsenic (As) and lead (Pb), and for mercury (Hg) electrodeless discharge lamps are used as the radiation sources.
- The liquid sample solution is subjected to atomization where it is converted into a mist. This mist further mixes with fuel gas and oxidant in the mixing chamber, which gets atomized when it is subjected to flame.
- The produced free atoms absorb light from the hollow cathode lamp or electrodeless discharge lamps of the target element.

11.2.5 Challenges

AAS analysis of herbal drugs and formulations faces the following challenges [4].

- **Analysis limitations:** It has the capability to detect one element at a time. AAS is time consuming and also requires a large sample quantity because multiple elements cannot be analyzed simultaneously.
- **Matrix interference:** Chemical and spectral interferences arising from herbal sample matrices, such as organic matter or anions, reduce the efficiency of atomization and therefore affect absorption accuracy.
- **Sample preparation:** The technique is suitable if the sample is in liquid form, but for solids, they must first be solubilized. This sample preparation step is time consuming and may not be possible for samples that are insoluble or are refractive materials.
- **Detection limitations:** AAS generally has satisfactory sensitivity for most heavy metals that are regulated in herbal drug products, but it may be inadequate for detection at ultra-trace levels without any specialized preprocessing procedures.
- **Destructive analysis:** The atomization process completely destroys the sample, thus making sample recovery difficult after analysis.
- **Lack of differentiation between elemental species:** AAS only quantifies total elemental concentration present in a sample, but it is unable to differentiate between different oxidation states or chemical forms (e.g., organic vs. inorganic arsenic), which is significant for toxicity and regulatory concerns.

11.2.6 Applications to Herbal Drugs

AAS is a sophisticated analytical tool generally useful in the analysis of herbal drugs to confirm their safety and quality. It is mainly used to detect and quantify toxic (such as lead, cadmium, mercury, and arsenic) and useful trace heavy metals/elements (such as copper, zinc, iron, cobalt, and chromium) present in both medicinal plants and herbal drugs. The evaluation of these elements is critical since contamination from environmental sources tends to introduce harmful metals into herbal drug products, whereas maintaining an accurate level of necessary minerals is important for achieving safety and therapeutic efficacy of the herbal drugs.

In a recent study on *Zingiber officinale* (ginger) rhizomes, the presence of volatile heavy metals, that is, As and Hg, and non-volatile toxic heavy metals, that is, Pb and Cd, collected from different geographical locations in Himachal Pradesh and Uttarakhand, India, was investigated using AAS. A total of 25 rhizome and corresponding soil samples from upper and lower layers were collected and analyzed. In soils, Pb, As, and Hg exceeded permissible limits while Cd levels were low; Hg showed uneven distribution across the field. In rhizomes, Pb was the most abundant and present in all samples, Cd was consistently below limits, Hg was higher, and As was found in a few samples from Himachal Pradesh. Overall, Himachal Pradesh samples had higher volatile metals, whereas Uttarakhand samples had more non-volatile metals, with abundance in rhizomes following the order: As < Cd < Hg < Pb, and in soil Cd < As < Pb < Hg. The results confirm that most samples were within FDA limits except two from Himachal Pradesh, thus underscoring the importance of heavy metal monitoring in medicinal plants [5].

In another study, the identification and quantification of the major elemental composition of *Andrographis echioides* (L.) Nees. leaves were analyzed using AAS to check their pharmaceutical potential. The presence of a total of 12 important elements, such as calcium, cadmium, copper, chromium, iron, lead, magnesium, manganese, nickel, sodium, silver, and zinc, was quantified. The concentration of the elements was in the order Ca> Fe> Mg. These elements were important for maintaining physiological functions such as bone health, blood formation, heart rhythm, and energy metabolism. Other elements, namely, copper, chromium, manganese, nickel, sodium, silver, and zinc, were present in smaller quantities, whereas toxic elements such as cadmium and lead were present at low levels, that is, 1.3 mg/100 g and 0.3 mg/100 g, respectively. These findings suggest the importance of analyzing a plant's mineral profile, therefore validating its traditional use for the treatment of various disease conditions. It also highlights the therapeutic efficacy of *Andrographis echioides* (L.) Nees in addressing nutrient deficiencies and developing herbal medicines [6].

11.3 INDUCTIVELY COUPLED PLASMA MASS SPECTROMETRY

Inductively coupled plasma mass spectrometry (ICP-MS) was developed in the early 1980s by integrating the use of inductively coupled plasma produced by high-temperature ionization with a mass spectrometer for sensitive detection of elements. The first commercial instruments came around 1983 from Sciex, VG Elemental, and PerkinElmer. ICP-MS has advanced from laboratory equipment to a commercial instrument having widespread analytical applications.

At present, ICP-MS is the most versatile technique valued for specific detection of ultra-trace multi-elements and isotopic/isobaric analysis below ppt to ppm levels across various fields such as environmental, pharmaceutical, clinical, and geological fields, allowing analysis of the most abundant natural isotopes for each element. It can perform qualitative, semiquantitative, and quantitative analysis by mass-to-charge (m/z) ratios and also measure isotopic ratios, since it is equipped with a mass analyzer. This technique is so versatile that it can replace almost all conventional inorganic analytical tools, such as inductively coupled plasma atomic emission spectroscopy (ICP-AES), in its analytical performance.

11.3.1 Principle

ICP-MS is an advanced analytical tool used for analysis of ultra-trace elements, combining an inductively coupled plasma (ICP) as an ion source, hyphenated with a mass spectrometer (MS) for element separation and detection. ICP-MS works on the principle of generating an ion source, which is a high-temperature inductively coupled argon plasma, that effectively undergoes atomization followed by ionization of elements existing in a sample. The sample (generally in a liquid state) is nebulized to get a fine aerosol mist, which is then transported by an inert carrier gas (argon) into the plasma. The atoms present in the sample are first converted into singly charged ions at plasma temperatures between 6000 and 10,000 K. These charged ions are further extracted by an interface region into a mass spectrometer situated inside a high-vacuum chamber, where they are sorted based on their mass-to-charge ratio (m/z). The separated ions then get detected, and the resultant signals are further processed to estimate the elemental composition and concentration with high sensitivity and accuracy.

11.3.2 Advantages

The elemental analysis of herbal drugs by ICP-MS has the following advantages [7]:

- **Detection of ultra-trace elements**: ICP-MS is capable of detecting toxic heavy metals such as Pb, Cd, As, Hg and essential elements such as Fe, Zn, Cu, Mn in herbal materials at ppb levels, therefore confirming its safety and compliance with regulatory limits.
- **Simultaneous analysis of multiple elements**: ICP-MS reduces the analysis time required for complete profiling of both useful and detrimental elements present in complex herbal matrices in a single run.
- **Broad detection range**: It precisely quantifies elements from very low to high concentrations, significant for analyzing both trace contamination and higher-level mineral content.
- **Isotopic analysis:** It allows isotopic fingerprinting for authentication and adulteration detection in herbal raw materials.
- **Minimal sample requirement**: It is suitable for the analysis of minimal or expensive quantities of herbal samples.
- **Hyphenation with other techniques**: It can be hyphenated with separation techniques such as HPLC or GC, for elemental speciation (e.g., distinguishing between toxic inorganic arsenic and less harmful organic forms) in herbal extracts.
- **Rapid screening**: ICP-MS allows quick screening of large batches of herbal drugs or extracts during processing for quality control purposes.

11.3.3 Instrumentation

In general, ICP-MS instrumentation consists of a series of units working together, as shown in Figure 11.3 [7].

The ICP-MS consists of the following parts:

- **Sample introduction system**: It consists of a nebulizer and a spray chamber.
  a. **Nebulizer:** The nebulizer converts the liquid sample into fine liquid droplets through a peristaltic pump.
  b. **Spray chamber:** This fine liquid droplet is then introduced into a spray chamber. which filters out droplets based on diameter size (>10 mm). The spray chamber has two main functions:
     i. It filters large aerosol droplets coming out from the nebulizer. The produced aerosol must have a droplet size of ~10 µm before entering the system.
     ii. It is used even out any "pulses" generated due to the pumping of the solution.
- **Plasma generation system:** It consists of an ICP-torch and RF coil. ICP-torch is a series of concentric quartz tubes that generates a partially ionized gas, which is an electrically conducting gaseous mixture consisting of a mixture of a positive charge of argon ions and electrons known as plasma. The ICP-torch is positioned in the center of a copper RF coil, which produces a temperature of about 6000 to 10,000 K.
- **Vacuum-interface system:** The interface section of the instrument allows the ICP and MS parts to be coupled employing a differential pumping at a vacuum interface system. There is a water-cooled cone that extracts the generated elemental ions from the plasma atmospheric pressure (760 torr) into a high vacuum low-pressure region (typically 10^{-4} to

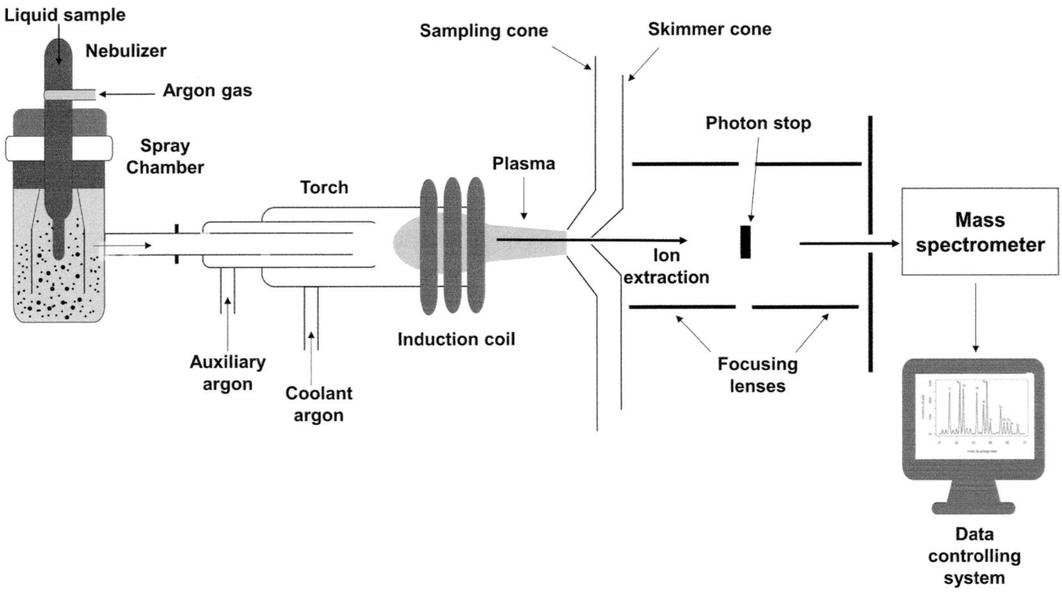

FIGURE 11.3 Instrumentation of ICP-MS

10^{-5} torr). The analyte ions are extracted through a pair of slots, known as the sampling cone and the skimmer cone, from atmospheric pressure to high vacuum.

- **Ion optics:** After the cones, through a series of lenses, "ion optics," the ion beam is focused into the mass analyzer by separating the neutral species and photons.
- **Interference removal system:** It consists of a collision or reaction cell (CCT or DRC) to minimize polyatomic molecular interferences. Either a quadrupole (with or without CCT or DRC) or a high-resolution magnetic sector mass spectrometer is used, which has a nominal m/z with an analyte. The collision/reaction cell operates with collision/reaction gases such as H_2, He, or reactive gases to either break down polyatomic species or shift interferences away from the analyte mass.
- **Mass spectrometer filtration system or mass analyzers**: The ions are separated according to their mass/charge ratio (m/z) and are detected with an electron multiplier. For routine elemental analysis, quadrupoles are most commonly used, offering fast and robust analysis; however, high-resolution magnetic-sector analyzers are used for resolving overlapping interferences. Additionally, time-of-flight and multi-collector analyzers are utilized for isotope differentiation or single-particle applications.
- **Detector and data controlling system:** Finally, ions exiting the mass analyzer are detected by electron multipliers that translate ion hits into electrical signals.

The data of all elemental masses are collected and stored through a computer interface, where the control software executes mass scanning, allows interference (isobaric/molecular) correction, brings about normalization of the internal standard, and performs routine calibration.

11.3.4 Procedure

- The herbal drug or extract is dried and ground into a fine homogeneous powder.
- Approximately 0.1 to 0.5 g of the powder is accurately weighed and transferred into reaction vessels for acid digestion to break down the organic matrix.
- Then a mixture of concentrated nitric acid (HNO_3) and hydrogen peroxide (H_2O_2) is used for digestion.

- Microwave-assisted digestion in a closed vessel system is mostly favored for the efficient and complete digestion of the sample under optimal conditions of temperature and pressure (generally ~200°C for 15 to 30 min is preferred).
- After complete digestion, the clear solution of the sample is further diluted using ultrapure water (about 50 mL).
- The resultant solution, if necessary, is filtered to remove any present residues.
- Internal standards are used to correct the matrix effects and instrumental variations.
- The standards of known element concentrations are prepared for quantification.
- The prepared sample solution is introduced into the ICP-MS instrument for analysis.
- The ICP torch then generates an argon plasma that ionizes the sample elements.
- The ionized elements then pass through the interface cones, followed by ion optics, and mass filter sorting ions based on the mass-to-charge ratio (m/z).
- The detector then counts elemental ions, and their concentrations are calculated by comparing them with the standard calibration curves.

11.3.5 Challenges

Following are the challenges of ICP-MS [7–10].

- **Mass interferences:** ICP-MS is often prone to polyatomic, isobaric (such as calcium, iron, and potassium), and doubly charged ion interferences. These interfere with the identification and quantification of accurate ion mass. Also, rare earth elements can cause interferences due to the presence of doubly charged ions and affect the detection of selected elements such as selenium and arsenic.
- **High cost and technical skill:** The instrument is expensive, leading to a higher cost per analyte concentration determination, and also requires an expert technician for data interpretation and maintenance.
- **Matrix effects:** Ion response in ICP-MS is significantly inhibited by matrix elements. Matrix interferences arise due to inconsistent atomization of different samples and between standards and samples.
  High concentration of total dissolved solids (approximately 0.3 to 0.5%) in test samples can also cause suppression of analyte signals.
- **Sensitivity limitations:** Detection limits vary from element to element. Some elements, such as fluorine, have too high ionization energies for effective plasma ionization, or where detection limits are insufficient for very dilute biological samples.
- **Memory or carryover effects:** The high sensitivity of ICP-MS can lead to carryover effects, where residual trace elements from previous analysis remain in the sample introduction system and carry over into successive analyses, thereby increasing the detection limits.
- **Spectral and isotope overlap:** The adjacent isotopes can lead to spectral overlapping, thus affecting the accuracy of detection. For example, abundant iron isotopes present in the blood samples often interfere with the analysis of manganese.

11.3.6 Application to Herbal Drugs

ICP-MS is extensively used in the analysis of herbal drugs owing to its ability to identify and detect multiple ultra-trace elements in a single run. Apart from the routine screening of heavy metals or elements, it allows isotopic fingerprinting for the authentication of medicinal plant origin and the detection of adulterants. When hyphenated with advanced separation techniques such as HPLC-ICP-MS, it permits elemental speciation, thereby distinguishing between toxic and non-toxic states of the respective elements. Its rapid throughput analysis makes it applicable for batch-to-batch quality control throughout the manufacturing process, whereas its high accuracy facilitates pharmacological research by linking it to the elemental profiles to claim therapeutic effectiveness or efficacy.

In a recent study, the heavy metal contamination in 80 samples of 10 commonly consumed herbal medicines from Hyderabad, India, was analyzed using LC-ICP-MS. The results showed that the samples were found to be contaminated with heavy metals; also 10% of them surpassed the limits of the Ayurvedic Pharmacopoeia of India (API). The most abundant contaminant was copper and chromium, and chromium content was high in the case of herbal crude drugs such as amla, brahmi, and lodhra bark.

Moreover, the health risk assessment showed Brahmi to be the greatest concern, with a hazard quotient of 1.8 and hazard index of 2.3, which was found to be above safer limits, thus suggesting probable non-cancer health risks. Its lifetime cancer risk was also greater than the range set by the US EPA. Monte Carlo simulations showed that Amla may also show greater non-cancer health risks [11].

In another study, by Kanrar et al., the elemental profiling along with the associated health risk of tea from north-east India (*Camellia sinensis*) was investigated by ICP-MS using a microwave digestion process. The elemental profile of about 321 black tea samples from 8 tea-growing areas of north-east India, namely Darjeeling, Terai, Dooars, North Bank, Upper Assam, South Bank, Cachar, and Tripura, was collected over three seasons from 2020 to 2021.

A total of 24 elements were profiled, and the macro-element levels showed the order Al > Mn > Rb > Ba > Zn > Cu > Sr > Cr > Ni > Sn, whereas Be, As, Ga, Tl, Li, Se, Cd, Ag, Cs, V, Co, and Pb were detected at trace levels. Mercury and uranium were found to be below detection limits. The elemental concentrations in these samples differed owing to regional variation arising from differences in soil composition, cultivation practices, and anthropogenic inputs.

However, risk assessment showed a hazard quotient and indices below 1 in all non-carcinogenic elements, thus demonstrating insignificant risk, whereas an increase in lifetime cancer risk was found to be within safer limits for elements like As, Cd, and Pb, but it was medium for Ni. As a result, black tea from north-east India was regarded as safe concerning elemental contamination [12].

11.4 CONCLUSION

The application of these advanced analytical tools for elemental analysis is indispensable in the field of quality control, standardization, research, and development of herbal drugs and formulations. The analytical tools discussed in this chapter give critical insights into the necessity of analyzing the elemental composition to confirm product safety and efficacy. The variations in elemental content, affected by factors such as geography, environment, and cultivation, support the requirement of routine monitoring. Overall, integration of robust analytical techniques adhering to strict regulatory guidelines, along with good agricultural and manufacturing practices, is critical to minimize contamination issues generally associated with herbal drugs and formulation. It ultimately safeguards public health and preserves the pharmacological as well as pharmaceutical integrity of herbal drugs.

REFERENCES

1. Zafar, M., et al., Elemental analysis of some medicinal plants used in traditional medicine by atomic absorption spectrophotometer (AAS). *Journal of Medicinal Plants Research*, 2010. 4(19): p. 1987–1990.
2. Uddin, A.H., et al., Comparative study of three digestion methods for elemental analysis in traditional medicine products using atomic absorption spectrometry. *Journal of Analytical Science and Technology*, 2016. 7(1): p. 6.
3. To, N.D., et al., Optimization of a procedure for the determination of some heavy metals in herbal medicines by atomic absorption spectroscopy. *Natural Product Communications*, 2025. 20(6): p. 1934578X251349924.
4. Seiler, H., A. Sigel, and H. Sigel, *Handbook on Metals in Clinical and Analytical Chemistry*. 1994, CRC Press.

5. Gupta, S., et al., Volatile (As and Hg) and non-volatile (Pb and Cd) toxic heavy metals analysis in rhizome of Zingiber officinale collected from different locations of North Western Himalayas by atomic absorption spectroscopy. *Food and Chemical Toxicology*, 2010. 48(10): p. 2966–2971.
6. PS, P. and A.K. AJ, Elemental analysis of Andrographis echioides (L.) Nees. leaves, a potential pharmaceutical plant using atomic absorption spectroscopy (AAS). *Asian Journal of Pharmaceutical Research and Development*, 2021. 9(3): p. 43–47.
7. Al-Hakkani, M.F., Guideline of inductively coupled plasma mass spectrometry "ICP – MS": fundamentals, practices, determination of the limits, quality control, and method validation parameters. *SN Applied Sciences*, 2019. 1(7): p. 791.
8. Hill, S.J., *Inductively Coupled Plasma Spectrometry and Its Applications*. 2008, John Wiley & Sons.
9. Soltanpour, P.N., et al., Inductively coupled plasma emission spectrometry and inductively coupled plasma-mass spectrometry. *Methods of Soil Analysis: Part 3 Chemical Methods*, 1996. 5: p. 91–139.
10. Schönbächler, M., Inductively coupled plasma mass spectrometry (ICP-MS), in *Encyclopedia of Geochemistry*. 2016, Springer. p. 1–6.
11. Validandi, V., et al., Heavy metal contamination in commonly consumed herbal medicines from local markets in Hyderabad, India, and associated health risks. *Food Additives & Contaminants: Part B*, 2025: p. 1–12.
12. Kanrar, B., S. Kundu, and P. Khan, Elemental profiling of north-east Indian tea (Camellia sinensis) by ICP-MS and assessment of associated health risk. *Biological Trace Element Research*, 2023. 201(8): p. 4114–4132.

Index